T0227406

SOCIETIES AND NATURE IN THE SAHEL

Societies and Nature in the Sahel provides an in-depth analysis of the diverse Sahelo-Sudanian environment. Drawing on a wealth of research and historical development, this book explores the links between environment and social systems in the Sahel, integrating ecological, demographic, economic, technical, social and cultural factors.

Examining the conditions for land occupation and natural resource use in the Sahel, this book offers a conceptual and practical approach to social organisation and environmental management in the face of rapid environmental change.

With a unique focus on the Francophone states, *Societies and Nature in the Sahel* presents important implications for future intervention strategies in order to address the crisis of desertification across sub-Saharan Africa.

Claude Raynaut is a Researcher at the National Centre for Scientific Research (CNRS), Bordeaux, France; **Philippe Lavigne Delville** is a Researcher at the Research and Study Group on Techniques (GRET), France; **Jean Koechlin** is a Professor of Biogeography, Michel Montaigne University, Bordeaux, France; **Pierre Janin** is a Researcher at the French Institute for Development in Cooperation (ORSTOM), France; **Emmanuel Grégoire** is a Researcher at the National Centre for Scientific Research (CNRS), Bordeaux, France.

ROUTLEDGE/SEI GLOBAL ENVIRONMENT
AND DEVELOPMENT SERIES
Series Editor: Arno Rosemarin

SOCIETIES AND NATURE IN THE SAHEL
Claude Raynaut
*with Emmanuel Grégoire, Pierre
Janin, Jean Koechlin, Philippe
Lavigne Delville*

SOCIETIES AND NATURE IN THE SAHEL

Claude Raynaut with
Emmanuel Grégoire, Pierre Janin, Jean Koechlin, Philippe Lavigne Delville

Assisted by Phil Bradley

Translated by Dominique Simon and Hillary Koziol

Routledge
Taylor & Francis Group

LONDON AND NEW YORK

First published 1997
by Routledge
2 Park Square, Milton Park, Abingdon, Oxfordshire OX14 4RN

Simultaneously published in the USA and Canada
by Routledge
711 Third Avenue, New York, NY 10017

First issued in paperback 2015

Routledge is an imprint of the Taylor and Francis Group, an informa business

© 1997 Stockholm Environment Institute (SEI)

Typeset in Baskerville by
Ponting–Green Publishing Services, Chesham,
Buckinghamshire

All rights reserved. No part of this book may be
reprinted or reproduced or utilised in any form or by
any electronic, mechanical, or other means, now
known or hereafter invented, including photocopying
and recording, or in any information storage or
retrieval system, without permission in writing
from the publishers.

British Library Cataloguing in Publication Data
A catalogue record for this book is available from the
British Library

Library of Congress Cataloging in Publication Data
A catalogue record for this book has been requested

ISBN 13: 978-1-138-88131-0 (pbk)
ISBN 13: 978-0-415-14102-4 (hbk)

CONTENTS

CONTENTS

CONTENTS

FIGURES

TABLES

CONTRIBUTORS

Emmanuel Grégoire Economic geographer, Researcher at the National Centre for Scientific Research (CNRS – France).
Pierre Janin Geographer, Researcher at the French Institute for the Development in Cooperation (ORSTOM – France).
Jean Koechlin Bio-geographer, Professor at Michel Montaigne University, Bordeaux (France).
Philippe Lavigne Delville Anthropologist and agronomist, Researcher at the Research and Study Group on Techniques (GRET – France).
Claude Raynaut Anthropologist, Researcher at the National Centre for Scientific Research (CNRS – France).

FOREWORD

The Sahelian crisis has been with us for many years now but, until the late 1960s, the outside world was largely ignorant of the turmoil that was going on in semi-arid West Africa, and of the tragedy that was about to unfold. The devastating drought of 1972 put an end to that. From that moment on, the Sahel has gripped the imagination of expert and layman alike. The terrible aftermath of that drought and the cumulative impact of those that were to follow has focused the attention of researchers, of aid agencies and of international political concerns. The growing awareness of the tragic consequences of these droughts, reflected in the collapse of pastoral systems, widespread impoverishment and destitution, the disruption of whole production systems and the threatening imagery of desertification, has drawn in the full force of international development assistance. Yet for all this effort, very little appears to have been achieved. By even the most generous estimation, the colossal amount of finance that has been dispensed through a multitude of aid programmes has yielded practically nothing. There has been much agonising over this failure of international goodwill, to the extent that we now see a much more intensive effort to understand better the realities of the situation, to grasp the complexities that underlie a tragedy that cannot be explained by simplistic references to desertification, overpopulation and misuse of the environment. Undoubtedly, one cause of the failure has been this misidentification of the problem, a too-easy acceptance of gross oversimplifications and explanations (and solutions) based on false paradigms of the North, which saw the problem as one of dysfunctioning bio-physical systems that could be remedied through technical improvements to environmental management. More recently, international efforts have been directed towards a more holistic approach to an understanding of Sahelian realities, in which both physical and social sciences play a role.

Nevertheless, even though the relevance of social concerns is now accepted, the many attempts at a synthesis of the Sahelian dilemma almost always start from the perspective of the physical environment: its unpredictable and changing climate, its impoverished soils, prone to both wind and water erosion, its fragile vegetation cover and its susceptibility to water deficits. In these descriptions, the environment is the primary focus of the discussion, the subject of analysis. Man becomes the object, the supplicant who, in order to endure, must play according to the rules imposed by such a harsh environment. Though often hidden, there is always a determinist streak running through these presentations. When the rules are not followed, the environment bites back. Land and its productive potential is degraded, food production declines, livestock perishes and man's hold on life and nature becomes increasingly tenuous. From this approach emerges the triad of climatic change, man's misuse of the environment and desertification. It is a paradigm that has dominated thinking about the Sahel for 30 years. Indeed its antecedents can be traced back to the earliest of European explorers.

The literature on the subject is not without alternative approaches. As the authors of *Societies and Nature in the Sahel* point out, Marxist interpretations place man as the victim of international economic and political forces beyond his control. By virtue of growing and continuing exactions, Sahelian societies are forced to exploit the resource base in non-sustainable ways. In these interpretations, environmental degradation is less to do with bio-physical limitations than with economic and social transformation; the integration of peasant economies into a capitalist global economy.

Within both these interpretations, we find the Sahel described in in-discriminate terms, oversimplified into either a series of internally homogen-ous latitudinal zones, corresponding to the major climatically determined biomes, or to inscribed nation states, within whose boundaries conditions are assumed uniform. In other words, through gross aggregation, the Sahel is reduced to a series of sweeping generalisations, a now-familiar lexicon of desertification, overpopulation, antiquated agriculture and irrational pastoralism.

For such an expanse, 500 km deep and 3,000 km wide (even in its western segment), this level of generalisation is clearly inadequate. We cannot account for the tragedy that is unfolding, nor seek to find workable solutions on such coarsely presented material. We still await a more reasoned, more sensitive analysis of the Sahel in all its human and physical complexities. In particular, we still seek a more refined differentiation of this vast region, one that supersedes the generalisations and aggregations which have singularly failed to illuminate the real conditions on the ground.

This text is an attempt to fill this gap in our knowledge, to break down the Sahel into its constituent parts. *Societies and Nature in the Sahel* neither presents the region purely in terms of environmental constraints nor does it position its peoples as victims of uncontrollable outside forces. Instead, it adopts a scale of analysis that allows for the derivation of a number of characteristic or 'type' situations. In a sense this is a typology which combines the most salient elements of the environment and of the societies which exploit it. It is, however, more than mere description. Throughout the text, there is a sustained attempt to explain and to understand the particular set of circumstances that determine these situations at the present time, and the historical processes through which they have evolved. Both the characteristics that define them and the situations themselves are subject to change. The same dynamics that have led to their formation, continue to exert pressures for further adaptation and change. These type situations are, therefore, not immutable. Nevertheless, they provide a reasoned point of departure for further analysis, which, hopefully, can yield more realistic and viable answers to the problems of sustainable development.

This book has been a long time in the making. What began in the early 1990s as an effort to work through the existing and voluminous literature on the Sahel, with the intention of preparing a report to act as a platform for further work, has evolved into a much more substantial text. It is rare indeed that time and resources can be assigned to review, analyse and ultimately synthesise such a wealth of material. We all know that the wheel is constantly being reinvented, that there is never the opportunity to research thoroughly the extent and depth of existing knowledge, that without such an effort the same errors, indeed the very same research programmes, are repeated, and the results of previous endeavours ignored. We all bemoan the lack of resources available for learning of and absorbing fully the lessons of the past. The rush to action research, to participatory appraisal and the jump to development action is all but unstoppable. Later reflection and regrets do little to brake the impetuous search for magic solutions.

In this case, however, resources and time have been made available to do just that. What is presented in *Societies and Nature in the Sahel* is the result of a thorough review of the literature, coupled to interpretative expertise and a strong theoretical underpinning. With active support from the Stockholm Environment Institute, Professor Raynaut was asked to bring together a team of researchers, whose task would be to uncover and access not only the full breadth of published literature, but also the 'grey' material hidden away in minor libraries and documentation centres. This material has been analysed in such a way as to yield a series of maps, representing the many facets of Sahelian production systems: the environment itself, the major forms of agriculture and pastoralism, the tools and techniques deployed by the different farming societies, demographic characteristics and land use systems. By overlaying these maps, the authors arrive at a synthesis, within which the Sahel has been divided into a limited number of 'type situations', defined by the conjuncture of a number of key parameters. From this point, the authors search for explanations that hinge on the interplay of both social and material imperatives and which incorporate fully the legacy of history and modern dynamics.

This is not yet another book about the Sahel, to be distinguished from the many others merely by its arrangement of material. On the contrary, it is in fact a completely fresh approach to the subject; perhaps the first that genuinely and successfully integrates new thinking about the nature of societies and their relationship to the environment. It is constructed on the basis of empirical fact, yet at the same time blends its material within a powerful theoretical foundation. Its most significant contribution is almost certainly its explanatory power, through which real insights into the logic, the rationale of peasant production systems can be gained.

Pious words are too frequently voiced about the merits of multi-disciplinary research, yet, as the authors affirm, it is a difficult enterprise to undertake successfully. Merely assembling information brought to the fore by different disciplines rarely yields a truly combined description and interpretation of events. In this case, it is clear that the thinking of the authors is trans-disciplinary. They have reached out beyond their individual disciplines to incorporate the thinking and methods of others. A genuine discourse unfolds, based upon a common conception of the nature of society and its relations with the environment. It is this theme that pervades the book, giving it a power that has thus far eluded other attempts to encapsulate the complexities and diversities of the Sahel.

Societies and Nature in the Sahel combines history, sociology and anthropology to come to a clearer understanding of the nature and functioning of societies in the region. This interpretation of social reality is linked to the environment, where natural resources are seen as socially constructed, not as absolute bio-physical determinants. Throughout Chapters 5, 6 and 7 we clearly see that such a construction denies the determinist explanation of current events. Individual environments are more or less habitable, depending on the nature and function of the societies that occupy them and utilise their productive capacity. While physical conditions such as rainfall or soil properties may limit modes of exploitation, they nowhere determine in advance the possibilities for human occupancy. Simply by looking carefully at the demographic data available to any researcher, in Chapter 3 Professor Raynaut demonstrates that the distribution of people in the Sahel cannot be explained by reference to the innate capacity of the environment to sustain human life. The truth is that the modern day pattern of population distribution is as much a result of social and political history as of any notion of 'carrying capacity'. If this analysis shows nothing else, it must surely force us to dispense with crude judgements about overpopulation and birth rates that are deemed to be too high. The fact of long-distance labour migrations from the Sahel is known, but perhaps their magnitude and the impact they have on residual populations and the functioning of rural societies is insufficiently appreciated. Here, in Chapters 3 and 4 their true significance is demonstrated. More importantly, the logic of labour emigration – as part of a range of strategies that have evolved to deal with the exactitudes of the modern world – becomes clear. As with many other key elements that are emphasised in the book, we cannot understand and interpret this haemorrhage of labour power without reference to the broader perspectives of social and material reproduction.

These chapters on demography reveal that the Sahel and its diverse societies are constantly changing. We see the dynamism of the external world, driving economic, political and social change, but as the authors make clear, this external pressure is mediated by internal adaptation – in trading patterns that are often hidden and informal, and which function both alongside and within the more formal externally orientated economies of the region. Clearly, crude Marxist interpretations cannot suffice here. The extensive and vibrant commercial networks that characterise the region are not the mere products of external control. They reflect an internal capacity to adapt to new forces, just as they have done for centuries, from the decay of the trans-Saharan trade systems, through the gradual absorption of Islam, and from the slave and colonial eras to the modern day.

To portray rural production systems in the Sahel in terms of a duality between pastoral and agricultural realms is misguided. We see that there are subtle intergradations between them, that the modern pattern of agricultural and livestock enterprises is not simply a continuation of past patterns. In fact it reflects the introduction of fundamentally new elements such as cash cropping in agriculture and livestock accumulation by non-pastoralists. These modern developments are compounded by strategies that deploy land resources in new ways. Paradoxically, development interventions that have focused on new tools and techniques (introduced to promote a move towards the intensification of agriculture) have often had the opposite effect. By making labour more productive, opportunities have been seized to extensify agricultural practices; but this is not uniform. In some regions, there has indeed been a gradual progression towards more intensive forms of cultivation – in Southern Mali and in the Senegalese Groundnut Basin, for example – but in others, the various factors of production combine to favour an extensification of effort. There are subtle combinations of forces at work here, that cannot be accounted for through a simple cost–benefit exercise.

A classification of Sahelian agriculture that relies on an assumed polarity between cash and food crop production, or on the dominance of certain crop types, tells us little. Chapters 6 and 7 reveal a complexity and a dynamism that seems to defy simple categorisation. From one region to another, from one village to the next and even from one family 'farm' to another within the same village, we find enormous variation. This variability is noted not just across space and in the relative emphasis given to different aspects of production, but also in time as, from one year to another, farmers adopt different strategies to maximise the forces at their disposal. As the authors take pains to elaborate, this is not to say that every situation is unique. While the specific outcomes of individual decisions may yield a unique, if temporary situation, what is of greater interest is the context within which such decisions are made: the extent to which history bears down on production decisions through its social and cultural legacy, the social relations between households in a village and the degree to which they are rooted in social reproduction strategies that have their origins in an earlier age. Alongside these aspects of social relations, we must also consider economic opportunity, varying as it does with fluctuations in commodity prices (including that of labour), access to markets and exchange rates. Even further, there are patterns of land availability (and rules over access to it), of changing methods of soil fertility maintenance and of changing perceptions of the environment itself. All of these define the scope within which production decisions are made. Rather than concentrating exclusively on their outcome (the type and quantity of production) we would do well to dwell on the interplay of these underlying factors. We observe, then, an extremely diversified agriculture – in crop mixes, in labour allocation, in tools and in land management. There is no 'Sahelian agriculture', there are many agricultures located within the Sahelian region.

The undercurrents to this diversity provide the means for the identification of 'type situations', which the authors have mapped. Seventeen categories and subcategories are identified, each defined by a certain combination of elements: historical, social, production and environmental. But as the authors insist, the spatial location of these situations is approximate, with a core area surrounded by a somewhat indeterminate fading away of the central characteristics. More important than cartographic exactitude are the intricate and changing relations between man and environment that they signify. The combination of these different elements is a means to an end, the end being the identification, the weighting and the elaboration of the interrelations among the principal forcing factors. In this, the book clearly demonstrates that a singular recourse to the environment, or to population pressure, or to 'outmoded' agricultural practices offers no explanation of the diversity of Sahelian production systems, nor to their current fragility.

Social systems lie at the heart of any explanation – but as Professor Raynaut cogently argues in Chapter 9, ethnicity is a poor tool for describing these systems. Instead, the proposal is to work on a social differentiation based on three axes of variation: the degree of centralisation of power, the nature of social stratification and the circulation and distribution of goods and wealth. Along these axes of variation, the authors highlight three major forms of social organisation: major trading states (archetypally the Hausa states), warrior aristocracies (such as the Soninké in the Malian, Mauritanian, Senegalese triangle) and lineage-based peasantries (for example the Sérer of Senegal).

Into this complex of environment, production systems and social organisation, the text moves to examine closely two fundamental transformations which impact on the relations between societies and the environment: the changing role of 'land' and the dramatic mutation of labour relations – the emancipation of the workforce. These two themes, in which we see a fundamental shift from social strategies that have historically sought to manage the labour force to those that now focus more and more on the control over land as the primary factor of production, represent a fundamental transformation of Sahelian and, indeed, African production systems.

In the concluding chapter, we find all these complex threads woven together in a synthesis which locates and assesses the major dynamic forces at work: environmental change, the metamorphosis of land and labour relations and the progressive integration of a global economy. It is clear from this overview and the main body of the text itself that the Sahelian situation, as we see it today, is a reflection as much of social as of material priorities. In returning to the opening statements of the introduction, the authors affirm that we cannot isolate the latter within a positivist scientific tradition that focuses exclusively on technical matters. Much more, we need to reflect on the fact that man's relation to nature is founded not only on material needs, but on the immaterial: the embodiment of culture, a mental interpretation of nature, or as Godelier (1984) succinctly tells us, the '*idéel*'. Without absorbing this new paradigm, we will continue to blunder in the dark, to fail in our attempts to understand the complexities and realities of the Sahelian predicament.

P. N. Bradley
REISAM, Hull University
17 April 1996

PREFACE

AIMS AND LIMITATIONS OF THE APPROACH

It must be admitted that we are embarking on a hazardous undertaking in writing this book. In only a few hundred pages we aim to encompass a vast portion of the African continent in order to draw out some of the major threads that guide the dynamics of relations between the agricultural and pastoral societies which people it, and the natural environment in which they lead their existence. At the same time, we wish to show clearly how diverse are the local situations, and how vigorous are the changes that are constantly taking place.

The conflict with which we are faced is practically impossible to resolve. In order to produce a coherent synthesis we are forced to generalise, but at the same time, in order to explore the diversity and variability of conditions, we must focus on detail. In fact, we have adopted an intermediate position, thus running the risk of being caught in the crossfire of criticism both from those who would have preferred a more bold and straightforward approach in tracing the major explanatory axes and from those who would have preferred more local detail and precision. We accept this risk because this middle path perfectly suits our scientific approach. Although it may not always be apparent, given that we do on some occasions go into a degree of descriptive detail, our aim is that of theoretical and methodological reflection on the question rather than a straightforward factual account. There is naturally an abundance of maps, tables, quantitative and qualitative information, references to concrete situations, etc., but all this is not simply presented to demonstrate a point, even if we can draw conclusions. The purpose of this information is, above all, to provide food for thought, to raise questions that might lead beyond the conventional explanations for the 'desertification' of the Sahel, and to suggest new avenues to explore.

The second challenge, very closely linked to the first, is that of interdisciplinarity. In order to avoid hasty simplifications, in an area which has certainly suffered in this respect, we need to approach the subject by comparing scientific approaches and viewpoints from several disciplines.

This is indispensable. Of course, the anthropological aim and the Utopian ideas on which it is founded, i.e., the never-ending search for a 'whole' that brings together the material and immaterial aspects of reality, is to the fore in the orientation of our approach. This could never have taken shape, however, without the support and stimulation of knowledge and critiques brought in by other disciplines, such as agronomy, geography and economy. Here too we have made no attempt to be exhaustive by considering all the many forms of collaboration possible. We could well have called upon the demographer and the historian, for instance, but we simply relied on the results of their work. We preferred to ensure coherence, by limiting ourselves to a small team created around one project.

Although its virtues are often extolled, as a scientific practice inter-disciplinarity is far from simple to apply. In order to overcome conceptual differences and language variations between specialists, it is necessary to apply a methodology that enables the comparison of a series of points of view and the integration of extensive and heterogeneous information. In this respect, maps are a most useful basis for bringing data together. In the first part of the book, we work through a series of cartographic stages which, as they follow on one from another and combine together, gradually produce a synthetic stratification that integrates data from a number of different disciplines. Although much space may be devoted to the maps, it was not our intention to produce a thematic atlas of the Sahel. For us and the reader alike, the graphics are simply an aid to reflection. They simplify, they lack detail and precision and there may even be inaccuracies, because they are based on heterogeneous and sometimes incomplete documents. Nevertheless, they do reveal the diversity of Sahelian reality and thus lead us to reflect on the underlying dynamics of which this spatial heterogeneity is the visible expression. This is their purpose. They are therefore inseparable from the text on which they are based and which gives them their meaning and determines their relevance.

Claude Raynaut

ACKNOWLEDGEMENTS

Five authors have put their names to this work. They assume collective responsibility for the scientific content, even though individual authors took charge specifically of writing one or several chapters. To the writer of these acknowledgements fell the special task of supervising the scientific content of the work as a whole, as well as overseeing the writing process. He is the one responsible for any deficiencies in any of the disciplines covered since, as scientific editor, he had the last word in decisions over worries expressed by the other writers about simplifications that they were forced to accept, given the method used in the book as a whole.

I would like to thank two people who have made an invaluable contribution: Charles Cheung who was responsible for producing all the maps, often from data which was very imprecise and ill-matched, and Bernadette Raynaut, who had the heavy responsibility of preparing the French version of this text for the various stages of publication.

Thanks must also go to Ralph Faulkingham for his most helpful comments on the first draft of the work, as well as for useful bibliographic details and personal information, especially in relation to the Hausa society. Thanks too to Claude Herry for some useful observations on demography.

I must not finish without acknowledging how much this work owes to Phil Bradley, who encouraged the project and ensured that financial support was forthcoming from the Stockholm Environment Institute. I am therefore particularly pleased that it is he who has prefaced the book. I should add that he has been very much involved directly and personally in the actual production of the book, as it is his skill that enabled us to transcribe on to computer all the data for the graphics presented here. He has also actively participated in the editing of the English version of the text.

Claude Raynaut

INTRODUCTION:
THE 'DESERTIFICATION' OF
THE SAHEL: A SYMBOL OF
THE ENVIRONMENTAL CRISIS

Claude Raynaut

As the last two decades have steadily unfolded, the mounting crisis in the Sahel[1] has forced itself on to the front page of the environmental debate. Well before the greenhouse effect or acid rain became public concerns or captured the attention of science, the desperate situation in the Sahel had become headline news. To western countries it represented the quintessence of a major environmental emergency. In this region of the world the crisis has been evident for a long time. Its most striking manifestation is a chronic deficit in plant production, that periodically reaches such a critical level that famine strikes humans and decimates livestock. The years 1973 and 1974, when the West finally became aware of the problem, were particularly acute.

The evidence of sustained environmental disturbance is constantly increasing: trees are dying, dunes systems are being reactivated, vast tracts of terrain have become sterile and are no longer fit for agriculture, and wells and marshes are drying up. Even the flow of some major rivers has been reduced to a trickle, as was the case with the Niger in 1984. Undeniable evidence such as this has caused much debate since the 1970s, resulting in innumerable articles, books and international conferences. The formulation of recommendations on arid zones at the United Nations Conference on Environment and Development in 1992 highlighted, yet again, the situation in the area.

In response to the problems that have stricken this part of Africa, from Senegal to Ethiopia, the mobilisation of public and private international assistance has been unprecedented. Dramatic predictions of the relentless advance of the Sahara and talk of 'desertification' led to the launch of major programmes to block the desert's advance (most notably the planting of tree barriers at its edge). Special organisations were also created to gather information and seek solutions, such as Comité Inter-Etats de Lutte contre la Sécheresse du Sahel (CILSS) (International Committee Against the

1

Sahelian Drought), Club Du Sahel (The Sahel Club) and Observatoire Sahara–Sahel (the Sahara–Sahel Observatory).

The urgency of the situation in the Sahel demanded immediate action, action that sometimes resulted in destructive effects (as in the case of food aid, the effects of which only became apparent much later). At the same time, however, it was soon recognised that an analytical effort was needed in order to understand the events leading up to the crisis, as well as to provide the means for conceiving long-term interventions. In response, several European countries, as well as the USA and Canada, financed research on the Sahel. As early as 1975 in France, the Direction Générale de la Recherche Scientifique et Technique du Ministère de la Recherche (the Research Ministry's Division of Scientific and Technical Research) asked for research proposals under its theme of 'The Battle Against Drought in Tropical Areas' which generated many field studies.

THE DIFFICULT ROUTE TO INTERDISCIPLINARITY

From the outset, this search for explanations of the Sahelian dilemma created competition between two opposing doctrines. The first of these (the 'naturalist' approach) is founded in the bio-physical sciences, and sees the pronounced and prolonged decline in rainfall as the fundamental reason for the observed ecological disturbances. In this view, the only possible action was to study ways of minimising its consequences. The other doctrine was socio-political and emphasised the disruption of peasant and pastoral societies as a consequence of colonial and post-colonial domination. This alternative viewpoint suggested that the principal causes of famine were the unequal distribution of food resources due to the spread of commercial agriculture and the weakness in global demand for the agricultural products of the Sahel. From this perspective, the drought was merely a symptom of a much deeper structural malaise.[2]

The futility of such a sharp conceptual divide soon became evident. Supporters of the naturalist approach could not ignore for long the role of human action in creating and continuing environmental degradation, and the critics of imperialism failed to explain how the imperial domination that actually extends over a large part of the world had provoked such a particular catastrophe in the Sahel. Furthermore, the explanations of these critics were so general that they did not include any consideration of the many different ways that Sahelian societies were actually responding to the crisis.

As the years passed and the drought became more severe, it was soon obvious that this continuing aridity was the effect of a long-lasting climatic change that could not simply be dismissed as an epiphenomenon. Researchers gradually realised that in order to explore the ways that multiple factors combined to create and perpetuate the crisis they needed to go beyond linear explanatory models. Thus, in France, between the

2

mid-1970s and the early 1980s, the Sahel became an early testing ground for interdisciplinary work between the natural and human sciences in the environmental domain, through the creation of several teams made up of different specialised disciplines. The experience was far from successful. It did, however, lead to one conclusion: an interdisciplinary effort must be methodologically constructed, it cannot be decreed. Without the appropriate methodological and theoretical framework, dialogue between the disciplines quickly comes to an end. Each remains impervious to the others' scientific approach and only expects from them the occasional data that might answer its own questions.

The natural sciences already had concepts at their disposal that enabled them to introduce humans into their explanatory models, but only as consumers placed at the top of the food chain where their actions (by the type and intensity of an impact or input) transform the ecosystem in which they take place. The process is known as anthropisation. Once humans are considered as one living species among others, the contribution of the social sciences is useful only to the degree that it helps to describe more accurately their material practices and to measure their intensity. According to this approach nothing is added to our understanding of the dynamics of the ecosystem by looking at the cultural dimensions of these practices. At best, culture is seen as the packaging of adaptive behaviour, that is the combined effort of imagination and ignorance. The need to take these dimensions into account becomes apparent only when beliefs appear to slip towards the irrational and create behaviours that are harmful. The goal then becomes one of eradicating these bizarre actions. The anthropologist is, thus, free to study the social representations and forms of organisation that govern human practices, but the naturalist is not prepared to incorporate such data into his or her own analytical model. It was with some impatience that members of interdisciplinary teams working on the Sahel sometimes saw their social science colleagues spending the majority of their time on areas of research that seemed to them far removed from what they considered to be the really important questions: that is their own.

For their part, the social sciences have always had difficulty making room for nature in their own approach. Whether nature is seen as a framework for life, support for resources or object of thought, its intrinsic properties remain confined to the periphery of the explanatory systems of the social sciences. Above all, social scientists see in nature, to use Descola's phrase, an 'animate reflection of society' (Descola, 1986: 10) and it is this collective image that makes up their true object of study. Perhaps only a geographer's landscape might attribute a more central place to the physical world and hence, perhaps, the easier collaboration with botanist, pedologist or geologist. Generally, social science researchers, and especially anthropologists, who experimented with interdisciplinary collaboration in the field in the Sahel during the 1970s lacked, more than their partners, the

3

conceptual tools for using the data gathered by those same partners. Irritated by the questions they were asked (questions they often deemed simplistic) they felt they were being used simply as 'service providers' and, with lasting bitterness, retreated from their attempt at collaboration (a collaboration that often had not been spontaneous but rather imposed by their home institutions). It is, in fact, specialisations such as agronomy and agro-economy (and the related fields concerned with the problems of livestock raising) that became the regular interlocutors of the natural sciences. By approaching human material practice in terms of instrumental rationality, that is as a system of means oriented towards goals of production and consumption, these fields were better able to meet the expectations of the natural sciences.

While the Sahelian experiment in interdisciplinarity did not achieve success, it did leave behind a considerable store of experience and set the stage for many useful reflections.[3] Early research was extended into the 1980s by a number of field studies, notably by l'Office de Recherche Scientifique et Technique Outre-Mer (ORSTOM).[4] As the environmental movement gained ground, the need for collaboration between the natural and social sciences took on a larger dimension. In France, the Centre National de la Recherche Scientifique (CNRS) initiated the Programme Interdisciplinaire de Recherche sur l'Environnement (PIREN) (the Inter-disciplinary Programme for Environmental Research) in 1979. Its mission was to promote the study of the interactions between ecological and social systems. Here again, an evaluation made at the end of a decade reported this collaboration between the two major disciplinary fields to be of limited success (Dobremez et al., 1990). Finally, during the preparatory phase of the United Nations Conference on Environment and Development of 1992, the notion of sustainable development (Bruntland, 1967) took shape. This was itself an offshoot of the idea of ecodevelopment introduced at the Conference of the Environment held in Stockholm in 1972 and widely promoted by Sachs (1980). This approach emphasises one fundamental requirement: the simultaneous consideration of both the human and the natural dimensions of development (although its very general nature leaves it open to many interpretations). Once more, we see the question of interdisciplinary study at the heart of these deliberations, not as a methodo-logical necessity but as an essential epistemological requirement for contemporary societies in conceiving their relationship with the world. In truth, it is no longer a need expressed simply by the research culture itself, but has become an axiom inspiring broader scientific and political institu-tions. Hence, in 1991, the United Nations Educational Scientific and Cultural Organization (UNESCO) organised a colloquium devoted en-tirely to the discussion of the challenges, obstacles and outcomes of interdisciplinary efforts (Portella, 1992). Even the development organ-isations, such as the World Bank and the United Nations Development

Programme (UNDP), are increasingly requiring an interdisciplinary approach in their actions.

Does this mean that concrete collaboration between the human and natural sciences has really progressed? This is not at all certain. With the example of the Sahel, we are forced to the conclusion that more ground has been lost than won. The majority of French and European current research efforts are studies of climatology or hydrology using highly technical means such as remote sensing devices.[5] In addition, major technical studies are devoted to the improvement of agropastoral production systems,[6] and to research on commercial networks.[7] Research programmes that include the social sciences in any real sense (especially anthropology) are now rare. The dominant development ideology, which links the principle of liberal economics with the principle of democratic politics, favours a mode of intervention that privileges local community initiative by development from the ground up. Within such a context, the scientific perspective, with its need for distance and concern with objectivity, finds itself somewhat disqualified in comparison to the participatory approach that gives voice to affected populations. Under the guise of a sociological approach, the use of rapid appraisal methods has increased. These methods are intended to lead pastoral and agricultural communities to identify their own needs and to become aware of barriers. With such an approach, the social sciences are asked to intervene only to facilitate a community's process of self-discovery. After the collapse of the large bureaucratic development structures of the 1970s and 1980s, smaller projects proliferated, often under the auspices of local or foreign non-governmental organisations (NGOs). With respect to the environment, which had become the primary concern of aid organisations, notions such as resource conservation and land management became the touchstone of these interventions.

In 1989, a colloquium held in Segu, Mali, assembled participants from various social circles of different CILSS member countries: public authorities, aid organisations, non-governmental organisations, peasant and pastoral communities (Shaikh, 1990). The colloquium was to represent a high point in the effort to elaborate a new 'participatory' doctrine of development and the fight against 'desertification'. The 'grass roots' participants in this dialogue were, in fact, mostly notables who were long-time partners in development interventions, such as representatives of peasant associations, co-operative movements and so on, who in the end were under the control of the political and administrative authorities of their countries. This situation may change in future with the transition to democracy and the fall of older authoritarian powers. This is particularly striking in Mali where, through the intermediary of their union structure, southern cotton producers confront the government with their demands. Local protest movements have certainly existed in Africa's past. They have even taken organised form, such as in the 1970s when the Soninke peasants from the Bakel region in Senegal tried to make their voices heard in the face of hydro-agricultural development along the river (Adams, 1977). The major difference in the current situation is that ideology and politics have come together in a way that now favours these popular movements.

Efforts to give voice to those forgotten during the first decades of development are perfectly legitimate. We are among those who have, for a long time, expressed this position (Raynaut, 1977b). Does this suggest, however, that from now on one should scrimp on efforts that are more objective and methodological in their description, analysis and comprehension of the way relations are structured between Sahelian societies and the environments they occupy and exploit? We think not, since a society's discourse about itself and its aspirations are, to a large extent, the product of social relations (which, notably, determines an individual's access to resources), cultural values (in which needs are rooted) and material necessities. None of these aspects is transparently clear, neither from within nor from without. If one is not to abdicate the will to understand the complex dynamics at work, beyond the explicit intentions of the protagonists, it is indispensable that a methodical effort is maintained in order to construct a reflection of a critical nature. This is needed not only to meet the goal of advancing knowledge, but also because this is the price of a development strategy that does not rely on a spontaneous harmony among particular aspirations but, instead, tries to anticipate the social and environmental consequences of conflicts between them.

AIMS AND METHODS OF THE STUDY

In an earlier work (Raynaut, 1989a) we suggested that because of the characteristically severe changes that accompany it and the diversity of domains that are affected by it, the Sahelian crisis can constitute an ideal model, or paradigm, in the light of which it might become possible to deepen and broaden a reflection of the relations between societies and their environment that is founded on fact. By methodically exploiting the data provided by the innumerable studies on the Sahel, we believed it was possible to identify the material and immaterial factors that are at work and interact when social systems and the physical conditions for their reproduction are simultaneously altered. It is a holistic goal, largely Utopian in its claim to embrace an entire reality, but nevertheless constructive if it helps tie together several threads from two opposing interpretations of the facts, i.e., one naturalist, the other social.

In order to accomplish this, we must yet cross the line, always perilous, between the formulation of a general theoretical goal and the concept of an operating framework for data collection, processing and analysis. In order to arrive at this we start from the empirical observation of diversity in the social and environmental situations encountered across the Sahelian landscape. There is not one but many Sahels, and this obvious variability bears witness to the different combinations of physical and natural features, the material conditions for resource exploitation and the forms of cultural, social and economic organisation. It is hoped that a description and analysis of this diversity will result in the identification of situations illustrating distinct states in the relations between society and nature. From here we can formulate a working hypothesis in which observation and comparison of these different situations may allow us to identify the various factors that intervene in the transition from one state to the next. Thus, in taking into account local situations, in all their manifestations, we should gain access to a more global level of analysis – that is to the operators of change themselves.

As the spatial dimension is fundamental to a grasp of this multiplicity of forms, maps have been valuable tools, supported by a detailed analysis of the complex and shifting realities that they represent, although only in a simplified and static form.

The paradox and limits of such an approach reside in the fact that, in order to highlight and compare the different local situations, it is necessary to use a common template built on a certain number of constant criteria of description. What this means is that we single out, from the beginning, those factors that we deem particularly pertinent to the study of variability in the relations between society and nature in the Sahel. The very act of listing these criteria is tantamount to formulating an implicit, if crude, model of the dynamics we intend to study. At this stage though, we are not trying to hypothesise on the nature of the interactions at work. The objective is only to define the fields in which to observe variation. To accomplish this, one does not start from a blank slate. Several in-depth, regional studies undertaken in the Sahel constitute, in a sense, the 'classic cases' for analysis of the Sahelian crisis. These are the Sine Saloum in Senegal, the Mossi Burkinabe plateau and the Maradi region of Niger (many of the findings are referenced later in the text). Several broad fields emerge, where phenomena likely to influence the relations between people and their environment can be situated:

- The physical and bioclimatic characteristics of arid and semi-arid environments.
- Population size and its impact on renewable resources.
- Economic constraints, at both the national and the peasant and pastoral community levels.
- The material and technical conditions of environmental exploitation.

At the beginning, then, our effort is one of describing the spatial variability that can be recorded in each of the major domains and expressing them in cartographic form. Beyond this descriptive and static approach we have also attempted to mount a more dynamic investigation of the origin of the different patterns of variability and their relevance for analysing the diverse conditions for resource exploitation. The first seven chapters of the book examine the different areas of analysis defined above as follows:

- ecological conditions (Chapter 1);
- population distribution and demographic dynamics (Chapters 2 and 3);
- major fluxes of economic trade (Chapter 4);
- multiple uses of resources (Chapter 5); and
- farming strategies and practices (Chapters 6 and 7).

We progressively combine the spatial divisions of each domain to arrive, in Chapter 8, at a proposed division of the area under study into separate zones. Thus several typical situations are identified and located in space. Each of these corresponds to a certain combination of the natural, demographic, economic and technical variables with which we have worked.

SOCIAL DYNAMICS AND USES OF RESOURCES

At the end of this first stage, then, we will have available a tool for the description of the diverse realities that are covered, but which are currently obscured and denied by images of the crisis in the Sahel that are too frequently homogeneous and simplistic. We will also have a preliminary framework for reflection on the complexity of the interactions that give rise to these local situations. This will enable us to build up a richer and more complex vision of reality, a vision which will include not only human beings but also the social systems by which they are organised.

In the first chapters, the description of resource use practices will lead us constantly to seek understanding well beyond the material necessities imposed by the physical environment, population size or technical constraints. In the course of this description we will repeatedly be confronted with the weight of history, the impact of conflicts over use and the diversity of strategies for utilising nature that are a reflection of a multiplicity of social and cultural systems. The three final chapters are thus devoted to a more in-depth analysis of the social logic of these systems and the dynamics to which they give rise.

On the basis of the facts and analyses accumulated at the end of this work, we conclude by attempting to achieve two goals. The first is to lay the theoretical groundwork that may serve to orientate a broadened approach to the relations between society and nature. The second is to try to draw from these theoretical reflections some lessons for the formulation of what might be a strategy for action. We are aware, of course, of falling well short of our initial ambition. Nevertheless, we will be satisfied if we have been able to help overcome a vision of the problems of the Sahel that is far too homogeneous and oversimplified. We also hope that, based on the examples we have considered here, we will have contributed to a better understanding of the role played by social realities that are, in their essence, immaterial (systems of representation, forms of organisation, power strategies, the often antagonistic dynamics of change) in the unfolding of the material processes studied by the natural sciences. In this sense, we hope to participate in the advancement of a truly interdisciplinary dialogue.

We close with several remarks on methodology. The first has to do with the delimitation of the area under study. For reasons mainly to do with the accessibility of the documentation, and taking account of the means at our disposal, we have restricted our study to the French-speaking Sahel. Moreover, the trouble in Chad over the last twenty years has not been conducive to the development of research, nor even to the collection of up-to-date information. There were too many gaps in the information available to us to make it usable, especially in the maps. Cape Verde, being an island, seemed to us to be too dependent on very particular dynamics to be included in a work of synthesis dominated by the situation of continental countries bordering the Sahara. Our study, therefore, covers five continental CILSS member states, namely Burkina Faso, Mali, Mauritania, Niger and Senegal. In all cases, the framework for data collection and representation is the nation state and its borders, even when these countries incorporate areas that are far removed from the natural conditions found in the Sahelian and Sahelo-Sudanian zones (the Senegalese Casamance, southern Mali and the south-west of Burkina, for example). In fact, as we shall see, from a human and economic point of view, these regions are closely linked to the national entities of which they are a part. Conversely, the northern edge of Nigeria has not been integrated into our analysis, even though it is in many ways an extension of the southern part of Niger.

Consequently, the choices we have made result in an incomplete picture of the Sahel. Yet, the area covered is sufficient to allow us to attain our primary objectives, namely to highlight the great diversity of local situations and to bring to bear elements of reflection about the dynamics that govern their variability.

The materials from which we worked caused us a major difficulty. The data provided by the many studies conducted in the Sahel over the last two decades are frequently quite localised, and, sometimes, are limited to a single village that was the object of research or the focus of a development project. In contrast, other studies are very superficial or very general, covering vast areas and describing only the average situation, which reflects poorly the diversity found on the ground. There is also a great deal of difference in the themes studied, the methodologies used and the social and spatial units within which data were collected. In-depth research programmes conducted from a multidisciplinary perspective, even with the inadequacies noted above, tend to concentrate on a limited number of regions, particularly the Nigerien and Senegalese groundnut basins, Mali's cotton zone, the Mossi plateau and its periphery in Burkina, and certain sectors of the Senegal River valley. Elsewhere, we encounter *real gaps*, areas virtually unknown to research, where we must content ourselves with fragmentary or uncertain information. This is particularly the case in eastern Senegal, western Mali and eastern Niger.

When attempting to map the distribution of such data, it becomes necessary to fill in the gaps in the documentation and to smooth over their heterogeneity. This can only be done by extrapolation, in an attempt to generalise to a greater area the few specific and reliable data available. To do this, we have resorted to the principle of non-contradiction. Information has been cross-checked, for example, concerning the natural environment, the human environment or development actions, that could contradict or modify the extent to which a particular phenomenon may be generalised. Where they have been available, we have also made use of existing synthetic maps. With regards to traditional technological practices, we finally decided to use the geographic distribution of ethnic groups as a framework for generalisation. Of course, the notion of ethnic classification is itself open to criticism (see Chapter 9), yet at the general level of analysis at which we are operating it does appear to offer an acceptable framework for approximation. Naturally, modifications were introduced whenever available information made them possible. Throughout the text, more detail is given about the sources used and the methodological approach to each specific theme.

Considering the many limitations that have been described, we do not claim to be either exhaustive nor beyond reproach with regard to the accuracy of the data. The precision of each datum entered into the maps is less pertinent than the relevance of the entire image that is created, highlighting dominant situations and notable discontinuities. We must stress the fact that the necessarily reductive act of generalisation, represented in the maps and other figures presented here, is inseparable from the nuances presented in the text in the form of more detailed analysis and close attention to ongoing developments. The creation of these maps constitutes only one of the tools we have used. The maps, nevertheless, furnish the concrete basis necessary for describing the diversity of Sahelian realities, that is an understanding with the essential objective (as we indicated above) of contributing to an interdisciplinary consideration of the dynamic of the relations between society and nature. Have we succeeded? The answer is not at all certain. If informed discussion has been opened on this critical issue, then our effort will not have been in vain.

NOTES

1 The precise geographical framework of this study is defined below. To avoid any possible confusion, unless otherwise stated, the currently accepted definitions

of Sahel and Sahelian are used, i.e., the countries on the southern border of the Sahara that, even if parts of them are situated in non-Sahelian bioclimatic zones, are affected by the problems of the use of arid and semi-arid regions.

2 Works such as *Qui se nourrit de la famine au Sahel* (Comité Information Sahel, 1974), *Sécheresses et famines du Sahel* (Copans, 1975) and *Seeds of Famine* (Franke and Chasin, 1980) are quite representative of this second trend. Conversely, when a French scientific periodical uses the term 'drought' it indicates a naturalist approach used to interpret the problems in arid regions.

3 Hence a seminar was held in Bordeaux, in December 1981, to discuss methodologies acquired through recent experience.

4 See especially Claude *et al.* (1991) and the research carried out by the ORSTOM Research Group, entitled Dynamiques des Systèmes de Production.

5 An important work has recently been completed by the Institut d'Elavage et de Médecine Vétérinaire des Pays Tropicaux (Institute for Breeding and Veterinary Medicine in Tropical Countries) that maps the vegetal resources of the Sahel (IEMVT/CTA, 1988), and an international organisation has been formed to gather and process hydrological data.

6 On this issue, see the works by Yung and Bosc (1992) on agricultural development in the Sahel.

7 In particular, efforts by the Organisation for Economic Co-operation and Development (OECD), the CILSS and the Club du Sahel on the Economic Regional Areas of West Africa, i.e., the seminar held in Lomé in 1989 on the cereal-growing regions and the 1993 meeting in Cotonou on international economic co-operation. We should also mention the network 'Knowledge of Cereal Markets' organised by the same groups.

1

ECOLOGICAL CONDITIONS AND DEGRADATION FACTORS IN THE SAHEL

Jean Koechlin

The Sahelian area is primarily identified by a set of physical and natural features. The principal criteria for a delineation and description of its characteristics are drawn from the observation of climate, soils and vegetation. The combination of these different variables creates a wide heterogeneity within local sites. The unequal impact of human activity brings with it an additional element of diversification, creating transformation processes which, in their most extreme manifestations, can lead to the degradation of environments which is then difficult to reverse.

It is the different factors associated with the variability in local situations that will be examined in this chapter.

CLIMATIC CONDITIONS

In the zone under consideration, the availability of water (more precisely the possibility of exploiting rainwater) represents the fundamental limiting factor for primary production. In this zone, therefore, primary production is dependent upon the seasonal distribution of precipitation and its annual variation even more than the annual, global amount of rainwater.

The climate of the Sahel is determined by two alternating aerial fluxes. The first is a flux of dry air, flowing north-east to east, which results from the Saharan high pressures of the Azores anticyclone. This dry air is known as the Harmattan. It is often dust-laden and blows from December to March. The second is a flux of humid air, flowing south-west to west, which results from the oceanic high pressures of the Saint Helen anticyclone. A zone of separation between these two fluxes, known as the Intertropical Front (ITF), oscillates during the year between the Gulf of Guinea in January and the 25°N parallel in August. This oscillation determines the rhythm of the humid and dry seasons.

The rains generally begin in the south in April and progress towards the north, arriving in June. The maximum rainfall is almost always reached in August. The rainy season ends at the beginning of September in the north, and in October further south. The downpours are often heavy and short at the beginning of the season, which can bring a considerable increase in run-off and erosion of soils which have dried out and are, therefore, difficult to infiltrate. In August the rains are less intense, allowing the soil (already moist) to absorb the rainwater better. It is, therefore, possible to observe a pronounced gradient of average rainfall from north to south, across the whole of the Sahel. There also exists, although less pronounced, a gradient of aridity in the shape of a crescent, from west to east, which results in a slight incline in isohyets. The shift southward of the 500 mm isohyet between Senegal and Chad is in the 200 km range. In reality, however, the situation is susceptible to considerable change from year to year.

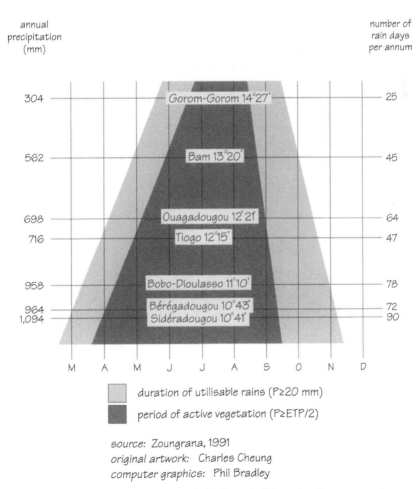

Figure 1.1 The relationship between latitude, rainfall and rainy season characteristics in Burkina Faso (1979–88)

In addition to the classic meteorological data, an ecologically important feature is the length of time available for active vegetation growth, defined as the period during which the rains are equal to or greater than half of the potential evapotranspiration. This will directly determine the vegetal productivity and the possibility of plant cultivation of shorter or longer cycles. This period decreases from south to north. In Burkina Faso, for example, the period of active vegetation is in the region of five months in Ouagadougou (12° 21' N) for 698 mm of annual rainfall, but only lasts two and a half months in Gorom-Gorom (14° 27' N) for 304 mm of annual rainfall (Figure 1.1).

According to these various elements, in this area a bioclimatic zoning is usually established, composed of different sectors which can also be identified by their vegetation:

- *Sub-desert sector*: annual rainfall less than 200–250 mm and a dry season of ten or more months.
- *Sahelian sector*: annual rainfall from 200–250 mm to 550 mm, dry season of eight to ten months, steppe vegetation.

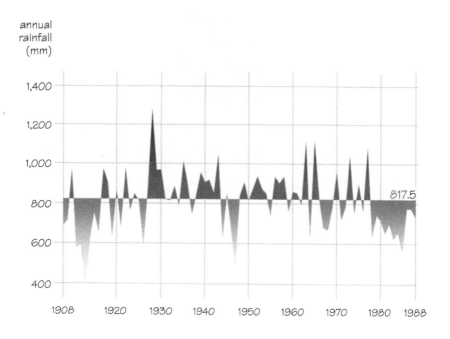

source: Zoungrana, 1991
original artwork: Charles Cheung
computer graphics: Phil Bradley

Figure 1.2 Long-term rainfall variability in the Ouagadougou region (1908–88)

14

- *Sub-Sahelian sector*: annual rainfall from 550 to 750 mm, dry season of seven to eight months, transitional steppe to savanna vegetation.
- *North Sudanese sector*: annual rainfall between 750 and 1,000 mm, dry season of six to seven months, typical Sudanese savanna vegetation.

The variation in precipitation is a normal component of arid zone climates, and of the Sahelian climate in particular. The droughts which the entire Sahel has experienced in recent decades, while perhaps exceptional in their duration, fit perfectly in this context. Figure 1.2 clearly shows this characteristic with, since the beginning of the century, alternating years and varying periods of dry and humid years.

The droughts of 1910 to 1916 and 1941 to 1945 are comparable in their intensity, if not their duration, to those that are now being experienced. Yet the 1950 to 1965 period is characterised by an extremely large number of years with surplus rainfall. Since about 1965, however, the rainfall deficits have been almost chronic in the countries of the Sahel and have had a tendency towards being more pronounced, with absolute minimums being recorded at many stations in 1983–4. The example of Burkina Faso (Figure 1.3) is absolutely typical in this regard, showing a regular tendency towards diminishing precipitation since the 1960s.

The rainfall deficits calculated for the period 1961 to 1985 for the Sahel (in comparison to the normal years of 1931 to 1960) are in the order of 25 to 30 per cent in the northern regions and from 20 to 25 per cent further south. During the same period, the isohyets have shifted towards the south, a shift that is as strong as the precipitation is weak, i.e., 200 to 300 km for isohyet 100 and 100 to 150 km for isohyet 500 (Figure 1.4).

Unfortunately, the short duration of these observations does not allow us to detect a periodicity in this phenomenon. It is difficult to know whether the low rainfall totals recorded since the end of the 1960s reflect an irreversible modification of the climate or a simple temporary accident.

A general tendency towards drought is perceivable since the beginning of the century as indicated below by the data for St Louis (Le Borgne, 1990):

Years	Average annual rainfall
1901–30	409.6 mm
1931–60	341.7 mm
1961–85	262.9 mm

This poses a problem regarding the choice of a standard (normal or deficient) precipitation situation for the period under study. Beginning in 1991, the member countries of the World Meteorological Office adopted 1961–90 as the standard, instead of 1931–60. Therefore, what could have been considered a deficit must now be considered normal. This, of course,

15

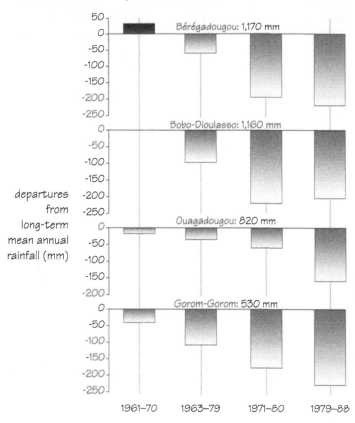

departures
from
long-term
mean annual
rainfall (mm)

source: Zoungrana, 1991
original artwork: Charles Cheung
computer graphics: Phil Bradley

Figure 1.3 Rainfall variability at four locations in Burkina Faso

does not change the actual rainfall received, but its implications will affect everything concerned with the problems of agropastoral development, because when the 1931–60 averages are used they give a much too optimistic view of the situation.

The repercussions of this aridification phenomenon are considerable in many areas. Nevertheless, within this tendency, the considerable variation of the rains in time and space must again be emphasised. The ecological significance of this variation is not easy to visualise since it is related not only to the importance of the rains but also to the length of the seasons, to

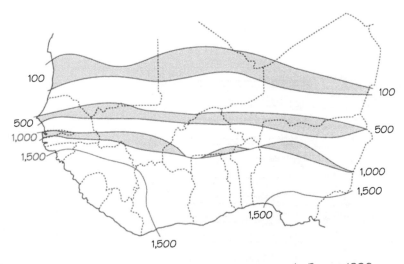

source: Le Borgne, 1990
original artwork: Charles Cheung
computer graphics: Phil Bradley

Figure 1.4 The southward displacement of the 100, 500 and 1,000 mm isohyets in the western Sahel during the period 1961–85

the date of their onset, to the number of days of rain and to the frequency of humid and dry periods during the course of the rainy season. Figure 1.5 illustrates some of these points, and clearly shows the wide range of inter-annual variation for these parameters.

The concept of aridity is complex and has to do with a number of factors. Besides their precipitation regime, the Sahelian countries are characterised by high temperatures, with absolute maximums reaching more than 45°C, with low atmospheric humidity for most of the year. Potential evaporation values are high and the water balance (actual evapotranspiration (AET)/ potential evapotranspiration (PET)) is less than one for much, if not all, of the year. For example, in Mopti, precipitation (P) equals 552 mm, while potential evapotranspiration is 1,984 mm, with a positive balance for only one month. In Gao the situation is even more severe, with precipitation at 261 mm and potential evapotranspiration as high as 2,255 mm and no months with a positive balance.

This emphasises the fact that groundwater reserves barely come into the picture and that the water imbalance (AET/PET), the principal factor of vegetation productivity, is almost equivalent to the rainfall deficit (P/PET).

17

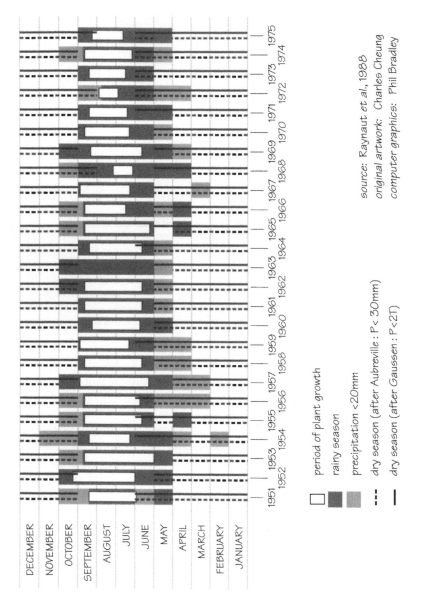

Figure 1.5 Inter-annual variability of rainfall characteristics at Maradi, Niger (1951–75)

period of plant growth
rainy season
precipitation <20mm
dry season (after Aubreville : P < 30mm)
dry season (after Gaussen : P<2T)

source: Raynaut et al, 1988
original artwork: Charles Cheung
computer graphics: Phil Bradley

EDAPHIC CONDITIONS

In the context of such aridity, how soils perform in relation to water (depth, texture, structure) is fundamental, especially as precipitation is low and uncertain. As a consequence agriculturalists may give soil moisture performance more significance than soil fertility.

Within the framework of traditional techniques, sandy soils make better use of precipitation because of their permeability, their capacity to retain moisture to a sufficient depth, the availability of this water to deep-rooted plants and low evaporation. Compact soils are much less favourable because of their low permeability which leads to run-off. The soil's compactness makes root penetration difficult. More importantly, water retention occurs at a shallow depth and, although the water is strongly retained in the soil, it is inefficiently used by roots, yet easily evaporated. These compact soils are also much more difficult to work, which is hardly compatible with the land use strategies on which peasants' technical systems are often based, as will be seen later in Chapter 5. Deep ploughing, with the aid of draft animal or mechanised cultivation, allows these constraints to be overcome to a certain extent, and puts more fertile clay soils, traditionally abandoned to pasture, within reach of agriculturalists.

Contrary to what is observed in tropical humid zones, where the deep alteration of rocks leads to very deep and relatively uniform soils, arid climates are responsible for a much greater pedological diversity, generally with shallow soils and where the influence of the bedrock remains very pronounced. In addition, the legacy of the recent climatic fluctuations of the Quaternary period translates into large areas of wind-polished hardpan and gravely soils and in the expansion of ancient dunal massif formations.

A certain number of soil types cover a vast geographic expanse:

- Dune soils, with a good moisture capacity, are chemically impoverished. Although sensitive to aeolian erosion, they suffer little from water erosion. Cultivation leads quickly to a structural modification of the surface layers of soil and a drop in fertility. These dunal formations are most widespread in Senegal, Niger and all of the southern Saharan margin.
- Soils over granite bedrock or metamorphic platforms and over sandstone are richer in fine particles and are of variable thickness. Their potential development is essentially based on their physical capacity, their saturation during the rainy season and their compactability during the dry season. Such soils cover only limited areas, and their value is often reduced by the presence of hardpan or graveley elements.
- Soils on hardpan, which are often rich in gravel, are usually very thin, with a minimal moisture reserve, which limits their suitability for agricultural use and makes them more appropriate for livestock production.

19

Such soils cover considerable areas (Senegal, Mali, Burkina Faso, west of the Niger).

• Other types of soils offer great potential, particularly alluvial and vertisol soils, but cover only small areas.

With certain exceptions (tropical black clays and particular alluvial soils), all these soils have very low chemical fertility (Boulet, 1964), especially the sandy soils because the lack of fine particles reduces the exchange capacity and leads to low base levels. As a result, restoration practices appear to be necessary, such as animal manure, mineral fertiliser (especially to fight certain deficiencies in N, P or K) or fallowing. Where restoration practices in heavily cultivated areas are lacking, the fertility of the land shows pronounced reductions.

The maintenance of fertility is the result of a complex process involving interactions between both mineral and organic soil composition and cultivation practices. In addition, the fertilisation systems must be perfectly adapted to local conditions in order to achieve their full effectiveness. Thus the fallow period will only have a positive effect on the crop yield if a large quantity of biomass is produced, that is on soils which are still relatively fertile. Furthermore, rotation systems of different crops on the same tract of land enable the potential fertility to be most effectively realised. As for inorganic fertilisation, it leads to an increased yield and a reduction in inter-annual variations which are tied to climatic uncertainties. Contrary to what is sometimes thought, inorganic fertilisation does not accelerate the loss of organic matter. Nevertheless, it is evident that combined organic and mineral fertilisation represents the only totally efficacious practice for increasing and stabilising yields and maintaining soil structure (Piéri, 1989).

NATURAL VEGETATION AND ECOLOGICAL CONDITIONS

Like agricultural production, natural vegetation is dependent on local conditions and on the perfect integration of the different parameters. The vegetation could, therefore, be considered as an indicator of ecological conditions and of the potential of the locality, and serves as a basis for bioclimatic zoning.

For the area under consideration, the scheme already proposed for classifying climate can be used. From north to south, the following formations can be schematically classified (the order is determined essentially by climate):

• *Sub-desert sector*: Semi-desert vegetation; herbaceous or shrub steppe (*Acacia, Panicum turgidum, Aristida* spp.). Precipitation less than 250 mm a^{-1}.

• *Sahelian sector*: Sahelian steppe, more or less shrubby or wooded (*Acacia,*

Commiphora, Balanites, ground cover of *Aristida* and *Cenchrus*). Annual precipitation between 250 and 550 mm.

- *Sub-Sahelian sector*: Forms the transition between Sahelian steppes and Sudanese savannas; progressive enrichment of flora by Sudanese elements (*Andropogon gayanus, Parkia, Anogeissus*). Annual precipitation between 550 and 750 mm.
- *North Sudanese sector*: Typically Sudanese shrubby or wooded savannas; enrichment by hardy grass species and increasing density of woody flora. Annual precipitation between 750 and 1,000 mm. This continues into the Sudano-Guinean sparse forests and dense, dry forests, with precipitation of more than 1,000 mm a^{-1}.

The climatic limits indicated above are very approximate and the moisture capacity of the soil can, to a large extent, lessen or aggravate climatic aridity. In Niger (Department of Maradi), for example, within the climatic context of a Sudano-Sahelian transition zone (between approximately 600 and 350 mm of annual rainfall), it can be stated that it is the edaphic qualities that are responsible for the organisation of the major features of the vegetation, more than the climate. The distinction between the sandy soils of the dunal formations and the more compact soils, as discussed above, remains fundamental. In this way, from north to south, it is possible to find the same major types of vegetation in the same types of soil, for example, a steppe shrub vegetation with *Combretum micranthum* on compact soil or a steppe shrub with *Combretum glutinosum* on sandy dunal soil. Of course, secondary floristic modifications will be tied to the climatic gradient with certain species disappearing, or being replaced by others, as the rainfall conditions vary latitudinally.

Although completely adapted to its environment, the vegetation has nevertheless been affected by the drought and its transformations make it possible to measure the impact of droughts on the region. However, added to this impact is the increasingly pronounced anthropic pressure.

The migration of the average isohyet position towards the south has naturally brought about modifications to the limits of distribution of certain species. Some, where located in an area of increased moisture deficit, have disappeared and have been replaced by other species which are better adapted to the new local conditions. In Chad, for example, Gaston (1981) observed a displacement of the edge of the desert, and of the Sahel, of approximately 50 km towards the south when comparing 1976 to the period 1964–5. This 'advance' of the desert resulted in the disappearance of woody species and hardy grasses such that only rare annuals survived. Conversely, Saharan species, such as *Cornulaca monacantha*, have appeared further south. In Niger, at the edge of the Sahelian and sub-Sahelian sectors, hardy grasses such as *Andropogon gayanus* have clearly regressed, to the benefit of typical Sahelian annuals (*Aristida, Cenchrus*).

In all of the Sahelian region, a large number of woody species, some with Sudanese affinities such as *Anogeissus* and *Prosopis*, others with shallow root systems such as *Acacia senegal* or *Commiophora africana*, were eliminated due to the lowering of the groundwater table. They are often replaced by *Balanites aegyptiaca*, which is better adapted to the new conditions as a result of its deep root system. In the northern Ferlo (Senegal), it was possible to trace the almost complete disappearance of trees and shrubs, except in the low-lying inter-dunal depressions (Michel, 1990). With such an evolution, it is not only the landscape that is transformed, but also a large part of the natural resources that disappear.

The tree, in particular, plays a fundamental role in natural and cultivated ecosystems in the Sahel (see Chapter 7). It is first and foremost a source of wood and numerous products used by man and beast. Its role is essential in protecting the soil, recycling water and producing organic matter. A species like *Acacia albida*, with its deep root system, plays a particularly fundamental role in maintaining fertility, first by supplying nitrogen, and, second, by recycling mineral elements. The fact that it loses its leaves during the rainy season allows it to be preserved on cultivated lands.

The tree represents a long-term investment, but is very endangered in crisis situations when immediate survival takes precedence. In synergy with drought, the increasingly pronounced human impact contributes equally to the degradation of vegetation. These impacts arise from exploitation of trees for firewood and charcoal production, the felling and pruning of trees by pastoralists, overgrazing (preventing the regeneration of ground cover), the sterilisation and denudation of soils through trampling and the destruction of vegetation by agriculturalists, whether directly or through a loss in fertility.

Although the regeneration of herbaceous ground cover may seem possible in the event of more humid periods or a better management of natural resources (thanks to the ability of some species to extend the germination of their seeds over several years), the disappearance of certain woody species may be irreversible, at least in the short term (due to the disappearance of seed plants or the destruction of seedlings by animals).

The drought has a clear, direct impact on the vegetation but the destruction of the ground cover by man can, in turn, also aggravate aridity in two ways: by modifying the soil moisture balance through increased run-off and a decreased atmospheric recycling of rainwater, and also by an increase in the earth's albedo. This results in a modification of the thermal balance of the terrestrial surface–atmosphere system and a reduction of the thermo-convective processes which create the rains. The cyclic system thus engaged will repeat itself, perpetuating or accentuating conditions favourable to drought (Courel, 1983).

VULNERABILITY AND RISKS OF LAND DEGRADATION

In such an ecological context, the land (its soils and its vegetation) is extremely fragile. The risks of degradation, indeed desertification, are elevated by the combined action of climatic uncertainties and human pressures.

In arid areas, the degradation of woody vegetation results in the exploitation of a resource which becomes more acute with demographic expansion, the growth of towns and agricultural development (Raynaut, 1980a; Bellot, 1982b). These demographic phenomena will be discussed later in greater detail (see Chapters 2 and 3). The rainfall deficit and the lowering of groundwater tables has also provoked a marked contraction of woody Sahelian areas between 1972 and 1982 which are easily visible on satellite images (Courel, 1983). In the short term, such a situation is difficult to reverse due to difficulties with regeneration of certain species, the systematic destruction of seedlings and shoots by livestock and the rapid growth of other species (*Guiera senegalensis, Balanites egyptiaca*) which are liable to engender a new vegetative landscape (Gaston and Dulieu, 1975; Grouzis, 1988).

In the Sahelian steppes, the majority of the ground cover is made up of annuals, the population of which can vary considerably from one year to the next depending on precipitation. A climatic accident or overgrazing can, for several years, compromise the regrowth of pastures and leave the ground bare by limiting seed formation. This problem has been studied in depth at Oursi Pool, using an ecosystem approach (Grouzis, 1988). Similarly, a study in Chad (Manière, 1990) describes the mechanisms of desertification very well. As everywhere in the Sahel, these mechanisms are linked to a failure of the rains, to demographic expansion, to increased numbers of livestock and also to the poor management of natural resources (i.e., the growth of herds overloads the pastures, and the expansion of cultivation leads to a decrease in grazing land). The equilibrium between the ecological potential of the local area and the social and economic demands of the populations is thus broken. This study was based on vegetation maps prepared by the IEMVT (Institut d'Elevage et de Médecine Vétérinaire des Pays Tropicaux) after the 1973 drought (Gaston, 1981), and on another one prepared in 1985 after the 1984 drought (IEMVT/CTA, 1988). The study complements these maps with SPOT (Système Probatoire pour Observation de la Terre) image findings, together with a close study of the vegetation carried out in the field. The phenomena of vegetative degradation and desertification are thus clearly evident. On the sandy plateaux and the large-scale dune systems, for example, the processes of desertification are marked by a large number of dead trees, especially among the oldest specimens. Certain species, such as *Acacia senegal* and *Commiphora africana*, are sometimes completely eliminated. As to the

23

herbaceous layer, the decrease in grazing potential may be considerable, but there are also a number of modifications which most often lead to a reduction in the quality of pastures, such as the disappearance of hardy species, the reduction of species that need more water (like *Schoenfeldia gracilis*) and their replacement by more resistant grasses, such as *Aristida mutabilis* or *Cenchrus biflorus*, and the expansion of certain plant communities (i.e., *Chrozophora brocchiana, Leptadenia pyrotechnica*). The degradation of the ground cover can, as a result, be accompanied by serious phenomena of aeolian deflation. Certain observations show, however, that the seed potential necessary for assuring the regeneration of vegetation in a more favourable precipitation context is not always exhausted. Such precise and objective studies clearly constitute an indispensable preliminary step in any planning attempt made in these regions of the Sahel.

Pedologically little altered, poor in organic matter, often badly protected by vegetation, the soils of the arid and sub-arid zones are, generally speaking, very susceptible to aeolian or water erosion. These phenomena can be significantly exacerbated when agricultural or pastoral pressure is combined with a rainfall deficit. The processes of degradation can therefore be significantly augmented, and can go as far as desertification; that is, dune systems begin to move again, gully erosion is reactivated, hardpans are exposed through denudation and soils compacted and 'sterilised'.

These risks are vastly increased by the more intensive agricultural practices (in particular animal-aided and mechanised agriculture) and by cultivation on slopes or on lands with unfavourable soil texture (UNESCO, 1991). In the Sudanese region bushfires certainly represent a considerable secondary erosion factor by eliminating ground cover (Monnier, 1981).

Despite the extent of the repercussions of the droughts over the last two decades, amplified by a considerable human and pastoral burden, it must not be forgotten that these are normally part of the Sahel's climatic landscape. Before the last war, French and English foresters were already preoccupied with the Sahara's movement southward (Patterson *et al.*, 1973). This issue is now more pertinent than ever, with the combined effects of aridification and human pressure. As we shall see the risks of degradation can be defined more precisely according to the diversity of the local environment.

THE PRECIPITATION GRADIENT

1. In the northern Sahelian sector the rainfall, which is less than 250–300mm with a dry period of approximately ten months, prohibits all regular agricultural activity. Grazing productivity, which is made very uncertain by the variability of rains, together with widely dispersed water sources, forces livestock to be mobile – leading to nomadism and transhumance. The sub-desert soils, little altered and lacking virtually any organic matter, or the arid brown soils (excluding zones of rocky

outcrops) are often formed by windblown deposits in interdunal depressions. Barely covered by vegetation and usually very finely textured, they are particularly sensitive to wind erosion. The vegetation which is found in these ecologically limiting conditions is also very fragile. All these zones are thus very exposed to the processes of degradation and desertification (which were widely manifested during the last periods of drought).

Livestock mobility is an ancient response by Sahelian populations to these conditions. In order to exploit their situation fully, most of the Sahelian countries are engaged in well-drilling programmes that give access to sectors previously without available water resources. Too often, unfortunately, as will be discussed later (see Chapter 5), the very high concentrations of livestock that result have caused an intense degradation of all the neighbouring grazing lands (Zoungrana, 1991).

In the past decades, the risks of epizootic diseases have diminished and the number of livestock has increased considerably. The overloaded pasture ecosystems have been the object of increasingly pronounced degradation resulting from the death of trees (eaten by animals or pruned by herders), the reduction of ground cover, soil denudation and wind and water erosion. The loss of fine particles through wind erosion and the phenomenon of leaching have led to a reduction in fertility, in groundwater reserves and, consequently, in biomass productivity resulting in accentuated degradation. It is estimated that fodder production in the Burkinabe Sahel was reduced by between 20 and 25 per cent between 1955 and 1975 (Toutain and Dewispelaere, 1978). Nevertheless, some research and experimentation suggests that, under certain conditions, these Sahelian localities are far from deficient in their capacity to regenerate (Grouzis, 1988). In particular, a grazing ban has had very positive effects on the structure and dynamism of the vegetation, herbaceous as well as woody, and on its floristic composition.

2. In the 250–300 to 500 mm rainfall zone, agricultural activities rendered more uncertain by the climatic variability coexist with pastoral activities linked to transhumance. During the dry season, the livestock found there are in areas untouched by trypanosomiasis. Increasingly, in the transitional zones today, it is in terms of superimposition and competition that agriculture and herding relate. The social aspects of this process will be discussed later (see Chapter 5), although the magnitude of its impact on the environment should be noted now, particularly the cultivation of wooded zones, sloping lands or thin soils on hardpans; the exhaustion of soil fertility and a compensatory recourse to extensivity; and overgrazing and rejection of herding in the most vulnerable sectors. Vegetation, soils and the water balance will thus be directly affected – in an environment that continues to be extremely fragile (Leprun, 1989).

These zones therefore appear particularly exposed to degradation due to a combination of many factors. The need for management is

25

particularly critical. Taking into consideration the different elements allows a country's degradation zones to be delimited. This is what Grouzis has done (Grouzis, 1984, cited in Grouzis, 1988) for Burkina Faso, taking into account the following elements: annual rainfall, population density, the level of land occupation, and bovine and small ruminant density (Figure 1.6). The state of degradation is all the more accentuated as aridity becomes pronounced or as the rates of occupation by man and livestock increase. Thus, in the central zone of the country the pressure on soils and grazing lands is such that the ecological equilibrium is destroyed, reinforcing the outward emigration of people which, as we will see later (p. 53), this region has experienced for a long time. This idea of a state of degradation should not be confused with that of the risk of degradation. The latter can be linked to the attractiveness of the area and the arrival of populations from more degraded zones, as would occur in the western part of the country. In the northern zones, the risk is linked instead to climatic conditions and to the fragility of the soils and vegetation.

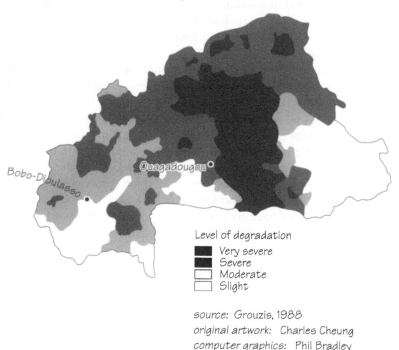

Level of degradation
■ Very severe
■ Severe
□ Moderate
□ Slight

source: Grouzis, 1988
original artwork: Charles Cheung
computer graphics: Phil Bradley

Figure 1.6 The different zones of degradation in Burkina Faso

3. In the 500 to 900 mm rainfall zone, with a dry period of between eight–nine and five–six months, the vegetation changes from Sahelian steppe to Sudanese savanna and open woodland, with a denser and more

26

diversified woody stratum and a ground cover enriched with hardy species. This is still an area of tropical, ferruginous soils, but the textures are quite varied. Rocky outcrops and especially hardpans occupy a very large area, which despite possessing unweathered or little-altered mineral-rich soils has a reduced agricultural potential.

On these slightly permeable soils, subject to more abundant and intense rains, water erosion becomes more pronounced than in more arid regions where aeolian phenomena often predominate. Run-off can lead to sheetwash and strip the slopes of soil or take a more linear form with more or less deep gullying. These phenomena are often linked to the degradation of vegetation and can have spectacular effects.

In this respect, from the start of the rainy season, the bushfires which leave the ground bare just before (often violent) storms play an important role. This is particularly relevant in the savannas where these fires, started for diverse reasons (such as hunting, pasture management and preparation for cultivation) become systematic. In addition to their role as an indirect factor in erosion, they are also responsible for a selective impoverishment of the flora and fauna, as well as disturbing the functioning of ecosystems by destroying the majority of the aerial organic matter.

SOIL DIVERSITY AND HUMAN USE

The risk of degradation occurring in any given area is also a function of the soil type, combined with the specific impacts linked to different production systems. Increased population densities, leading to high land occupation, can result in serious impacts on the natural environment, in terms of stress on the ecosystem's resources, disruption of its equilibrium and soil degradation. We will return later (pp. 40–1) to the complex relationship which can be established between population density and environmental impacts. Risks and impacts will differ as a function of edaphic conditions and mode of exploitation, whether cultivation is manual or mechanised, dependence on inputs, pastoral practices, etc.

A further critical point is that soil structure is specifically endangered. Less protected by vegetation, soils are alternately subject to wind and water erosion. The removal of fine particles leads to a decline in fertility and a lowering of soil water retention capacity, and, ultimately, to a reduction of natural or cultivated biomass and plant productivity. These degradation risks are much greater with mechanised agriculture because the physical impact on the soils is much more pronounced.

Other factors can equally affect vegetation resources, particularly the degradation of woody vegetation due to clearance for agriculture, stresses linked to herding and meeting the population's need for construction material and firewood (Bellot, 1982b). Besides the shortage of an essential resource, this destruction of vegetation will only accentuate the general

JEAN KOECHLIN

degradation of the area and the susceptibility of the soils to erosion (Toutain and Dewispelaere, 1978; Dewispelaere, 1980; Dewispelaere and Toutain, 1981; Dewispelaere *et al.*, 1983).

As already noted, a light, sandy soil is preferred by agriculturalists because of its favourable soil water properties and the ease with which it can be worked. However, its low fertility leads to mediocre yields and, in a traditional setting, favours more extensive production systems. The fundamental problem that needs to be resolved continues to be that of renewing fertility, which is made difficult by the reduced possibilities to fallow and insufficient forage (which is a limiting factor for livestock and therefore for the production of manure). The processes of degradation result in the denudation of the soil, which follows the reduction in plant cover. These processes are increasingly obvious with rises in land occupancy, which acts in synergy with climatic causes. The impact of cash crop agriculture, especially the groundnut, will be much more pronounced, especially with respect to the physical degradation of the land. In this way, the traditional system in the Senegal Groundnut Basin (which allowed for the maintenance of soil fertility) is now, more often than not, overstressed. The expansion of cultivation eliminates any possibilities for long-term fallow and for the integration of agriculture and livestock and, therefore, for the production of manure. A similar situation occurs in the Department of Maradi in Niger where the agricultural occupation of sandy soils which are favourable for cultivation is more than 90 per cent (Raynaut *et al.*, 1988).

In zones characterised by tropical, ferruginous, sandy-clay soils, with many sandstone outcrops or hardpans carrying fresh or little-altered mineral soils, the detailed pedological situation is much more complex, and the map at 1:5,000,000 scale can represent only the predominant edaphic type in each cartographic unit. The agricultural occupation produces then the same mosaic character as edaphic conditions, with high concentrations of agriculture in the most favourable situations, i.e., alluvial, light soils and sufficiently thick soil on hardpans. Health constraints (particularly onchocerciasis) can also limit human settlement in the valleys, as we will see in Chapter 3.

The traditional systems of subsistence cultivation permitted, by their diversification, a good adaptation to local conditions and the renewal of an area's resources. However, population increases, followed by the establishment of new agricultural techniques and the development of cash crop production such as cotton and maize have profoundly upset these equilibria. The resulting danger is from the relief and the slightly permeable soils that favour erosion, which is especially serious where hardpans exist close to the soil surface. It is further accentuated by animal-aided cultivation and by mechanised agriculture which make it easier to extend cultivation on to marginal lands and to suppress the regeneration of woody cover.

28

This is the case in the south of Mali, where relatively favourable soil and rainfall conditions have allowed for the large-scale development of cotton and maize cultivation, with a considerable increase in animal-aided cultivation and the use of fertilisers and other crop treatments. It must be noted, however, that increased cotton production is linked more to extensification (80 per cent) than to intensification (20 per cent). As for subsistence cultivation (millet and sorghum), recourse to technological inputs is much more limited although, here also, the land surface under cultivation has greatly increased with demographic growth. Equally, the development of animal-aided cultivation has led to lowered stock mobility and an increase in the number of animals. Reduction in fallow, declines in soil fertility, overgrazing, degradation of vegetation and erosion are all consequences of this situation (Bosc *et al.*, 1990).

The most unfavourable zones, on plateaux or wind polished hardpan, are often covered with a type of vegetation known as *brousse tigrée*. These are the ones used by livestock, although pasture productivity is very poor. Because of this these environments are naturally very fragile. Even in the absence of any accentuated human pressure, the combination of variable environmental factors (rain, geomorphology, pedology, bushfires) is sufficient to provoke and maintain serious degradation. For example, hardpan plateaux, almost bare of ground cover and with a sparse shrubby vegetation, together with some very eroded, wind polished areas, represent the extreme state of such situations and are at the very point of desertification.

In such environments, vegetation, and particularly woody vegetation, has suffered enormously from drought, from the exhaustion of groundwater reserves, and from the lowering of water-tables which have led to high vegetation mortality. Such situations, unfortunately, affect large areas of West Africa, although not in a completely continuous fashion. They occur particularly in wind polished or hardpan plateaux, sandstone tablelands and in areas which are at times locally covered by sandy deposits, and are separated by depressions where agricultural activities are possible. This is the case for all of eastern Senegal, contiguous with western Mali. Such conditions also prevail over most of northern Burkina Faso and west of the Niger River.

In depressions and large valleys (Senegal, Niger), on recent alluvial soils that are more or less hydromorphic, and on tropical black soils (vertisols), problems are raised by irrigated agriculture. These soils often have a high fertility, especially the vertisols. However, their cultivation can be limited by unfavourable physical properties, such as compactness, deoxygenation and shrinkage cracks, which require a great deal of management. A possible danger arising from irrigation is salinisation, which is a particularly serious phenomenon because it is difficult to reverse, especially in heavy soils. This phenomenon is linked to high evapotranspiration, the use of more or less brackish water and poorly managed water flow under irrigation (UNESCO, 1977; Florent and Pontanier, 1982).

Thus, policies which are sensitive to environmental problems appear to be indispensable for safeguarding and improving the area's potential. They must be based on a precise understanding of the area's capacities and limitations. It is imperative that the protection of flora and fauna is respected. A full awareness is necessary of the traditional practices of extensive agriculture (techniques and means of cultivation, processes for restoring fertility) that are no longer compatible with either the present rates of land occupation or the present imperatives of agricultural production which require the association of food sufficiency and the development of cash crop agriculture. The latter, too often, impose on the area what is in effect a surcharge, incompatible with the regenerative capacities of the environment.

We will return later (Chapters 6 and 7) to the problems which result from the evolution of agricultural and pastoral production systems. It is sufficient to note here that the present context is extremely propitious to a serious degradation of the environment. This can take diverse forms and intensities depending on local conditions. In addition, the Sahelian environment's susceptibility is such that situations which are difficult to reverse with respect to vegetation as well as soils can very rapidly be reached. The vegetation represents, by its specific structure and composition, a very precise indicator which may be particularly suited to the use of remote sensing as a diagnostic method. In effect, the recourse to satellite data is an important contribution to the understanding of vegetation and its mapping, provided there is a good field knowledge of the actual land. Yet, results appraising the state of the vegetation, of its biomass and crop yield are still very variable (Pons, 1988b).

LARGE ECOLOGICAL ZONES

From an ecological point of view, the map of the physical environment (Figure 1.7) depicts a spatially complex situation, linked essentially to the variety of edaphic conditions. At the heart of the zone situated between the southern limit of desert vegetation (with very scattered or contracted vegetation in sectors with a favourable water supply and very pronounced xerophytic adaptations of the plants) and the northern limit of sparse Sudanese woodlands, the following principal types of pedological situations have been taken into account:

• Brown arid soils on ancient ergs, where annual rainfall is less than 350 mm, or tropical ferruginous soils on aeolian deposits, with higher precipitation with or without ferric hardpans or rocky outcrops of shallow depth.

- Tropical ferruginous, sandy-clayey to clayey-sandy soils, more or less deep on diverse geological substrata or on hardpans.
- Fresh, mineral-rich soils or little altered soils on hardpans or rock outcrops – the mountainous massif represents a specific environment.
- Hydromorphic soils in depressions or floodplain zones, which can present health problems.

Climatic zoning is simpler, at least at the scale of this map (which is too large to reflect local peculiarities). The zoning is characterised by an aridity gradient which increases from south to north, with the continental effect, i.e., a southern downward shift of the climatic limits as the distance eastwards from the Atlantic coast increases. In the same way, vegetation limits are clearly inclined in an east-south-east direction, through the contact between Sahelian and Sudanese zones which is situated about 15°N latitude in western Mali but is at approximately 13°N to the south of Niamey.

The combination of climate, vegetation and soil factors delineates a certain number of major zones as described below.

The sub-desert margin

The rainfall, less than 350 mm with a dry period of about ten months, prohibits all regular agricultural activity. Vegetation is Saharo-Sahelian to Sahelian steppe. Forage productivity (made very uncertain by rainfall variability), together with the widely dispersed water sources, limit herding to nomadism and transhumance. Soils are sub-desert, little altered and practically bereft of organic matter, or arid brown (isohumic soils found in steppes). Except for the zones of rocky outcrops, they are most often formed by aeolian deposits in dune systems. They are sensitive to wind erosion because they are barely covered by vegetation and often finely textured. The vegetation, which is found in ecologically limiting conditions, is equally fragile. As a result, all these zones are very exposed to the processes of degradation and desertification, which were widely manifested during these last periods of drought (UNESCO, 1991).

Forage production is essentially based on annual herbaceous species. The biomass produced is a direct expression of the quantity and distribution of the rains and is, consequently, very variable. In addition, the floristic composition and the productivity of the herbaceous layer are highly dependent on seed production and their germination capacities (Grouzis, 1988). The real carrying capacity of these pastures could therefore be very different from one year to the next. From north of the Senegalese valley this margin extends, with varying breadth, continuously to Lake Chad and beyond. It is in Niger that it occupies the largest area, although it is divided by the Aïr massif and its southern extensions.

31

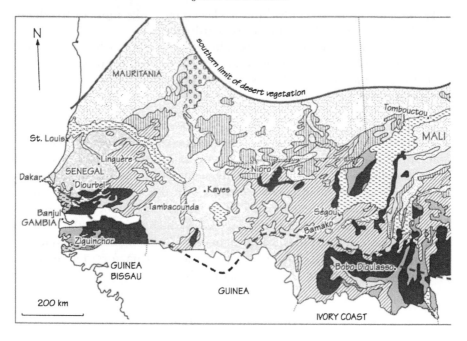

Brown arid soils on ancient ergs. P < 350 mm.
Herbaceous or shrubby steppe. Uncertain agriculture,
suitable for extensive transhumant pastoralism.

Tropical ferruginous soils on aeolian deposits. Good soil
moisture characteristics, but low fertility. P = 350-650
mm. Steppe progressively changing to savanna. Suitable
for agriculture, with extensive herding.

Discontinuous pockets of sand overlying hardpans or rock
outcrops.

Tropical ferruginous clayey-sand to sandy-clay soils.
Subsistence cropping, with cotton and pasturage.

As above, with localised hardpans or ferruginous gravel,
reducing agricultural potential.

Hardpans or rocky outcrops with localised colluvium.
Tropical ferruginous soils and fresh mineral soils.
Extensive herding or agriculture on deeper soils.

Figure 1.7 Bio-physical environmental zones of the western Sahel
(continued opposite)

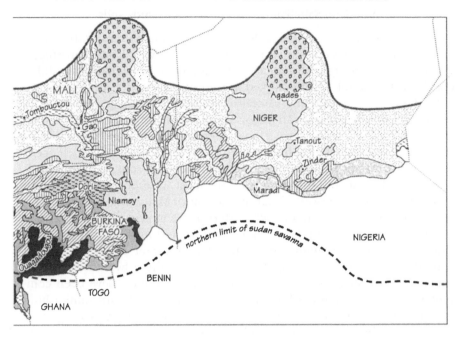

Hardpans or rocky outcrops. Fresh mineral and barely developed soils. Valley colluvial deposits are usable.

Hydromorphic soils more-or-less prone to flooding, and vertisols. Flood recession agriculture, rice cultivation, pasturage.

As above, with salinity problems.

Mountains

Southern limit of sub-desert vegetation: change to semi-desert herbaceous or shrubby steppe.

Northern limit of sudan *Isoberlinia* savanna and secondary forest-savanna mosaics (Senegal); source White, 1986.

original artwork: Jean Koechlin, Charles Cheung
computer graphics: Phil Bradley

A Sudano-Sahelian transition zone on sandy soils

These are tropical ferruginous soils on ancient dune systems, with a mean annual rainfall between 350 and 650 mm and a typical Sahelian steppe vegetation, where agricultural activities affected by climatic variability coexist with pastoral activities linked to transhumance.

In these very constraining rainfall conditions the soil's water capacity represents, for the peasants, an important criterion for its selection. Despite their low fertility, these sandy, loose, permeable soils are cultivated in preference to soils richer in fine elements and more compact, which occur in interdunal areas abandoned to pasture. These sandy soils, in the southern margin of the Saharan dune systems, are most widespread in Niger and western Senegal. In the west of Mali and southern Mauritania discontinuous sandy patches on hardpans or rocky outcrops are most common.

Today, the relationship between agriculture and herding in these transition zones is increasingly competitive (see Chapter 5). Since the last droughts, the terms of the equilibrium between these two activities (i.e., transhumance, sharing of lands, manure contracts, etc.) have been completely disturbed. Herders find themselves confronted with diminishing forage resources, not only due to drought and the resulting ground cover degradation, but also due to the cultivators' increasing annexation of the land.

The environmental impact of these changes is serious and includes the cultivation of wooded zones, sloping land or thin soils on hardpans, the exhaustion of soil fertility and compensation through intensification, and the overgrazing and abandonment resulting from herding in the most vulnerable sectors.

A very diverse mosaic of edapho-climatic conditions

This is climatically the domain of tropical ferruginous soils, in Sahelo-Sudanese conditions. The pedological textures are quite varied, from sandy-clayey to clayey-sandy, although sandstone massif and iron hardpans often reduce the soil depth and agricultural potential. Sudanese savanna vegetation predominates.

In western Niger, alternating fossil dunes, sedimentary basins and hardpan plateaux give rise to a mosaic of ferruginous soils that are more or less washed out, and little-altered soils on sandstone, hardpan or gravel. Soils from tropical ferruginous sand to clay–sand dominate in Burkina. These soils are poor, badly structured, often compact and are very susceptible to water erosion and poor agricultural practices, such as too great a reduction in the period of fallow. Sloping land and high population densities are serious aggravating factors. The geomorphology of central and southern Mali is characterised by alternating sandstone plateaux (Manding and

Dogon plateaux) and low-lying regions with an extensive hardpan. The result is, once again, a mosaic of diverse tropical ferruginous soils, little-altered soils and fresh mineral-rich soils. In southern Senegal, these tropical ferruginous soils, which are more or less sandy, rest upon old continental sandstone or hardpans. Given such conditions, it is important to underline the great vulnerability of agricultural potentials, which are often very limited, and the exceptional susceptibility of the area to extreme climatic events or to poorly managed exploitation.

Vast zones of hardpans or rocky outcrops

The distinction between this and the preceding zone is difficult to make. It is mainly a question of the proportion of fresh mineral-rich soils, or less altered soils on hardpans or rocky outcrops, which dominate and the more altered tropical ferruginous soils of greater depth which occur in this zone. The flora is typically Sudanese or Sudano-Sahelian as a result of the climatic conditions, but the vegetation often takes the very characteristic form of *brousse tigrée*.

These conditions are unfavourable to widespread agriculture because, in order to establish itself, such agriculture would have to profit from all the edaphic microsituations allowing the formation of deeper, more altered soils with a better moisture economy. The most unfavourable zones are used for grazing, but the productivity of the pastures is very low. These are, therefore, situations that are naturally very sensitive. Even in the absence of all accentuated human pressure, the combination of different environmental parameters, i.e., rains, geomorphology, pedology and bushfires, are sufficient to provoke and maintain serious degradation. The hardpan plateaux, almost denuded of ground cover and with meagre shrub vegetation, or certain very eroded wind polished hardpans, are on the edge of full desertification and represent the extreme state of such situations.

Such zones occupy large areas in eastern Senegal, all of western Mali and eastern Burkina Faso, the latter extending to Niger.

On the mountainous massifs of Mauritania, the Adrar and the Aïr, the previously discussed edaphic conditions (fresh mineral-rich soils or little-altered soils) are accentuated by the sub-desert climate and relief. Life is concentrated in the valleys and around water sources.

Humid zones

In humid zones, the soils can be quite varied (diversified alluvium, vertisols) with an equally large diversity for potential development. This is particularly the case for the Inland Delta of the Niger and the large river valleys. Saline vegetation (mangrove, tanne) is located in Senegalese zones subject to tides, in Casamance and the Saloum valleys.

Different agricultural schemes have had serious repercussions in this region. In the large valleys traditional systems were based on recession agriculture, but the reduction of river levels has considerably restricted this practice, and it has been replaced or is practised along with irrigation schemes (see Chapter 5).

Here again, work on the large irrigated perimeters, combined with the drought, has considerably altered the environment. In saline areas, seasonally flooded depressions have long been developed as rice fields. However, the salinity of the water-tables and the reduction of river levels have provoked a serious decline in the mangroves and led to the abandonment of rice cultivation in certain sectors (Michel, 1990).

CONCLUSION

We can see finally the degree of diversity in the Sahelian environment. It is due to differences in climate, soil, vegetation and development, although a common point is the extreme precariousness of the area to all impacts, whether natural or human-induced. This explains the seriousness, often dramatic, of the present crisis which is linked as much to a persistent drought as to an extreme and excess pressure on the ecosystems. As mechanisms interact and processes are driven forward, the environment is increasingly undermined and attains levels of near-desertification that are difficult to reverse with the present systems of exploitation. These systems, based on traditional practices and balanced by natural mechanisms of regulation, are no longer compatible within present climatic, economic and social contexts. Therefore, new forms of natural resource management must be found and put in place, but in each case these must be adapted to the diversity of local conditions, of which a complete understanding is an indispensable precondition.

2

THE DEMOGRAPHIC ISSUE IN THE WESTERN SAHEL: FROM THE GLOBAL TO THE LOCAL SCALE

Claude Raynaut (with the collaboration of Pierre Janin)

Of all the phenomena invoked to explain the environmental problems occurring on the African continent, and in particular the countries of the Sahel, the continued and rapid increase in population is the one most often cited:

> When examining the feedback effects between populations, agriculture and environment, we are speaking of Africa's survival. . . .
> First feature: the population explosion, the most striking phenomenon presently and in the future of the African continent. . . .
> The impact of this explosion? It is already terrible today and predictions suggest catastrophe if nothing is done.
>
> (Falloux, 1992: 1–2)

How do the countries in this study fare? Table 2.1 summarises the actual and estimated demographic data currently available on the subject. It reveals that over a forty-year period populations have increased, on average, two and a half times. This table does not give us the full picture, however, for during the same period, the average annual rate of growth has also continued to increase. As Table 2.2 shows, this configuration of growth can be labelled as 'over-exponential' (Meyer, 1985). Such an evolution is not limited to the countries cited here. In all of Africa the rate of demographic growth is more rapid than in the rest of the developing world (IUCN, 1989). Some estimates predict that by the year 2020 the total population of the continent will reach 1.5 billion inhabitants, or 20 per cent of the world's population as compared with 12 per cent at the current time (Loriaux, 1991).

This acceleration in total growth rates is fundamentally due to a demographic dynamic which characterises the first phase of the process of change known as the 'demographic transition', in which the mortality rate decreases in an appreciable way, with lifespan from birth going from about 33 years during the 1950s to more than 45 years in the 1980s (IUCN, 1989; J. Schwartz, 1992), while fertility tends to increase. The next stage is an

Table 2.1 Population growth in the western Sahel (millions)

Year	Senegal	Niger	Mauritania	Mali	Burkina Faso	Total
1950	2.5	2.9	0.8	3.9	3.7	13.8
1960	3.0	3.2	1.0	4.6	4.3	16.1
1970	4.0	4.1	1.2	5.7	5.1	20.1
1980	5.7	5.3	1.6	7.0	6.2	25.8
1990	7.4	7.1	2.2	9.4	7.9	34.0

Table 2.2 Mean annual population growth rate for the western Sahel (by decade)

Decade	1950–60	1960–70	1970–80	1980–90
Annual growth rate (%)	1.6	2.5	2.8	3.2

Source: FNUAP, 1988.

adjustment to a new demographic stability, characterised by a decrease in both mortality and fertility, which has yet to begin in the western Sahel (Figure 2.1). Some models anticipate this stage during the current decade, yet there is little evidence of this being seriously confirmed. Hypothetically, and taking into account the effect exercised by the 'demographic momentum' (growth generated by the entry of youths, currently less than 15 years old, into their period of fertility), natural increase of the Sahelian population should follow an accelerated rate well into the twenty-first century.

DEMOGRAPHY AND ENVIRONMENT, A CONTROVERSIAL ISSUE

In the countries covered by this study, past and future population explosions are, therefore, an incontrovertible fact. However, the consequences that this evolution could have in terms of economic development and environmental dynamics remain subjects for debate.

Proponents of an ecosystems approach insist that an uncontrolled growth of population presents a severe challenge at the global level and to the forces that govern the biosphere. This is, in particular, the thesis supported in the introduction of the 1988 United Nations Population Fund (UNFPA) report on the state of the world population:

> By his activities, man submits nature to constraints which translate into an ever heavier toll on the natural resources essential for all life: water, air and land. In the developing world, a slowing of the growth and a more balanced distribution of the population would alleviate the economic pressures occurring on agricultural lands, energy resources, catchment basins and forests. . . .

(FNUAP, 1988)

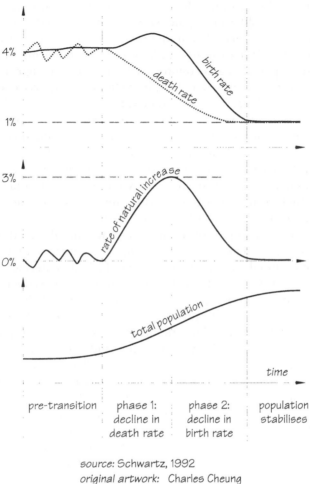

source: Schwartz, 1992
original artwork: Charles Cheung
computer graphics: Phil Bradley

Figure 2.1 The demographic transition

Such reasoning seems to apply even more to the Sahelian and Sahelo-Sudanian areas, where primary production is low and where the balance between demand and the availability of resources is susceptible to disruption faster than elsewhere. A constant increase in the demands on these natural systems would drive them inevitably past their internal regenerative capacities and begin the process of degradation. The key notion here is that of 'carrying capacity' which postulates the existence of an objective extraction threshold beyond which the intrinsic balance of the ecosystem is put in danger. In the Sahel, precipitation is the principal factor limiting primary production, and the dry phase of the last twenty years has again lowered this threshold, causing the baseline for crisis to shift downward.

Whatever the evidence, this model reveals its limits when submitted to close inspection. If the existence, at least in heuristic terms, of a tolerable threshold of 'exports' for an ecosystem is difficult to challenge, the question should be asked whether it is legitimate to link exports exclusively to population. Actually, at least two other variables also carry a determining influence on the intensity of stresses a society exerts on its environment.

First, there are the technical conditions for the exploitation of resources. There is no need to dwell on this point here since it will be discussed again later (Chapters 6 and 7), but it should simply be remembered that certain practices, notably in this case the poorly controlled introduction of new crops and tools, can appreciably modify the intensity of the impositions exerted on the environment (Bosc et al., 1990).

Second, there is that part of production that is not destined to meet the direct or indirect basic needs of the population. This appears and develops in response to stimulations and constraints which are external to the social and physical reproductive needs of human groups, such as the introduction of new consumption goods, price fluctuations and speculative behaviour, as well as politico-administrative pressures. If, for example, in Niger in the late 1940s the sale of 20 kg of millet could pay the tax obligation for one person, by the beginning of the 1970s that person would need to sell 90 kg to meet the same obligation (Raynaut, 1977b: 161). A similar regression in the terms of exchange between agricultural prices and imported goods (incidentally, pursuing a trajectory much greater than the rate of demographic growth) is reported for Senegal (Bosc et al., 1990: 151). At identical population levels, the demand on primary production can vary considerably under the effects of such parameters. The average per capita need for material and energy is thus far from fixed, and the environmental impact of demographic growth is difficult to isolate from other factors with which it combines and reacts. A uni-factorial explanatory approach therefore risks a failure to comprehend the complexity of the concrete dynamics and diversity of local situations.

Another criticism can be levelled at this demographic model of ecological crisis. It addresses the fact that the population level is considered in exclusively negative terms, i.e., only as a function of the level of extraction that it supposedly causes. More realistically, it should also consider the reverse, and more positive, aspect of the size of a population, i.e., the potential workforce that it represents. An increase in population not only translates into an increase in needs that must be satisfied but it also translates into more able bodies. In this latter scenario the workforce can constitute a limiting resource with respect to the application of more intensive forms of exploitation, protection and environmental management (Mortimore, 1989: 209; Tiffen et al., 1994). Under certain conditions, then, high human densities can, and with certain limitations do, constitute a favourable factor in reaching agro-technical thresholds and establishing solid relations between a human group and its environment. Boserup developed this analysis in 1970 (Boserup, 1970). However, it should not be applied in a mechanical way any more than should the contrary theory. Its validity is linked to the existence of connected factors which favour technological change. Nevertheless, the examples of the Dogon in Mali, or that of the Sérer in Senegal, have long illustrated the fact that even in the same Sahelo-Sudanian milieu, high population densities are at the very least compatible with the existence of balanced production systems (Gallais, 1965; Lericollais, 1972).

Even if only for mobilising public opinion, we must still guard against retaining a model that is too mechanically linear in its description of the impact demographic factors have on the emerging Sahelian crisis. Certainly, at a continental or Sahelo-Sudanian scale, the general demographic tendencies are incontestable. Nevertheless, having taken this into account, there is little here to illuminate the realities on the ground or to elucidate the articulation between population and the environment. The demographic question, like many others, cannot be dealt with validly using global tendencies or averages – important and significant as these figures may be. The population of the Sahel is distributed throughout the region in a very unequal fashion. From one location to the next, both human densities and population growth rates can vary considerably. This variability, combined with that which affects natural conditions and technological practices, gives rise to a great diversity of local situations, where prognoses on environmental risks are far from even. To advance the debate and to progress in an analysis of the diversity found in the Sahel it is therefore necessary to go beyond generalities. We must show how the demographic push varies from one region to another. We must identify the core areas of high population density, along with those that are almost deserted. We must locate the centres of rapid growth, where a positive balance between in-migration and out-migration reinforces the trend of natural growth, as well as those sectors of little dynamism where, by contrast, a negative migration balance diminishes the impact of natural growth.

We now turn to the creation of a map which can illustrate this demographic diversity. It is a first step in an attempt to identify the diversity of local environmental situations.

VARIED POPULATION DENSITIES

The discontinuity in population distribution within the Sahelo-Sudanian zone is not a recent discovery. Observers have long noted the existence of alternating centres of marked human concentration and vast expanses which are sparsely populated. What is missing in support of this very general observation is, however, a map that shows this distributional contrast and that delineates the nuances and the transitional areas which separate these extreme situations. Such information is sometimes available for certain countries, but the year of data collection often varies, as do the ways in which the information is presented graphically. It therefore appeared necessary to construct a new document, based on demographic data which is as homogeneous and as recent as possible.[1] Fortunately, in the second half of the 1980s, the CILSS countries conducted national census campaigns in which data were gathered in a relatively standardised manner. It is these data that have been used to illustrate the distribution of the rural population in Figure 2.2.

Table 2.3 summarises the sources used for the different countries covered by the study. A first and immediate problem concerns the spatial unit of representation in these data. Those results which are easily accessible are aggregated according to the administrative divisions specific to each country. From one case to the next, the level of detail thus obtained is not identical, thereby introducing an element of discontinuity from one national border to the next. In certain cases, the area covered by one administrative unit is so vast that the figure for the average density in no way expresses the heterogeneity of the actual population distribution. This situation is particularly apparent in all the northern administrative divisions of the countries mapped, i.e., those which incorporate a large portion of the Sahara.

These biases have been overcome by means of both an interpretation and

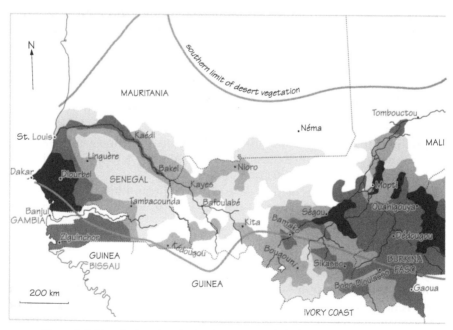

Figure 2.2 Population density in the western Sahel (continued opposite)

Table 2.3 Sources used for mapping rural population densities

Country	Year of census
Burkina Faso	1985
Mali	1987
Mauritania	1988
Niger	1988
Senegal	1988

Source: National censuses

an adaptation of the approximate image furnished by the spatial projection of the census data. On the one hand, regroupings and generalisations were carried out in those sectors of the map where the scale of coverage produced a finely detailed mosaic – which is difficult to compare with results obtained in other sectors (where the information base was less detailed).

On the other hand, in the sparsely populated sectors, the demarcation of boundaries was made more precise by taking into account complementary information about the localisation of habitation (the Institut de Géographie Nationale (IGN) maps in particular).

The results obtained through these efforts are presented in Figure 2.2. The classification of densities developed in this map was determined in a manner that highlights large homogeneous areas and, therefore, favours

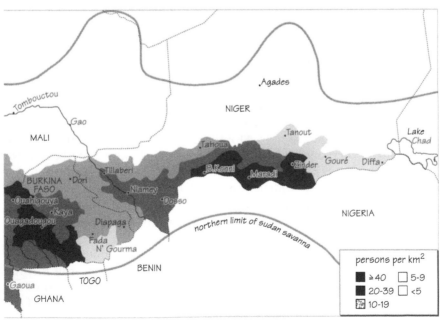

original artwork: Charles Cheung, Pierre Janin
computer graphics: Phil Bradley

the readability of the document. What is obtained is a picture of population distribution in the Sahel, from Lake Chad to the Atlantic Ocean. This map raises a number of issues.

Densely-populated sectors

First we will highlight the existence of three large population sectors which run from west to east and give the pattern to land occupation.

1 *The Senegalese sector* Centred around the Cape Verde peninsula and the groundnut basin, the Senegalese sector reveals a halo of decreasing population density bands, which faithfully reflects the historical progression of human occupation towards the east (Ferlo, eastern Senegal) and towards the south (Casamance). This has been stimulated by the growth of commercial agriculture and favoured by communication routes (in particular the railway, towards Tambacounda and eastern Senegal). For its part, the Senegal River spearheads an extension, which, while gradually tapering, greatly prolongs the high density zone towards the east.

2 *The Malian-Burkinabe sector* This is the most expansive sector. Located between the Niger River (Bamako to Mopti) and the eastern slope of the Mossi Burkinabe plateau (Ouahigouya to Kaya), it occupies almost all of the central part of the area mapped, creating a centre of demographic gravity. More than half the rural population of the five countries in this study is concentrated there. This demographic magnate has a largely bipolar disposition with two principal cores of population (the smaller of which extends south of Segou and the larger one on to the Mossi plateau) linked by a large zone where the densities, although lower, are always greater than 20 inhabitants per square kilometre. The structure of this section is, moreover, much less uniform than in Senegal. In Mali, several secondary concentrations are attached to this large sector: in the Central Delta of the Niger, around Mopti; on the Dogon plateau, near the Burkina border; and, although less pronounced, near Toumboctou (to the north) and Sikasso (to the south). In Burkina, however, there is yet again a halo effect, which is marked by densities that decrease in proportion to the distance from the Mossi heartland. This partitioning undoubtedly reflects a complex population dynamic resulting from the combination of a number of factors to which we will return in an analysis of what, at this point, only appears as a static spatial configuration.

3 *The Nigerien sector* It would, perhaps, be more accurate to call this the 'Nigero-Nigerian' sector, since what is found in Niger certainly represents only the northern extension of a zone of population concentration whose pole is located in Nigeria and is, therefore, not visible on the map (Ninnin, 1993). Nevertheless, in the truncated version presented here, the structure is similar to that found in Senegal, with bands of decreasing

44

density progressing out from a highly concentrated core of population wedged on the Nigero-Nigerian border. The latitudinal stratification of these bands illustrates the northern movement of sedentary agricultural populations (under the combined effects of demographic growth and the diffusion of groundnut cultivation). This tendency is most certainly combined with the return movements of pastoral–nomadic populations towards the south. We will return to an analysis of this dynamic later (p. 50). For the moment it should be noted that the Nigerien sector is sandwiched between two extremes: the desert barrier, which prevents the pursuit of a northward expansion in any form, and a major political and economic frontier to the south which presents a barrier to population mobility.

In the end, these population concentrations comprise only a modest part of the zone which extends south of the vegetation limit of the desert. Not only is this zone covered in its northern half with expanses only sparsely peopled with nomadic pastoralists, but further south there also exist vast areas of 'low pressure' demographics surrounding sectors of high concentration. These latter can be classified into two major categories: the 'demographic voids' (less than ten inhabitants per square kilometre), and the 'intermediary zones' (between 10 and 20 inhabitants per square kilometre).

The demographic voids

The first and largest of these stretches across 1,000 km from Senegal to Mali. Starting from the Ferlo, it covers a major part of eastern Senegal, the upper basin of the Senegal River (upstream from Kayes) and joins, in the east, the Interior Delta of the Niger River. Such a vast expanse includes, of course, a great variety of climatic and natural situations; a fact which raises the question of the historical causes which have let these areas stand aside from the demographic push which has otherwise populated this part of the African continent (see Chapter 3).

A second 'demographic void' covers the eastern third of Niger, to Lake Chad. A striking feature here is the sudden rupture which is evident at the eastern edge of the densely populated Nigerien sector where, without any transition, densities fall from more than 40 inhabitants per square kilometre to less than ten. Whatever the causes of this repellent force on human occupancy, they must be demonstrably strong to provoke such a dramatic fracture.

The intermediary zones

Strictly speaking, one of these intermediary bands is located outside the Sahel-Sudanian region because it extends, on both sides of the border which separates Mali from Burkina Faso, south of the ecological limit of the

Sudanian savanna. The internal solidarity among regions within national boundaries, however, prevents them from being excluded. Regardless of their specificities in the natural and human world, both southern Mali and southern Burkina are integral parts of larger territories. To separate them would be artificial. What characterises this geographic region is, in fact, the discontinuity of its population distribution (which the map can only partly reveal). Even when mean densities appear low, pockets of higher density may exist. This is particularly the case in the Sénoufo country, close to Sikasso.

The 'Burkinabe–Nigerien' sub-region stretches from the latitude of Dori to that of Fada N'Gourma and crosses diverse physical environments. In fact, what is found here, much more than an autonomous zone, is a transitional band linking the 'Malian–Burkinabe' and the 'Nigerien' sectors. This viewpoint will be confirmed when we explore the dynamics of growth.

Two major observations emerge from this discussion. The first concerns the diversity of local situations that are revealed and the importance of the contrasts that distinguish them. It should be clear from this that treating the demography of the Sahel and its impact on the environment at an aggregate and global level risks dangerous oversimplification. If the population concentrations observed in the 'population cores' can induce high levels of extraction, it cannot be the case as concerns the vast expanses which remain almost untouched by permanent occupation. Some observers conclude that these areas represent a development potential, one that must be exploited within the perspective of establishing a geographic re-equilibrium with respect to human pressure. Thus we find the *terres neuves* approach, which conceptualises the exploitation of Sahelian resources in global terms, i.e., in terms of interregional complementarity (Hunter, 1977; Rochette, 1986). This is not a new proposal (it inspired the direction that l'Office du Niger took in the 1930s – see Chapter 5) and it is no less important today. What remains, of course, is to assess the obstacles that must be overcome if the potential of the *terres neuves* is ever to be realised. This means that an effort must be made to understand the reasons behind the current population distribution and heterogeneity, as well as to take into account the large contemporary trends in population dynamics.

A second point affirms the general irrelevance of national boundaries with respect to population distribution. Several of the major geo-demographic groupings that have been identified straddle two countries and only gain their total significance when understood in this broader context. Thus the impact of high human densities is continuous across both sides of the Burkina Faso–Malian border, while the problematic of the *terres neuves* is as applicable to eastern Senegal as it is to the western edge of Mali. Beyond political partitioning (as significant as it has become in certain respects) this statement invites further study of the major determinants which have long guided the distribution of population, which the current state of land occupation represents and which, in the final analysis, is its legacy.

VARIABILITY OF POPULATION GROWTH

From our earlier observations it must now be clear that the examination of the spatial distribution of rural population remains incomplete without an understanding of geo-demographic dynamics. Documentary sources which have been used, and which concern migratory movements within Sahelian countries, are very incomplete and, on their own, make it difficult to draw a general panorama of the current situation. To obtain a better representation, an alternative approach has been adopted which rests, primarily, on a comparative analysis of demographic growth rates which can be considered as indicators. In areas where growth is higher than the average found in the Sahel as a whole, we have estimated that natural growth is being reinforced by external input. Conversely, in those areas where rates of growth clearly fall below the average, emigration is deduced. This reasoning implicitly sets natural growth as an overall constant in the areas covered by the study. Although this working hypothesis probably cannot be verified when comparing local situations with extreme characteristics (refugee nomads as opposed to agriculturalists or city dwellers) it is an acceptable approximation when addressing populations with relatively homogeneous living conditions. To confirm and complete the indications derived from this comparison of growth rates, available information concerning migratory movements between rural zones has been used. Although most often limited, they can help in the understanding of the dynamics to which the identified areas are subject.

This approach presents some difficulty in its application. Here again there is a problem of insufficient sources. National censuses have been infrequent in the Sahelian countries and the periods covered are usually not contemporaneous. It has, therefore, not been possible to draw on data which can uncover long-term trends. With some variation between countries, the period that has been used covers approximately the decade preceding the given census year (Table 2.4). Figure 2.3, therefore, does not reflect the genesis of the actual distribution of the population, but instead reflects the evolutions that currently accompany it.

In order to produce a simplified graphic representation, the rates have been grouped into three categories: the highest level (more than 3 per cent

Table 2.4 Periods for growth rate calculation

Country	Period
Burkina Faso	1975–85
Mali	1976–87
Mauritania	1977–88
Niger	1977–87
Senegal	1976–88

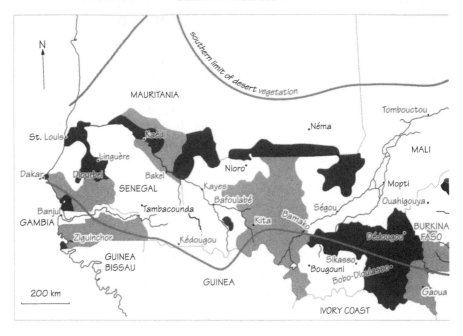

Figure 2.3 Demographic change in the western Sahel (continued opposite)

per annum) corresponds to what will be termed 'poles of attraction' and the lowest (less than 2 per cent per annum) corresponds to the zones of low growth or 'demographic depression'.[2] In the intermediate category (between 2 per cent and 3 per cent per annum) rates of growth balance out, representing neither migration gains nor losses in the population.

Poles of attraction

Several areas showing high rates of demographic growth can be identified in Figure 2.3. Two extensive areas of strong growth are immediately apparent, on both sides of the Mossi Burkinabe plateau.

The periphery of the Malian–Burkinabe sector

The centre of attraction which extends to the west of the plateau is well known and has been the subject of a number of studies undertaken during the 1970s (Lahuec and Marchal, 1979). In summary (and while certainly oversimplifying a complex demographic reality) this centre plays the role of an outlet for the zone of high demographic pressure established in the Mossi country. Historically, this movement was supported by the conquering forces of the Mossi kingdom. After the colonial conquest, it was

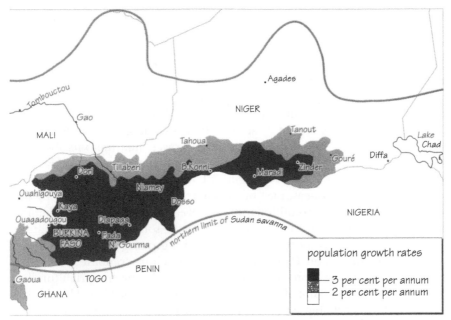

original artwork: Charles Cheung, Pierre Janin
computer graphics: Phil Bradley

amplified by those desiring to escape the constraints exercised by the new power, such as taxation and requisitions (Marchal, 1986). Since independence, it has been strongly accentuated and has made gains towards the west and south as the regions of origin have become saturated and their agricultural potential degraded. Efforts made by the national authorities to create receiving areas (such as the Kou valley) have, so far, only had a limited impact on the size and direction of these migratory flows because, for the most part, they are not reached by these planned interventions. Data from the 1985 census suggest the emergence of a movement of settlement of migrants towards the south, i.e., beyond Bobo-Dioulasso. As Rochette has observed, the effort of the Bobo and Dagari autochthonous populations can no longer contain the arrival and settlement of these Mossi immigrants; to use his expression 'the dam has given way' (Rochette, 1986: 13). In reality, the vast area of high growth that has been described here is not limited to the Burkinabe territories but in fact transgresses across the border and extends towards the central part of southern Mali. Nevertheless, despite this geographic continuity, the migratory movements that affect this region of Mali are, in nature and origin, different from those just described. Precise data are lacking, but the available information suggests that they constitute a withdrawal of populations from the northern regions which have been hit hardest by drought. Such people are moving south in search of more

49

favourable climatic conditions and are responding to the attraction of the cotton zone (Raynaut, 1991a).

Further to the east, there is a second area of strong demographic growth, beginning in Burkina and extending into Niger. This dynamic arises from the progressive in-filling of those areas of low population density which separate the Malian–Burkinabe and the Niger population sectors. The long-standing and sparsely populated areas between Mossi country and the Niger River are slowly being settled by agriculturalists coming from both the east and the west. Here, the movement is much slower than the one observed on the Malian side of the western border separating Mali and Burkina. The migration from the eastern margins of the Mossi country is ancient, especially in the north-east part of Burkina, but it has been slowed by the unfavourable climatic conditions which have prevailed in the Burkinabe 'Sahel', especially over the last twenty years. In the south-west, this migration continues hesitantly, especially towards the rice growing zones located east of Fada N'Gourma and Diapaga. These fluctuations combine with other currents, originating from the east, which drive Nigerien agriculturalists towards the western bank of the Niger River and into Burkina. This double movement has probably accelerated since the census on which this study is based, due to the completion of the Niamey-Ouagadougou paved road, which has contributed to the opening up of this region.

Central Niger

In the central part of Niger, the demographic progression at work is perceptibly different. It rests on the succession and then the merging of several factors. During the colonial period, and for reasons approaching those described in the case of the western Mossi plateau (i.e., demographic pressure, the desire to escape colonial control), agricultural populations expanded northwards (as noted above and which translated into a series of bands of decreasing density). This movement was encouraged by favourable climatic episodes. Paradoxically, it was strengthened locally by a French administration that sought to contain Tuareg disturbances. These zones are exposed to high climatic risk and have therefore been badly hit by the repeated droughts of the last 20 years. As a result, the population that had settled in the region was forced into a progressive outward and southward migration. In addition to this, the attraction posed by the Nigero-Nigerian border, with the multifarious 'traffic' that it generates, has created a dynamic parallel economy, while at the same time hindering the southern expansion of a population through the turning back of citizens of the Niger Republic, or even the closure of border posts.

If the major trends described here continue (the growth of high demographic pressures in the centre of Niger, an encroachment on to sparsely populated areas which straddle the Burkina/Niger border, and increasing densities of agricultural occupation in western Burkina and southern Mali) one can anticipate the future merger of the two major highly-populated sectors and the creation of an enormous zone of high demographic concentration extending from Bamako to Zinder (Niger).

The margins of the Senegalese Groundnut Basin

Compared with the situations which have just been described, the far western Sahel appears quite different. Despite the ancient and intense demographic pressure that is found in the Senegalese Groundnut Basin, the increase that is currently observable on its margins is far from being as strong as those that exist elsewhere in similar situations. After a long history of rapid and continued expansion it appears now that a slowing trend can be observed. Figure 2.4 illustrates well the past evolution of the groundnut basin and the intensity of its progression east for half a century.[3]

Even today, the concentric bands of population density still bear witness to such a movement. Nevertheless, a comparison of the 1976 and 1988 censuses suggests that the rate of expansion is currently slowing. Even if the drive is being maintained in the Linguère region, further south (from Diourbel to Tambacounda) we discern a stagnation or even a fall in growth. It seems that informal settlement in the *terres neuves* of eastern Senegal is setting the pace, in spite of official efforts to organise it. In fact, the demographic dynamic of the rural zones of western Senegal cannot be understood in isolation because, perhaps more so here than elsewhere in the Sahel, it must be linked to another major phenomenon to which we will return later (p. 77) – that of urban growth.

Outside of these major centres, several very localised phenomena are observable which have little relevance if considered on a zonal scale. The data obtained for Mauritania are so unreliable that they should be

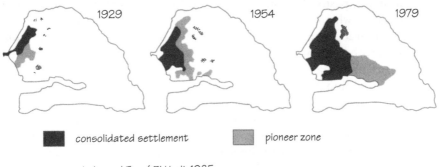

consolidated settlement   pioneer zone

source: Lake and Touré El Hadj, 1985
original artwork: Charles Cheung
computer graphics: Phil Bradley

Figure 2.4 The expansion of the Senegalese Groundnut Basin

interpreted very cautiously. The band of strong growth which extends the length of the Senegal River, on both sides of Kaédi, may illustrate the attraction of the irrigation schemes developed in the 1970s, while, further east, the Mauritanian–Malian border seems to create a draw and retention effect which is similar to the one noted for Niger, along the Nigerian border. In Mali, the pocket of growth observable south of Bafoulabé is indicative, in this case, of the presence of the Manantali Dam.

Zones of demographic depression

In contrast to the large centres of attraction that have just been identified, there are vast expanses which reveal comparatively low growth rates. This growth rate differential has diverse, possible interpretations, according to particular situations. Nevertheless, it is a sign of an overall recomposition of the entire population distribution and reveals large population movements that are at work.

The Sahelian band

A first observation concerns the relative depopulation of the strictly Sahelian band. The coarseness of the growth categories adopted in the map do not enable us to grasp the real differences that are observable between growth rates. In the majority of cases, annual population growth in the northern regions of Mauritania, Niger and Mali is equal or less than 0.5 per cent, which amounts almost to a situation of stagnation. Care should be taken in interpreting these evaluations since they concern regions which are very sparsely populated from the outset. Nevertheless, the tendency seems real. Several causes can be suggested. First of all there are the effects of the successive droughts of the last twenty years on the birth and mortality rates of nomadic Fulani and Tuareg populations.[4] Furthermore, pastoralists and agropastoralists have migrated towards the south; those who have lost their livestock and can no longer pursue traditional practices as well as those who have been successful in migrating far south with the few heads of cattle they were able to save (hence the Nigerien Fulani that are now found in Nigeria and Central Africa). Also gone are the pioneer agriculturalists who, as we noted with the situation in Niger, had earlier sought opportunities in sectors of high climatic risks. This movement often occurs in the direction of sparsely populated agricultural zones, contributing to the high rate of demographic growth identified earlier (notably in the south of Mali), but it also poses serious problems for immigrant populations who must adapt to new environments and learn different production practices (see Chapter 5). This exodus also ends in the cities, where floods of uprooted people converge and try to settle. Added to this phenomenon, during the 1980s, was the impact of political tensions between the Malian and Nigerien powers and their Tuareg minorities, resulting in some young men leaving for Algeria and Libya (Bourgeot, 1990).

The Mossi plateau

A very different situation is unfolding in north-west Burkina Faso, on the Mossi plateau. Here, weak growth rates confirm a demographic saturation, as mentioned earlier, which is particularly acute in the northern departments, especially in the Yatenga around Ouahigouya, where the rate is estimated to be around 0.5 per cent. Here, low densities are the counterpart of the expansion observable at the plateau's periphery. Moreover, this stagnation is also due to an exodus that, for several generations, has drained young men from the region towards the plantations and port cities of the coastal countries. These absences can last several years and their effect reverberates throughout the home populations, especially because the returning migrants do not entirely fill the gap left by new departures (for some migrants do not return), while those who do return have a tendency to settle in the large cities (Sirven, 1987). Other geographic areas are also experiencing the phenomenon of decreasing demographic pressure in some ways similar to that of the Mossi plateau. This is the case in Mali, in certain sectors of the Interior Delta (upstream from Mopti), as well as in the Dogon country (between Mopti and the border with Burkina Faso). It is also the case in Senegal, in a large section of the groundnut basin.

The Upper Senegal basin

The vast area of demographic depression which extends throughout eastern Senegal and western Mali constitutes a third major example of low growth. Contrary to what is occurring on the eastern edge of the Malian–Burkinabe population sector, these empty spaces exert no attraction. On the contrary, the gap relative to the areas of high population continues to widen. There is here a phenomenon of resistance which, especially in Mali, can be related to the existence of strong natural constraints, as will be shown later (p. 58), but which indicates above all the presence of socio-economic obstacles which we will identify below (p. 67).

The south of the large Malian cotton basin (in the region of Bougouni, Sikasso) is the last major example of demographic depression. The reality that unfolds here is particularly complex. First of all, we note the impact of onchocerciasis which, for a long time, has proliferated along the region's rivers (see Chapter 3). Despite favourable results over the past few years battling against the simulids (the vectors of this parasite), the ravages caused by this disease among the local population continue to discourage many potential immigrants (Raynaut, 1991a). Although slowed by this constraint, a settlement movement is nevertheless occurring which originates particularly from the Bend in the Niger and the Dogon country. This settlement is an extension of the movement, the effects of which were noted in the cotton basin extending immediately to the north. This human influx might have favoured a strong demographic growth had it not been counterbalanced by a simultaneous flux of departures which perpetuate, in the autochthonous population, an ancient tradition of migration (see Chapter 3). As in the case of the Mossi country, these are temporary migrations for which the rate of return does not fully compensate for new departures. In the end, the migratory flows, inward and outward, tend to balance out – resulting in a modest growth rate. This geographic area has a strong, hidden demographic dynamic that, if current factors were to be modified, could be the starting-point for a rapid growth in population. Projected efforts to develop resources and infrastructure in this region by the Compagnie Malienne de Développement des Textiles (CMDT) could very well be a triggering factor at this point.

CONCLUSION

The preceding analyses highlight well the fact that establishing a sharp opposition between 'centres of attraction' and 'zones of demographic depression' could lead, in many cases, to disassociating two sides of a single reality. Under the effects of a certain number of pressures and stimulations, among which are those exercised by the harsh climatic conditions of the past two decades, a reshaping of the population distribution in the Sahel–Sudanian region is occurring. It is only in localised areas that this movement resembles a gradual diffusion, evening out density differences and producing a more uniform land occupation in the surrounding areas. If considered in its totality what is apparent, on the contrary, is that this movement is subject to discontinuities, i.e., it privileges certain territories and leaves others on the margins. Highly differentiated local situations emerge and, in order to analyse them, explanations must be drawn which are not always of the same nature. A complex dynamic is at work, where actions by a variety of factors combine. It is to the principal of these factors that we now turn, with the goal of illustrating the importance and limits of their influence.

NOTES

1 Since the completion of this work a document has been published on population density distribution throughout West Africa (Ninnin, 1993). It confirms our own conclusions.
2 By 'depression' we mean a growth which is obviously lower than that of the surrounding areas.

3 For a detailed study of the most recent phase in the expansion of the groundnut basin (1954–79) see Lake and Touré el Hadj (1985).
4 We know of no specific study on the impact of drought on birth rates of nomadic populations. However, its effects on the birth rates of sedentary populations have been analysed (Faulkingham, 1977a, 1977b, 1980).

3

POPULATIONS AND LAND: MULTIPLE DYNAMICS AND CONTRASTING REALITIES

Claude Raynaut

It would be impossible to reduce the variety of specific situations described earlier to several simple factors. Thus, the purpose of this chapter is simply to emphasise the role played by the most prominent among them. Nevertheless, we will certainly fall short of giving, in a satisfactory manner, an account of all the local situations which can be encountered in the Sahelian zone. At least we will attempt to introduce a little comprehension to what has remained, until now, a simple statement of diversity.

THE NATURAL ENVIRONMENT: ONLY A PARTIAL EXPLANATION

When one attempts to grasp the principles which rule over the distribution of people in space, there is a strong temptation to search for a single interpretative theme among the many constraints and incentives of the environment. This is even more true when, as is the case with Sahelian societies, we are faced with technical systems that have few methods for transforming the environment and are, therefore, forced to adapt to it. Analysis only partially confirms this hypothesis. If the physical and natural environment influences the patterns of human settlement, it is far from ruling over it in any mechanical fashion. In the Sahelo-Sudanese context this is verified by the three major domains where environmental factors can weigh heavily: climate, soils and topography, and parasitic endemic diseases.

Climatic zones

It is clear that the average precipitation increase that occurs with increasing distance from the Sahara accounts for the major contrast between the Sahelian belt, with its extremely low population densities, and the southern section where the majority of the population is concentrated. In the same way, the worsening climatic conditions of the last 20 years explain, at least in part, the demographic decline in the northern zones. The agricultural populations who ventured into these zones around the middle of the century are now returning to the south in search of land, while the many pastoral nomads, who saw their flocks decimated several times over, no longer have the means to pursue their old activities. More locally, the aridity that characterises regions such as the Senegalese Ferlo, the Burkinabe Sahel and far eastern Niger explains to a large extent their long-lasting under-population as well as the resistance which currently prohibits a more sustained occupation.

However, if we overlook the wide divergence between zones where agriculture is possible without inordinate risk (thus allowing concentration of population) and the vast territories dedicated to extensive herding (where human occupation is very sparse) and if we try to apply the climatic explanation to a more detailed level of analysis, it rapidly becomes apparent that it fails to account for local diversity. It is clear that the long-term presence of the immense demographic depression that covers eastern Senegal and western Mali cannot be interpreted solely on the basis of climatic considerations. For example, the aridity of the Kayes (Mali) region is no more severe than that of the Yatenga Burkinabe. In a general manner, therefore, the alternation between densely occupied zones and under-populated ones (encountered along the same latitudes from Senegal to Chad) cannot be explained by climatic considerations. While climate permits or prohibits certain modes of exploitation in the area, it alone does not determine the presence or absence of human occupation.

Soils, topography and geology

When considered over the whole of the zone relevant to our study, edaphic conditions bear witness to a heterogeneity that Figure 1.7 (pp. 32–3) reveals. Let us review its major features. Soils offering very good potential for agriculture are rare, i.e., vertisols of the large valleys and brown eutrophic soils on the alluvium. Brown arid soils on ancient ergs cover most of the northern zone devoted to pastoralism. In the agricultural band, poor tropical ferruginous clayey-sandy soils dominate, which are poorly struc-tured and sensitive to erosion. In a very localised manner (central and east of the Niger, west of the Senegal) these soils rest upon aeolian deposits. Everywhere else, they cover a lateritic hardpan. Where soils are very thin or where the hardpan outcrops at the surface, agricultural exploitation can become impossible. In this case, as in that of central and southern Mali's great sandstone plateaux, only those sectors where colluvial deposits are formed, as well as the valleys themselves, are suitable for agriculture.

In the entire zone under consideration, not only could the agricultural potential never be very great, but specific constraints can sometimes be paralysing. However, human occupation only partially reflects this diversity and it is not always possible to establish a line between the concentration of people in some areas or their relative absence in others, and the variations in edaphic conditions. If the densely populated Senegalese and Nigerien groundnut basins (Maradi/Zinder region) correspond to areas of soils on aeolian deposits (only slightly fertile yet with good hydrological capacity), this correspondence is only approximate, since the patterns of population density are far from coincident with those of soil types. The exceptionally high human concentrations on the Mossi plateau, for their part, do not in any way reflect particularly favourable conditions for agricultural exploitation even if, locally and especially in its northern section, more fertile soils are found. This observation is even more true concerning the sandstone massifs in the Dogon country. As to the contrast between sparsely occupied southern Mali and the high densities observed in eastern and south-eastern Segou, it can hardly be explained by the existence of contrasting edaphic potentials. Such remarks do not suggest that Sahelo-Sudanese agriculturalists are indifferent to the characteristics of the soils which they work. Just the opposite – they are sensitive to the fine nuances which disappear at the generalised level depicted by a map, such as the one upon which our analysis rests. They exploit local micro-environments, linked to topography and hydrology, which allow them to settle where conditions are otherwise unfavourable. Within a regional framework, Marchal has clearly shown that in the Yatenga Burkinabe there is a close correspondence between the nature of the soils and population distribution (Marchal, 1983: 682). A very similar finding was made in the Maradi region of Niger (Raynaut *et al.*, 1988). In western Burkinabe, the massive influx of Mossi agriculturalists which is taking place today is essentially along the length of the Black Volta valley and not the interfluve, where agricultural potential is lower (Collectif, 1975: fasc. 1, map 7). However, similar possibilities elsewhere are not exploited in the same way. Thus, even if part of the sparsely populated areas of east Mali can be explained by the infertile heights of the Manding plateau (along the Guinean border) and of the Kaarta massif (north of Bafoulabé), it is much less plausible to use agro-ecological arguments to account for the relative emptiness of the large plain that extends from Kayes to Nioro, as well as in the Senegal River valleys and tributaries. A similar observation can be formulated for eastern Senegal, whose physical characteristics are insufficient to explain the low occupation or the moderate rate of demographic growth (Moussa Soumah, 1980: 180–4).

To understand fully the influence the physical environment can exert on settlement, the question of water access must not be neglected since, in this region, an insufficient or irregular supply of water can severely limit the potential for development. It is a factor which can exercise a marked local influence on human occupation. This is especially the case in the geographic sector that covers south-east Burkina and extends into south-west Niger. Here the presence of crystalline bedrock prohibits the existence of water-tables. In the absence of wells, water supply is insufficient to sustain large-scale settlement. Once this limiting factor is alleviated, the effect on the population dynamic is immediate. Thus, in the Linguère region of Senegal, the demographic advance towards the Ferlo reflects an increase in well-drilling.

> It can be concluded that water points increasingly constitute 'rescue' centres for many families haunted by the drought of the last several years. The consequences of an increase of men and their herds around wells pose problems and can be very substantial and ambiguous.
>
> (Ba Cheikh, 1980: 140)

Topography seems to play an ambivalent role and, depending on the specific situation, creates either an attractive or a repulsive effect. In Mali, in the Dogon country and in the Sénoufo region around Sikasso, as in Niger in the Ader massif (between Birni-N'Konni and Tahoua), the mountainous massif and escarpments have, during the last century, offered a refuge for populations which were historically menaced by a general insecurity. In eastern Mali, on the other hand, particularly towards the Guinean border, the Mandigue mountains remain sparsely populated.

As in the case of climate, topography and pedology determine a general framework of obstacles and opportunities for human exploitation, but it is according to other criteria that social groups are able to overcome the former and to take advantage of the latter. Hence, the ill-defined relationship which has been observed between the physical characteristics of the environment and the distribution of population. In certain situations, soils and topography can provide a valid explanation of population distribution but, elsewhere, they do not offer an appropriate interpretation.

Endemic parasitic diseases

Borrowing from P. Gourou's observation, Y. Lacoste notes that the poor use of valleys is one of the major features of the geography of settlement in Africa – a feature that contrasts sharply with the structure of land occupation observed in Asia (Lacoste, 1984). This observation is not so true for the Niger and Senegal River valleys, but is certainly verified in all instances when travelling south in the direction of the Sudanese margin, which accounts for a significant portion of the Sahel. Unhealthy conditions are most frequently invoked to explain this phenomenon. This argument, as we will see, only partially explains the large variation in population distribution, given the broad scale of this assessment.

The development of certain endemic parasitic diseases, characteristic of tropical environments, is highly dependent on the existence of localised ecological conditions. This is especially the case for schistosomiasis and onchocerciasis, both of which are linked to the presence of surface water – stagnant in the case of the former and free flowing for the latter. The

59

aquatic larvae of bilharzia penetrate the human body through the skin when it is in direct contact with water. The simulids, black flies which transmit the filariae responsible for onchocerciasis, have a flight range of no more than 10 km (at least for the females which are the only vectors) and the danger they present is restricted to the proximity to waterways. Schistosomiasis causes serious lesions of the urinary tract, while onchocerciasis causes a series of skin problems, often leading to blindness in its terminal phase. Thus both are very disabling and represent a potential obstacle for the occupation and development of infested areas. Figure 3.1 shows the location of the principal areas of endemism for these two parasites in the Sahel. A juxtaposition with the demographic maps (Figures 2.2 and 2.3, pp. 42–5 and 48–9) allows us to draw several conclusions about the relationship between the distribution of risk and human settlement patterns.

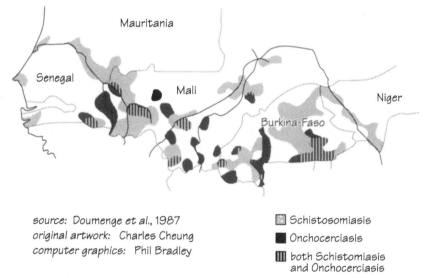

source: Doumenge et al., 1987
original artwork: Charles Cheung
computer graphics: Phil Bradley

Schistosomiasis
Onchocerciasis
both Schistomiasis and Onchocerciasis

Figure 3.1 The distribution of schistosomiasis and onchocerciasis in the western Sahel

First of all, with respect to schistosomiasis, there is the frequent co-incidence between areas of hyper-endemism and sectors of human concentration, particularly certain segments along the banks of large rivers and their tributaries. This is not surprising considering the zonal distribution of these areas. They cover all of the northern margin of the Sahelo-Sudanese zone. The alternation between a short rainy season and a longer dry season favours the formation of stagnant water reserves (pools, dried-up river branches). These stagnant waters are favourable to the development of bilharzia but, in a semi-arid context, they also constitute a precious resource where people concentrate to practise recession agriculture, fishing and herding. The connection between the appearance of the symptoms of schistosomiasis and contact with water is far from always being understood by the populations concerned (Villenave, 1983). The attraction of these sources of water is not, therefore, mitigated by the danger they may pose. However, while there is often a coincidence of densely populated areas and bilharzia infestations, this generalisation does not always hold. For instance, numerous centres of hyper-endemism are situated in sectors of low population density, for example, eastern Burkina Faso and Mauritania. In this case, the hydrological conditions are favourable for the development of vectors and parasites, but other factors limit the attraction of the water to humans, for example, Sahelian pools which are almost exclusively used for watering livestock. In the end, what these different observations show is that the geographic distribution of bilharzia does not exercise a measurable influence on land occupation, at least not at the macro-regional scale with which we are dealing.

In the case of onchocerciasis, the reality is much more complex. The underpopulation of entire regions in the southernmost parts of Senegal, Mali, Burkina Faso and Niger seems to be related to the intensity of onchocerotic infestation. It is largely due to the ravages caused by this illness that settlement attempts by Mossi immigrants in the White and Red Volta valleys have long ended in failure:

> Because of the interrelated economic, social and demographic consequences, which add to the purely clinical manifestations of the illness, within a hyper-endemic zone, it results in a progressive erosion of the workforce, in a threat to the reproductive potential of family groups and finally to their decampment or extinction on site.
>
> (Lahuec and Marchal, 1979: 83)

A similar commentary can be applied to the region that extends west from Sikasso in Mali and, perhaps, also to the Kedougou area in Senegal, although few case studies are available. Several elements, however, lead us to qualify the link between onchocerotic infestation and underpopulation. Not only does the existence of archaeological remains bear witness to an ancient occupation of the Volta valleys (Marchal, 1983), but recent history also shows that they were populated during the colonial period. In fact, their current depopulation is at least partly due to the forced evacuation imposed in the course of a campaign against trypanosomiasis. Specific examples demonstrate that, under certain conditions, life in the valleys is possible without paying too high a price in illness (Lacoste, 1984).

A comparison between the current distribution of population and the centres of endemic diseases is, however, difficult. Since the early 1970s, a vast inter-African programme to destroy the simulids' breeding grounds has been organised across eight countries (Benin, Burkina Faso, the Ivory Coast, Ghana, Guinea, Mali, Niger, Togo) with very positive results. In

Burkina the exploitation of the valleys of the Volta, finally freed from this constraint, is accelerating. In the Sikasso region of Mali, a development project in the zones from which onchocerciasis has been eradicated is currently under study. Likewise, rural development efforts are in progress in south-west Niger. The observation that the valleys are being depopulated is hardly applicable today. Nevertheless, several observations confirm that, even in the past, high infestation levels did not always exercise the same dissuasive effect on settlement patterns. This is especially the case in Burkina, concerning the Yatenga and, further west, the valleys of the Black Volta and its tributaries. In the first example, the existence of an area of major endemic disease coincides with ancient and high human densities. This is undoubtedly due, in part, to the fact that the increase in small surface water management works during the 1950s favoured the development of the simulids' breeding grounds subsequent to settlement (Lacoste, 1984: 220). Such an example, however, suggests that under certain conditions a hyper-endemic situation is compatible with high human densities. The second example is situated at the front line of Mossi immigration. While settlement has recently been accelerated by the success of the anti-onchocercotic battle, the movement is, in fact, much older. These counter-examples illustrate a second aspect of the epidemiological model of onchocerciasis.

> The more dense the population, the less serious the consequences of onchocerciasis overall and the more the settlement front can stabilise in proximity to waterways. The depopulation process which affects human groups installed beyond the protective barrier of a very densely structured settlement front is thus far from being a process without end which would rid a given region of its population.
>
> (Lacoste, 1984: 83)

The obstacle that onchocerciasis presents is thus not insurmountable, and the relationship between risk and population density is not linear. If scattered populations are subjected to a level of individual exposure which condemns them to extinction or withdrawal, the risk is lessened at a certain threshold of population density and settlement concentration. In such high density situations, its collective effects become sufficiently unimportant to become socially tolerable (the gravity of the symptoms and the progression of the disease being proportional to the number of bites received). Whereas a hesitant and sporadic settlement would be condemned to failure, a massive and sustainable pioneering front could, thus, be established and result in a complete occupation of the area.

In the end, the intensity of the onchocerciasis threat is mainly linked to the structure of the land occupation and to the modes in which nature is exploited. Production systems which lead to the clearing of large areas (thus breaking the lines of tree cover which are favourable to simulid development) and where work organisation favours a grouped human presence in the fields, help reduce the risks of exposure when compared with extensive systems where fields and men are widely dispersed. The health risk which blocked the settlement of the humid Sudanese valleys, far from being the pure and simple effect of an ecological constraint, would thus be the historical product of an interaction between its physical characteristics and an evolution in the relationship between social groups and their environment. This is the thesis which Hervouët convincingly defends regarding the White Volta valley:

> It is probable that the simulids' breeding grounds have been allowed to multiply since the beginning of the century while humans became dispersed over land, exposing themselves differentially to the simulids' aggression. They thus certainly created favourable conditions for the transmission and diffusion of river blindness.
>
> (Hervouët, 1992: 160)

Once again, we find that the link between ecological constraints and human development is not linear. Rather, it exists within the context of a global dynamic which associates it with co-factors that allow this link to be effective or not and which modify its consequences. These factors, of a cultural, social and economic nature, orient the choices that human groups make and the strategies they use within a framework of constraints and incentives found in their environment. Among these human factors, certain appear more prominent than others.

THE ENDURING HERITAGE OF THE PAST

This is a domain that can only be briefly reviewed here. Nevertheless, certain aspects of human occupation in the Sahelo-Sudanese zone cannot be clarified without mentioning, although hastily and summarily, some of the historical features of societies found in this region of Africa.

From the Atlantic to Lake Chad, large centralised states have successively appeared and disappeared over the course of past centuries. To cite only a few examples, political entities such as Ghana, Mali, Songhai empires, the Dyolof and Yatenga states, and the city-states of the Hausa world were highly organised and powerfully armed and policed. They tightly controlled the activity, and sometimes even the settlement patterns, of the populations under their control, such as in the planned settlement policies put in place by the Mossi chieftainships in the Yatenga (Marchal, 1983). The peaceful areas which their existence brought about, as well as the hubs of activity to which they gave life, favoured the concentration of populations. By contrast, the insecurity created by their conflicts with rivals or neighbours, whom they wanted to conquer, could empty entire regions of their inhabitants.[1] In different forms and for a long time, political divisions have thus forcefully contributed to moulding the geographic distribution of populations. Could this influence confer upon this distribution a certain flexibility in relation to the natural environment's incentives and constraints? It appears that it could, for a number of reasons which we will now examine.

As we will see in Chapter 4, the existence of these states rested upon trade. Crossroads between the large trans-Saharan caravan routes and the multitude of passageways which drained riches from the heart of black Africa constitute the geographic position and function around which these states developed and prospered. Salt, iron, north African and European manufactured goods, and trinkets coming from the north were all traded for gold, ivory, copper, ostrich feathers, leather and slaves (Mauny, 1967). Cowrie shells served as the main currency on account and for exchange. This large international commerce also had an internal ripple effect, engendering an entire internal exchange network in the sub-Saharan zone throughout which local products circulated, including millet, livestock, dried fish, cloth and tools (Mauny, 1967; Meillassoux, 1971; Thornton, 1992).

The distinctive feature of such an economic system rested in its capacity to give rise, at the local level, to an abundance of service and artisanal activities not directly related to primary production. The development of large trading cities throughout this part of Africa is the most evident manifestation of the existence of such an economic structure. If Jenne, Mopti, Tombouctou and Gao prospered along the Niger River, it is only partly due to the resources drawn from their immediate environment (even if the contribution from fishing was substantial). It was mostly due to their location along the route of a major waterway and their role as an off-loading site between trans-Saharan traffic and circulation routes into the interior of the continent. In the same way, the prosperity of large Hausa cities such as Katsina and Kano owed much less to the richness of their hinterland than to the monopoly which they were able to establish, by political and military means, on international trade in this part of Africa. This means that the establishment of large centres of power, areas of intense production of riches and intellectual centres (such as Tombouctou as a centre of Islam) responded to different stimuli than those on which an exclusively agro-pastoral civilisation depended. By their attraction, and benefiting from a division of labour which attracted foodstuffs produced over a vast territory, such centres were able to become areas of population convergence. Thus, the high population densities which we currently observe at certain points along the Interior Delta of the Niger River, notably between Mopti and Jenne, as well as further north in Tombouctou, are partly the heritage of this history. The same holds true for the population locus of central Niger, because its existence is inextricably linked to the presence of large Hausa markets further south. Here the reality is, nevertheless, more complex. In addition to the effects of such economic interactions, this zone experienced the political troubles of the last century and a withdrawal towards the north of the Hausa populations fleeing the Fulani 'holy war' led by Ousman Dan Fodiyo.

More valuable than speculating on a distant past, useful lessons for understanding the present can be drawn by looking at the political situation as it was in the nineteenth century. We have tried succinctly to summarise it in Figure 3.2. Although the history of this region has always been troubled, the two centuries preceding the present one witnessed a profound reorganisation – a prelude to colonial penetration. In brief, beginning in the seventeenth century, the crumbling of the Songhai empire gave way to a mosaic of small autonomous states which were fraught with constant turmoil. This condition of great political and military instability certainly reflected the concurrent ambitions of local princes but, more fundamentally, it was the result of an upheaval of the economic base upon which this part of Africa had for so long depended. As the great trans-Saharan trade system declined with the rise of commerce centred on the European coastal trading posts, not only was there a reorientation towards the south of exchange channels, but also a staggering increase in the slave trade. Not that this trade was a novel occurrence, since it had existed with north Africa for centuries, but it took on a new quantitive dimension in the seventeenth and eighteenth centuries, under the impetus of the Caribbean and American slave trade.[2] Thus, the whole of western Africa, to the limits of the Sahel, found itself in an unprecedented whirlwind of problems. From then on the military confrontations related less to the pursuit of political objectives than to the pure and simple hunt for slaves. The state of Segou, which was entirely organised around war, is a perfect illustration of this

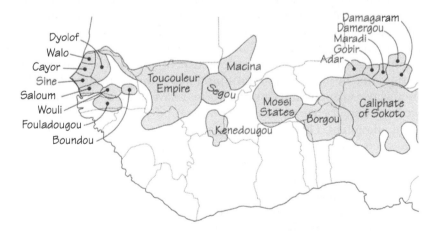

source: Coquery-Vidrovitch, 1988
original artwork: Charles Cheung
computer graphics: Phil Bradley

Figure 3.2 The states of the western Sahel in the mid-nineteenth century

situation (Meillassoux, 1975: 87). At the same time the trade system was profoundly destabilised by the enormous quantity of cowries that European houses of commerce dumped on the coastal trading posts. These shells were used as an instrument of payment which they procured cheaply from their Indian Ocean territories (Johnson, 1970).

In reaction to this general political, economic and social decay, and under the influence of important charismatic Fulani individuals, there exploded a series of movements aimed at establishing, by fire and the sword, theocracies ruled by Islamic law. These were the Sokoto caliphate which built principally upon the ruins of the Hausa states, Macina on the Interior Delta of the Niger and the Toucouleur empire of El Hadj Omar. Whereas order more or less reigned over these territories, insecurity rose at their borders because these new powers threw themselves into the slave war against the 'heathens' (Meillassoux, 1975: 88–91). To this confusion was added the resistance movements provoked by their hegemonic political aims. This was the case, for example, with the Hausa and Hausa-speaking states of Maradi, Gobir and Damagaram, which refused to submit to Sokoto.

In a different historical context and far distant from the large federated empires which reigned from the Middle Ages to the seventeenth century (between Upper Senegal and the Bend of the Niger), the Senegambian region was also experiencing a profound political fissioning at the dawn of the colonial conquest. The proximity of merchant trading posts planted along the coast which carried out the colonial slave trade contributed to division, conflict and insecurity in an even more direct manner than in the rest of the Sahelian zone (Becker and Martin, 1982: 100ff.).

Regarding the problem with which we are particularly concerned, that of population distribution, such a historical background is rich in significance. In the turbulence which reigned almost everywhere, security constituted a major interest, the search for which often took precedence over the search for conditions favourable to the practice of productive activities. This security could be obtained by regrouping under the protection of a military power whose weapons and cavalry opposed slave-hunters' raids and, where necessary, were used to check more massive and better organised attacks of hegemonic intent. Often, this strong power was also the guarantor of an unimpeded movement of merchant caravans over its territory, even while charging for the rights to passage.[3] Even before the colonial presence, the Mossi Burkinabe plateau experienced high population concentrations.[4] This is better explained by such factors as those described above than by the presence in the area of particularly attractive natural conditions. In spite of the rivalries among them, the animist stronghold Mossi kingdoms were able to hold in check the aims of the great Islamic empires that surrounded them. In the same way in Mali the past existence of the now extinct states of Kenedougou (around Sikasso) and Segou and, to a lesser degree, Macina (Interior Delta of the Niger) can still be traced on the map of population density (Figure 2.2, pp. 42–3). In Niger, the outline of the population sector which extends from Konni to Zinder can be overlaid almost perfectly over the regional territories of the Hausa and Hausa-speaking states. Defensive criteria may also have influenced the choice of geographic location. In this respect, the most well-known case is that of the sandstone plateau where the Dogon people took refuge. In the same way, in Niger, the topography of the Ader massif and the shield of the Maradi valley played an important role in the resistance of Hausa-speaking populations of southern Niger to the hegemonic Sokoto state.

It is not only some of the dense population centres that can be explained, at least partially, by historical analysis; a similar approach can aid the understanding of some areas of low demographic density. This is especially the case in the Upper Senegal Basin (in Mali) and in the Upper Gambia (in Senegal). These two regions, following the decline of the medieval empire of Ghana, continued to constitute a confrontational zone between two rival powers, where no stable, enduring power established itself (with the short-lived exception of the Bambara state of Kaarta, where we again see the demographic 'line' from Kayes to Nioro). The ephemeral and turbulent existence of the Toucouleur empire of El Hadj Omar did not, for its part, create large population concentrations. In the end, the current demographic void in this region is certainly related to the insecure situation which reigned for many centuries and to the role the region played as a supplier of slaves for the states along the Senegalese coast (whether for their own use or for resale to nearby trading posts). The current sparse population observed in southern Mali in territories which Caillé had earlier described as prosperous and densely populated can be attributed to similar causes (the ravages of the nineteenth-century slave wars, aggravated by the brutal domination of the Samori armies) much more than to the threat of onchocerciasis (Caillé, 1980).

Obviously, precolonial history does not offer the ultimate explanation for occupation of space by people, but it does offer a framework which gives a better understanding of what kinds of choices have sometimes driven people to take liberties with the incentives and constraints exerted by their natural environment. If we are aware of the fact that the reproduction of local societies, which we today perceive as agricultural, depended as much on their trading activities and their strength in war as on their capacity to exploit the resources of their immediate environment, we can better interpret certain paradoxes apparent in their geographic settlement. It is the heritage of this past which the colonial power took over. The colonial power partly assured its continuity because it established itself where populations were already settled and, through the territorial divisions it instituted and the infrastructure it put in place, reinforced existing population patterns. Eventually, these patterns were intensified by changes to the prevailing demographic balance, which resulted from the end of the haemorrhage of slavery and the fight against endemic diseases.

History, meanwhile, hardly stopped with the arrival of colonial powers, no more than it stopped with the emergence of independent states. While certain forces worked towards permanence, others conversely pushed towards a redistribution of the population. Soon after the end of the resistance movements which accompanied the colonial conquest, the armed peace imposed by the new dominant powers allowed peasant populations to exploit lands which earlier had been too risky to settle. Simultaneously, the constraints imposed by the new powers (taxes, requisitions, various exactions) and the freed slaves' thirst for liberty caused many to flee to distant areas. In Burkina Faso, for example, such events resulted in the origin of the expansion movement on to the margins of the Mossi plateau, as well as the first attempts to reoccupy the Volta valleys (Marchal, 1986). They also resulted in the principle which presided over the reclamation dynamics in the Maradi region of Niger during the first third of this century (Raynaut, 1975). The colonial administration in its time, and later the governments of independent nations, gave themselves the task of organising these population settlements. From the 1930s onwards this was particularly the case with l'Office du Niger (Niger River Office) in Mali and its colossal project (see Chapter 5). The objectives of this project, the creation of a colonial centre and an outlet for the more populated areas of the south (reaching as far as Mossi country, which was then part of the colony of French Sudan), are still far from being reached. Nevertheless, despite many ups and downs, the programme continues to be pursued and developed around Nioro, resulting in a concentration of population which can be seen on the population density map (Figure 2.2, pp. 42–3). The current *terres neuves* project in eastern Senegal, the Development Project of the Valleys of the Voltas in Burkina and, to a certain extent, the onchocerciasis eradication programme all come from a similar procedure. Although their direct impact on land occupation remains modest, their unintended effects are not negligible because the spontaneous settlements on the periphery of the developed sectors are much larger than those under official control (Rochette, 1986).

In the end, the dynamic of close-range, rural migrations only partially reveals the changes in the structure of land occupation imposed upon the Sahelian countries since the beginning of the colonial era. It does not explain well some of its enduring features, which have been raised here, particularly the persistent marginalisation of certain demographically void areas. In reality, to understand the forces which have created population movements and which have moulded the physiography of this part of the African continent throughout this century, the analytical perspective must be broadened to make room for two major phenomena. The first of these is the large labour migrations which have displaced massive numbers of people from the interior of the country towards the coastal zones, while the second is the movement associated with the powerful growth of urban centres movement which, in ever-growing numbers, draws rural people towards the cities.

LABOUR MIGRATIONS AND THE RURAL EXODUS

As we have discussed, the eighteenth and nineteenth centuries saw the upheaval of the economy of the Sahelo-Sudanese societies, with the geographic redeployment of commercial trade routes and the hegemony of the slave trade. These were, in fact, only a prelude to even more radical changes which would arise after colonial partitioning took effect. Thus, the war and plunder economy in which the states of the region had come to specialise came to a sudden end. The large trans-Saharan trade, already weakened, eventually dried up and the commercial routes collapsed. Salt, minerals, artisanal products, cereals, kola nuts and livestock which had circulated, and assured complementarity, between different regional areas, all but disappeared (Meillassoux, 1971; Baier, 1980). The large agricultural exploitations that depended on slave labour (the Hausa version of which is described in the memoirs of Baba of Karo (Smith, 1954)), and which had expanded throughout western Africa after the end of the transatlantic trade, disappeared (Meillassoux, 1975). On the ruins of a secular social and economic structure was built an organisation whose objectives and principles were completely new.

Poles of attraction and labour reserves

The geographic reorientation of the trade routes, which had begun several centuries earlier with the establishment of European trading posts on the Atlantic coast, was completed when the interior of the continent fell under colonial control. West Africa was partitioned into several autonomous, sometimes rival, holdings, each of which was administered by a coastal centre, a large port city from which economic activity emanated. Between the continental regions and the coast a criss-crossing pattern of trade emerged where goods produced by the metropolitan industries were brought inland and agricultural commodities destined for export were transported to their shipping site (Suret-Canale, 1968–72).

Nevertheless, after a period of false starts and grand Utopian projects, the French colonists quickly discovered that the marginal Sahelo-Sudanese regions, which covered a large portion of the territories they now controlled, offered only limited economic potential. They harboured few riches of which to take advantage except, finally, the inhabitants themselves, that is as potential consumers of manufactured products and as a usable labour

69

force. The necessary precondition for the development of such resources was the introduction of productive activities whose output would be tradable in the metropoles and the international markets. During the initial phase of the colonial takeover, that is the phase devoted to the establishment of a basic infrastructure necessary for the exercise of power (especially public buildings and channels of communication), manpower was mobilised very directly in the form of forced labour. In a longer-term perspective, however, it was not possible to continue exclusively with a method that did not produce marketable goods or create demand and, as a result, blocked the creation of a market. Rapidly, the objective thus became the creation of conditions favourable to the development of a monetary economy, that is by creating the need for cash, by offering the means to acquire it and by proposing possibilities for its use.[5] The introduction of a poll tax, payable only in cash (in contrast to the earlier colonial period's goods and cowrie taxes), the promotion of commercial agriculture (principally groundnuts and cotton) and the abundant offerings of manufactured products competing with those of traditional artisans constituted, almost everywhere, the three pillars of a strategy intended to put in place the new extractive economy. This system was orchestrated by the colonial administration and implemented by the large trading houses which, through the exchange of local and imported products, disciplined local societies to the needs of metropolitan and international markets (Raynaut, 1977a; Gregoire, 1986).

Two constraints, however, inhibited the simple application of this general picture to the whole of the Sahelo-Sudanese zone. One was of natural origin because groundnut farming, like that of cotton, was subject to the limiting conditions of climate and soil. The other was of a commercial nature because the distribution of bulky and cumbersome agricultural products over long distances necessitated a communication infrastructure and a well-adapted means of transport. Railroads and roads were built, but a preoccupation with profit (made more critical by the low value of the transported commodities and by the colonial administration's lack of financial resources) limited such investments to a few privileged axes, which might be conceptualised as a number of large drainage canals oriented towards the coast. The evolution depicted in Figure 3.3 is very significant in this regard. The combination of these two imperatives gave rise to regional specialisations with commercial agriculture concentrated in geographic areas where the constraints of production and distribution could be minimised. Hence the two great groundnut basins were created: that of Senegal (the older and the more important of the two because it benefited from its immediate proximity to the port of Dakar) and that of central Niger (more recent and modest, contiguous with the large northern Nigeria production zone and serviced by railroad since the beginning of the century) (Péhaut, 1970). The privileged production zones for cotton agriculture were found in Sudan (in what later became independent Mali)

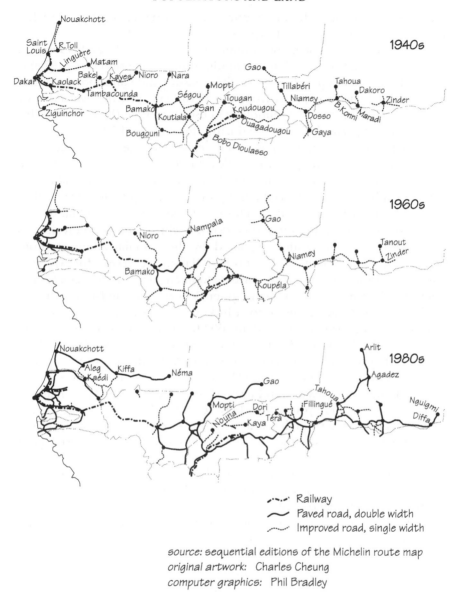

Figure 3.3 Growth of the transport network in the western Sahel

and in Upper Volta (future Burkina Faso), but, as we will see later, their history does not present the same continuity as those of the groundnut zones.

In the end, a major principle guided land management, one which found it more profitable to attract labour to a favourable and easily accessible production site than to allow it to function in its original 'ecological niche' and, afterwards, to go to great expense to reap its benefits.

The history of the Senegalese Groundnut Basin exemplifies this strategy. We have previously discussed the chronology of its expansion (see p. 51 and Figure 2.4). This was possible only with a massive and continuous input of temporary labour brought in from the periphery, i.e., the *nawetaan*, or *navetanes* according to the terminology in use at the time (Faye, 1980). This movement was particularly intense during the 1930s and 1940s. It reached some populations of the colony itself, especially the Toucouleur and Soninké from the Senegal River valley (a region marginalised by the construction of the Dakar–Bamako railroad) as well as the inhabitants of Upper Gambia. It also displaced immigrants in large quantities from other territories under French control. This exodus and the contractual service arrangement between worker and landowner which took place on site was systematically organised and regulated by the colonial administration, which assured a sort of exploitation of 'labour reserves', for which it intended no autonomous economic development. The two principal zones where this began were the Malian Upper Senegal and the Guinean Fouta Djallon. In this context, the railroad linking Dakar to Bamako played the role of 'drainage canal' and, of the 60,000 to 70,000 *navetanes* present annually in the Senegalese Groundnut Basin before the Second World War, 60 per cent were of Malian origin, of whom the large majority were Bambara from the Kayes, Bafoulabe, Kita and Nioro regions (later joined by migrants from around Segou, Sikasso and even Mossi country). The large majority of these *navetanes* were young men between 20 and 40 years of age and, as a consequence, western Mali found itself for several decades (in fact until independence) with a massive and continuous loss of its productive labour force.

Within the framework of this system of labour division to which, under French tutelage, West Africa was subjected, the Sahelo-Sudanese zone was given the role of a labour reserve. Apart from this Senegalese case, only l'Office du Niger, which was the source of grandiose colonial projects, and the Niger groundnut basin were exempt from this strategy. This system had its flaws, however, because as soon as the quest for money became the primary reason for migrant departures,[6] strict planning and control over the flow of labour became difficult. In fact, the attraction exercised by the large economic coastal centres flowed over the boundaries meant to separate the areas under rival administrative European powers. Thus, thanks to the dynamism of mining and cocoa cultivation, from the 1920s onwards the British colony of the Gold Coast (subsequently Ghana) became the destination of choice for migrants coming from Upper Volta, central and southern Sudan, as well as western Niger (Painter, 1985). This competition was strongly felt by those directing l'Office du Niger, who demanded labourers in order to realise their grand project, as well as by the Ivorian coffee and cacao planters who also needed labourers. The return to authoritarian labour requisitions only partially compensated for the flight of the workforce: an escape in the face of the exactions that accompanied such requisitions. In 1944 British authorities estimated the number of French subjects residing in the Gold Coast at 200,000, half of whom had become permanent residents. But, as Painter described well, such evaluations were very probably underestimates (Painter, 1985). The situation progressively evolved over the years with the development of coffee and cocoa production in the Ivory Coast in the late 1950s, favouring the creation of an employment centre and settlement area, the advantages of which began to rival those offered by its anglophone neighbour (Painter, 1985). After independence, there was a reversal of these tendencies. The Ghanaian economy rapidly declined, while that of the Ivory Coast enjoyed rapid growth and became the sole focus of a vast labour reserve extending from Niger to Mali and from Benin all the way to Guinea with its centre of gravity in what was then Upper Volta and, more specifically, the Mossi country (Marchal, 1986a).

Further east, a third large zone of influence, centred in Nigeria, also appeared quite early. Contrary to the two zones just mentioned, it was not along the coast that it developed but deep in the interior around what was, at the time of the colonial conquest, the commercial heart of West Africa, namely Hausa country and its market cities. The arrival of the railroad in this area in 1910, and then the early development of groundnut agriculture, rapidly created an intense hub of activities and exchanges. Quite naturally, its influence extended throughout the region's historical hinterland, particularly the Hausa and Hausa-speaking areas of Niger under French control. Contrary to what was occurring at the same time in Senegal, this zone of attraction did not stimulate a massive human exodus. We can find at least two reasons for this. First, Nigeria itself had access to a labour reserve within its own borders, making the need for large, external labour migrations unnecessary. Second, there was a rapid, though slightly later, development of a groundnut production area within Nigerien territory itself (Péhaut, 1970). We should also note that the mode of operation, and sometimes the commercial operators (being the same on both sides of the border), created relations based more on market exchanges than on workforce transfers.[7] More than the workers, it was the groundnut itself and the various trade items that crossed the border in one direction or another, depending on relative differences in prices (Collins, 1974; Egg and Igué, 1993). Having said this, we should not underestimate the attraction exercised by tin mining on the Joss plateau (Freund, 1981), although in Niger it does not seem to have created the sizeable population transfers that have taken place from Mali and Upper Volta to Senegal, Ghana and the Ivory Coast.

After independence, the erosion of the value of Nigerian money and the relative strength of the CFA franc[8] often made the practice of border trafficking more appealing than labour migration. Things were to evolve appreciably during the 1970s, at the same time that the Sahelian countries were strongly affected by drought. The large anglophone country thus entered an intense period of economic growth following the surge in petroleum prices. The increase in construction sites and the rapid growth of the port of Lagos created innumerable unskilled employment possibilities, attracting a flood of young men from all the countries along the Niger River. However, this migration was of a different sort to that illustrated by the preceding examples because the destination was urban, the departures were short term, often of short distance and limited to the dry season. These characteristics, as well as the later emergence of Nigeria's attraction, account for its weaker influence on the demography of the Sahelian countries, especially when compared with the influence of the Senegambian and Ivorio-Ghanaian centres (with respect to their own spheres of influence).

With the help of schematic maps, Figure 3.4 illustrates the geographic distribution of these centres of attraction, as well as their spatial progression over time. Today, the demographic heterogeneities which are encountered throughout the Sahelo-Sudanese zone, whether they concern the density of occupation or the rate of growth, still cannot be understood without reference to the division of roles thus instituted on a West African scale from the beginning of the colonial era and which, *mutatis mutandis*, has until now been perpetuated. These migratory fluctuations displaced considerable populations over the course of decades and, even though they generally involved only temporary displacements, in the long run they exerted a significant drain of the productive labour force from the interior. According to Amin (1974), they were the source of a demographic deficit whose cumulative effect reached 4.8 million people between 1920 and 1970. Such an evaluation is, of course, completely conjectural, but it furnishes an order of magnitude that seems reasonable. For Burkina Faso alone, the number of officially registered emigrants exceeded 700,000 people in 1985 (Sirven, 1987). Doubtless fewer were involved in Mali during the same period, but the effects would have been no less dramatic.[9]

As for western Mali and eastern Senegal, they have not ceased (since the beginning of colonisation but also after their independence) to be used as labour reserves, first within the Senegambian area, then towards Ghana and currently towards the Ivory Coast. More recently, since the 1960s, waves of emigration to Europe have begun. Thus, by the early 1970s, there were 35,000 Malians, 21,000 Senegalese and 10,000 Mauritanians in France, practically all of whom were ethnically Soninké, from all three countries of origin along the Upper Senegal valley (Colvin, 1981). The 1990 French national census counted 43,692 Senegalese, 37,693 Malians and 6,632 Mauritanians. These figures are certainly underestimates due to a high level

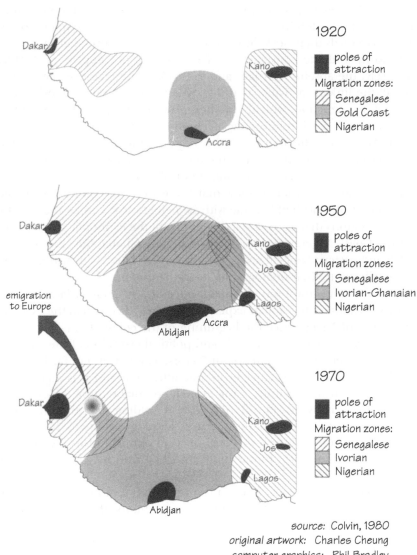

1920

poles of
attraction
Migration zones:
Senegalese
Gold Coast
Nigerian

1950

poles of
attraction
Migration zones:
Senegalese
Ivorian-Ghanaian
Nigerian

1970

poles of
attraction
Migration zones:
Senegalese
Ivorian
Nigerian

source: Colvin, 1980
original artwork: Charles Cheung
computer graphics: Phil Bradley

Figure 3.4 Labour reserves and poles of attraction in West Africa

of illegal immigrants. Undoubtedly these emigration-related population losses, following those of the slave trade, explain to a large extent the low densities and modest rates of growth which are today characteristic of this part of Sahelo-Sudanian Africa. It is interesting to compare this fact with the isolation of this same region with respect to the development of the inter-African road system, which was particularly intensive during the

1980s (see Figure 3.3). Here the principle of an interregional division of roles continues to dominate. As long as the goal remains the migration of labourers towards their work destination, the existing railroad fills the need satisfactorily and the construction of roads does not appear profitable. Conversely, the absence of roads and the lack of merchandise circulating (with all the linkage activities that this dynamism generates) deprives this region of an attraction, which, elsewhere, has been especially powerful.[10] In a way, this is a self-reinforcing mechanism that exacerbates marginalisation. Moreover, the nature of the region's immediate international neighbours most certainly contributes to its permanence. It is not necessary to emphasise Guinea's economic and political situation over the course of the past decades in order to show that there were no conditions for the development of a profitable trade with its neighbours. Unlike the Sahelo-Sudanese regions that were able to benefit from the economic dynamics of the Ivory Coast, Nigeria, and for a period of time Ghana, the extreme west of Mali and eastern Senegal are geographically isolated, which in no small way contributes to their demographic marginalisation.

In a more conjectural way, in the absence of any detailed facts, we can put forward a somewhat similar explanation for the situation in the extreme eastern part of Niger. Since the beginning of the colonial era, this region has remained outside the economic and political evolution of the country. This is explained by a conjunction of factors. The region lies far from the capital, it offers no satisfactory economic attraction and it is adjacent to Chad which, during the time of the French colonial empire, belonged to French Equatorial Africa (a different federal holding than that of Niger) and where, since independence, serious security problems have been recurring. It too found itself in a sort of dead end within its own nation. Eastern Niger, therefore, found itself absorbed into its southern neighbour's area of influence.[11]

The situation of southern Mali, in the Bougouni region, is also clarified when we consider its labour migration movements abroad. As already discussed above (pp. 49, 53), if the population movements from Dogon country, the Interior Delta and the Bend of the Niger (Mopti, Tombouctou) towards the southernmost part of the country do not provoke a more marked acceleration in the growth rates in the destination zone, it is because they are offset by departures that drive inhabitants onwards to the Ivory Coast and Ghana (in accordance with a tradition which has its explanation in the history just discussed).

The densely populated sectors, namely north of the Mossi plateau in Burkina, the Dogon country and the Segou region of Mali, illustrate a third type of situation. The short-range settlement migrations towards the Volta valleys do not explain sufficiently the weakness in observed demographic growth rates (see Chapter 2). No doubt, as we have just seen, the effects of the powerful and ancient currents of attraction exerted on these regions by the economic hubs on the coast have a more pronounced effect.

In a determining fashion, the interregional division of roles that appeared in western Africa following the colonial conquest and the geo-economic remodelling which it brought with it, certainly oriented the destiny of the Sahelo-Sudanese zone throughout the twentieth century. It should not be forgotten that although the landlocked position of most of these countries is a major constraint (Collectif, 1989), it is not the pure and simple expression of an inevitable geographic fate. It exists only in relation to a particular structure of land management that designates economic roles, defines exploitable resources and organises circulation routes. For centuries the Sahelian states have been situated at the heart of extensive commercial channels which nourished this part of the African continent. Resting on an economic base where services and crafts no doubt occupied a more important position than primary production, far from suffering from their position at the heart of the continent, they have, in fact, enjoyed a great advantage. The submission to new interests (under the historical conditions which we briefly reviewed above) led to an upheaval of the principles that organised the zone's economy. Resources were redefined, localities received new specialisations and communication resources were reoriented. The current distribution of population and the local variability of growth rates is due at least as much to the influence of this recent past as it is to the determinations of the natural environment or to the legacy of secular history.

Nevertheless, the emphasis placed on the past must not obscure the role subsequently played by new forces of change. If the effect of attraction exercised by several large regional centres of economic activity accounts for some of the demographic phenomena observed in the Sahel, it is a movement of another kind, urban growth, which may mould the demographic landscape of this region of Africa in the decades to come (following a pattern that is found throughout the developing world).

Acceleration of urban growth

Within a general trend of accelerated growth, two major characteristics underscore the demographic evolution of the Sahel: the particularly rapid rate at which the urban population has increased in this process, and the dominant role played in this increase by metropolitan capitals and a limited number of other large cities.

In spite of this general tendency, Table 3.1 shows that the urban population remains largely in the minority in Burkina, Mali and Niger. The situation, however, is very different in Mauritania and Senegal. In the first of these two countries, the urbanisation rate has reached 42 per cent. Nouakchott, with close to 530,000 inhabitants, is home to the overwhelming

Table 3.1 Urbanisation rates in 1991

Country	Urbanisation rate (%)
Burkina Faso	9
Mali	19
Mauritania	42
Niger	19
Senegal	38

Source: J. Schwartz, 1992

majority of the country's urban population. The level of urbanisation is slightly lower in Senegal, according to the source used above, but there is a certain lack of agreement between the data used here and those of the general 1988 census figures, which indicated that 40 per cent of the country's population lived in cities. Be that as it may, urbanisation is a massive phenomenon which affects not only Dakar and the Cape Verde peninsula, where a large megalopolis of 1.5 million inhabitants has formed, but also all of the western part of the groundnut basin, where a chain of urban concentrations of more than 50,000 inhabitants is found (Table 3.2). The role played by Dakar as the capital of the French West Africa Federation during the colonial period, its economic dynamism through its function as a port, as well as its position as the railroad's starting-point and the repercussions of the attraction exercised by the development of groundnut agriculture, are the principal factors from which this urbanisation phenomenon draws its exceptional intensity. Even today the urban coastal margin of Senegal attracts not only the inhabitants of its own hinterland, but also immigrants from the neighbouring countries of Guinea, Guinea-Bissau and Mauritania. If current trends continue at the same rate, half of the population residing in Senegal will soon live in cities.

In the other Sahelian countries the phenomenon is much less accentuated. Nevertheless, in each of these the population of the capital numbers several hundreds of thousands of inhabitants. Bamako officially counted close to 650,000 inhabitants in its 1987 census, Ouagadougou recorded 442,000 in 1985 and Niamey's population was 400,000 in 1987. Other urban centres rank far behind, except for Bobo-Dioulasso which, when compared with Ouagadougou, perhaps owes its competitive position to its role as former capital of Upper Volta. In spite of the extremely rapid growth experienced by some of these cities, the human loss from the rural environment to these secondary cities remains moderate.[12] Except in the case of Senegal, it thus appears that outside the geographic areas closest to the capitals and principal provincial centres, the exodus to the cities continues to exercise only a modest influence on settlement dynamics in the rural zones. This influence, however, continues to reinforce international migrations, since the same departure zones which support the

Table 3.2 Urban population growth in the western Sahel, in centres of more than 30,000 inhabitants

Country	Town	Population at last census (1980s)	Average annual growth rates (%)
Mauritania	Nouakchott	393,325	10.21
	Kaédi	30,515	3.66
Senegal	Dakar	1,500,459	3.96
	Saint Louis	115,372	2.18
	Louga	52,763	10.45
	Thies	175,465	3.58
	M'Bour	76,751	6.27
	Diourbel	77,548	2.75
	Kaolack	152,002	3.21
	Tambacounda	41,885	4.05
	Ziguinchor	124,283	4.90
	Kolda	34,337	5.05
	M'Backé	38,947	3.76
Mali	Bamako	646,163	4.01
	Kayes	48,216	0.00
	Sikasso	73,050	4.19
	Koutiala	48,010	5.32
	Ségou	88,877	2.82
	San	30,688	2.50
	Mopti	73,979	3.02
	Tombouctou	31,925	4.75
	Gao	54,874	5.37
Burkina Faso	Ouagadougou	442,223	9.86
	Bobo-Dioulasso	231,162	7.23
	Banfora	35,204	11.04
	Koudougou	51,670	3.44
	Ouhigouya	38,604	4.16
Niger	Niamey	398,265	5.32
	Dosso	27,092	9.55
	Tahoua	51,607	4.66
	Maradi	112,965	8.55
	Zinder	120,892	6.85
	Agadez	50,164	8.50

growth of Bamako, Ouagadougou or Niamey also support the flow of young men making an exodus to Abidjan or Lagos. The cities of the Sahelian countries (medium-sized agglomerations, capitals) are also intermediate links in a vast circulation system of men which encompasses all of western Africa. It is this system in its entirety which, in the end, contributes to a remoulding of the distribution of the region's population.

With this description of the principal factors which help us to understand the demographic heterogeneity of the Sahelo-Sudanese zone one conclusion stands out. That is the impossibility of building a single, satisfactory explanatory model, homogeneously applicable to the entire area under

CLAUDE RAYNAUT

consideration. Each local situation is the product of a singular combination of explanatory elements, which, though present everywhere, vary in their relative importance from one case to the next. As with any complex reality, we are thus confronted with a diversity of individual cases. In order to move forward, we cannot, however, leave it at that. An effort must be made to introduce order into this diversity, and this can only be done through some form of categorisation. An effort needs to be made to go beyond the mosaic of local examples to develop a typology that identifies the more important situations, concordant with the major trends which are at work today and which will impact on the future.

LARGE GEO-DEMOGRAPHIC COMPLEXES

With such a classificatory approach, a first stage consists of reducing the diversity of individual situations to a number of type cases. Overlaying the two preceding demographic maps (Figure 2.2, population density (pp. 42–3) and Figure 2.3, demographic change (pp. 48–9) yields a synthesis constructed from a grid that distinguishes seven major groupings (see Figure 3.6).

Among the geographic sectors where population density is more than 20 inhabitants per square kilometre, we can identify two very contrasting

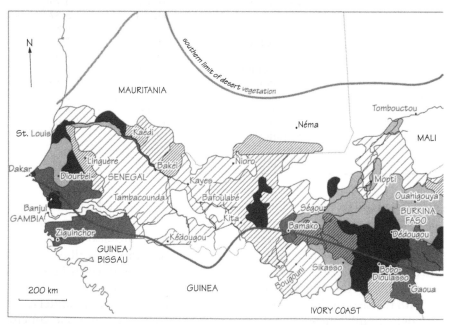

Figure 3.5 The combination of population density and demographic growth in the western Sahel (continued opposite)

situations, which are also the most significant from the point of view of dynamics.

In case A, the presence of a settlement pattern which is very dense at times (more than 40 inhabitants per square kilometre in Niger) combines with an effect of demographic attraction resulting in an annual growth rate of more than 3 per cent. In similar areas, which we term 'poles of attraction', the exploitation of natural resources is evidently the object of an intense dynamism which, sooner or later, will give rise to strong tensions.

The situation illustrated by case C can justifiably be interpreted as an example of such a 'demographic saturation'. Here, high densities accompany a low growth rate which, as we now know, is the consequence of strong migratory movements. These have a stabilising effect on the level of land occupation, which has become largely incompatible with the maintenance of conditions for sustainable production.

Likewise there are two strongly opposing cases within those areas marked by low densities:

- that of the 'pioneer zones' (case D) which until now were little populated, but for which the demographic dynamism indicates that they have begun to receive immigrant populations at an accelerated pace; and

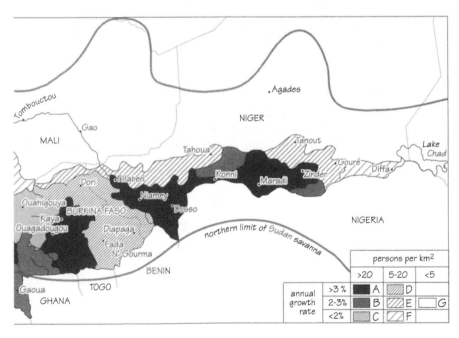

original artwork: Charles Cheung
computer graphics: Phil Bradley

81

- that of the 'peripheral zones' (case F), which not only remain sparsely occupied but in which populations continue to increase only slowly, leading to a progressive widening of the gap separating them from more dynamic zones.

One last category (case G) also deserves attention. With less than five inhabitants per square kilometre, these zones are practically deserted and, almost everywhere, have a growth rate of only about 0.5 per cent per year. Without necessarily becoming depopulated in an absolute sense, these areas experience a negative migratory balance which reflects the population's slow movement towards the exterior. In this context, we can therefore consider them as 'zones becoming abandoned'.

The two central categories of this grid (cases B and E) correspond to sectors whose demographic growth remains moderate. These are intermediary situations which can be found in densely populated as well as in sparsely occupied areas.

Even if it introduces more clarity into the referencing of local situations, such a classification is still too formal and analytical. As we have already realised, the reality of the region does not consist of land segments cordoned off. It is, on the contrary, composed of interconnected spaces which create large areas corresponding to these territorial complexes ordered by a dominant feature. The spatial importance of these geo-demographic complexes is represented schematically in Figure 3.6 in the form of contours whose intersections reflect the absence of clear boundaries and the existence of transitional areas subject to cross-influences.

I The Sahel's western extreme: an area of urban polarisation

Two dominant features characterise the western extremity of the Sahelo-Sudanese zone. The first is the weight of its historical heritage, namely that of the Senegalese Groundnut Basin. The centre of this territory corresponds

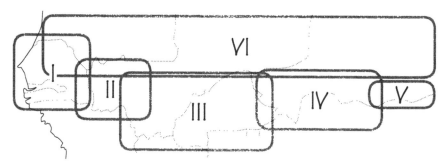

original artwork: Charles Cheung
computer graphics: Phil Bradley

Figure 3.6 Major demographic zones of the western Sahel

82

to lands devoted to this agriculture since the beginning of the century, and sometimes even earlier. Although always densely occupied, the lands' low demographic growth is indicative that a state of saturation has been reached. Nevertheless, even on their margins growth remains very moderate. In the direction of the Ferlo, demographic penetration, such as it is (especially around Linguère), has come up against a series of constraints, principally difficult access to water, unfavourable soils and the presence of sylvi-pastoral reserves. Towards Tambacounda, the conquest of the eastern *terres neuves* appears to be slowing down and the pioneering settlements that they are extending have nothing in common with the powerful out-migration movements that took place around the area between the 1940s and 1970s.

The second major feature, narrowly linked to the preceding one, is the intensity of urban growth and, more specifically, the demographic explosion being experienced by the Cape Verde peninsula. The hinterlands of Dakar and the whole of its environs act like a magnet (Kane Diop, 1989), relieving the demographic pressure that, under different circumstances, would no doubt exist at its rural margins. In Mauritania, although on a smaller scale, a similar attractive force is observable around Nouakchott. In this part of the Sahel, the demographic interactions between city and country attain a level that is without parallel elsewhere in the Sahel. The role of urbanisation constitutes an essential structural element in defining this territorial complex.

To this central core are attached two sub-areas which are closely linked: the Senegal valley to the north and Casamance to the south. Even if, at another level, the two might be conceptualised autonomously, especially concerning their modes of development, the migratory currents which have long bound them to the groundnut basin, through the *navetanat* system, as well as their contribution to Dakar's demographic growth, justify grouping them together.

In terms of the spatial distribution of population, the future of this territory will depend primarily on the balance established between two opposing forces: on the one hand, the capacity of rural regions to retain their inhabitants, and, on the other hand, the level of attraction exercised by the urban world.

II The Upper Senegal Basin: a labour export zone

This is the largest area of demographic low pressure in the whole of the western Sahel. The haemorrhage of the slave trade was succeeded by its role as a labour reserve for the profit of other regions of western Africa. Furthermore, it was marginally positioned with respect to transportation resources and market trade, and suffered even further from soil and topographic constraints. These multiple factors have contributed to the creation of the situation observable today. Whether in the direction of the coastal countries, or to Bamako, or even to Europe, this territory continues to export its able-bodied workers. However, it does not benefit from the attraction effect that such vast expanses might present to adjoining densely populated regions. The inhospitality of the environment, even if it is locally real, is insufficient to explain this territory's dominant centrifugal forces. In terms of a reallocation of land and its subsequent development the question to be asked is which obstacles can be eliminated in order to unblock the inertia under which all of western Mali is struggling?

III Between the Niger River and western Burkina: an area of high population mobility

This area is characterised by the combined interaction of several population dynamics. The central area of high population density, which extends from the Niger River, at the level of Segou, up to the Yatenga Burkinabe by way of the Dogon plateau, is experiencing a phenomenon of demographic saturation which, on its southern and eastern margins, results in a vigorous population push – either towards the 'poles of attraction' which are already densely occupied, or towards the 'pioneer zones' where development efforts have recently begun. On these migratory currents are superimposed others, originating from the northern margins, that is the Interior Delta and the Bend of the Niger, and Sahel Burkinabe. Their destinations are partly the same as those movements originating from the already saturated sectors, although they seem to be targeted specifically towards occupation of the southernmost regions, especially to the south of Bougouni in Mali. Finally, to these internal movements are added the ancient and powerful migratory currents which originate from the same regions but whose destinations are far beyond national borders.

To a large extent the geography of these internal population displacements coincides with the area's inherent incentives and constraints. One tendency which dominates is a movement towards the south in search of a more favourable rainfall regime. In a period of repeated climatic shocks this is a trend that has been the deciding factor for many of these departures. Nevertheless, the success of the international programme to eradicate onchocerciasis, even if it has not always managed to alleviate all fears among potential settlers, has constituted a necessary precondition for the development of zones where, until recently, the presence of simulids continued to pose a threat. When, as is the case in Burkina Faso, this public health action is accompanied by public hydro-agricultural management schemes, colonisation movements are clearly reinforced, although often in an indirect and unpredictable manner (Rochette, 1986). The phenomenon is very perceptible south of Ouagadougou, in the White and Red Volta valleys.

84

In all the territories corresponding to this geo-demographic complex, human mobility is a major and omnipresent phenomenon. Nothing can be understood of the relations between social groups and their environments if this is not taken into account. The land settlement patterns bequeathed by ancient and recent history undergo a profound recomposition. Experiencing emigration, the most populated areas stagnate, while those whose occupation was until now sparse develop rapidly. The result of this double movement is the progressive erosion of contrasts in interregional density and a more and more complete occupation of the land. Within this global dynamic are the international migratory fluctuations that act as regulatory mechanisms.

IV The Nigero-Nigerian border: a field of polarisation

In all of the region that covers central and western Niger the most pronounced feature is the attraction which, however densely populated, these spaces continue to exhibit. These 'poles of attraction' occupy a dominant position and, contrary to what we have observed everywhere else, show no signs of becoming 'saturated'. Even in the regions of central Niger where densities exceed 40 inhabitants per square kilometre and sometimes attain, as in the Maradi department (Raynaut *et al.*, 1988), values equal to or greater than 60 inhabitants per square kilometre, demographic growth is not waning. This situation is confirmed by the modest push of the 'pioneer fronts' (high demographic growth into weakly populated sectors) that is evident in the narrow band extending west of the Niger River. In this case it is as if, instead of an expansion movement away from the most heavily populated areas, there was a progressive accumulation of populations along the Nigerian border. As we have seen above (p. 50), several phenomena are simultaneously responsible for this particular dynamic. First of all, there is the southerly movement of agropastoral and agricultural populations driven by successive droughts. Second, we note the fact that migrations abroad, without being completely unknown in this part of the Sahelo-Sudanese zone (and which can be quite vigorous in the western part of the country (Painter, 1985)), do not appear to have attained proportions comparable to those observable in Mali and Burkina Faso. Third, in Niger there exist practically no areas which are sufficiently sparsely occupied to serve as receiving and settlement areas for the more populated sectors. Finally, the presence of the border also undoubtedly explains to a large extent this southerly population concentration. This may appear paradoxical in the sense that, from a social and cultural point of view, this dividing line is nothing more than an artefact. This is particularly evident regarding the vast Hausa and Hausa-speaking population area which overflows into the territory of Niger, but whose demographic and historic centre of gravity is found in Nigeria.[13] The underlying cultural unity here is quite real, but contemporary economic and political divisions also introduce a break in continuity that has an effect on the daily life of local populations and also, indirectly, on their spatial distribution. In the economic domain, the differences created by the meeting of two monetary and customs zones is the source of innumerable opportunities for speculation, which are made even more profitable by fluctuations in the relative value of the Nigerian naira and the CFA franc. Not only livestock and local agricultural produce but also products imported from overseas, continuously cross the border, supporting a small yet intensive border commerce as well as vast trading operations and an international transit from which very large fortunes are made (Grégoire, 1986). Thus, by its very existence the border generates intense economic activity and exercises a powerful draw for the population of Niger. As to its political dimension, it is far from negligible. At the whim of sudden changes in the internal situation of each of the two countries, and following the ups and downs of their mutual relations, obstacles sometimes arise to block the movement of their respective citizens. In the past these have gone as far as the return of Nigerien citizens to their native country and even to the complete closure of the border (Raynaut and Abba, 1990).

It is difficult to predict which course this unique demographic dynamic will take in the future. Much will depend on the adjustments made by Nigeria. If the economic and monetary situation stabilises, the border will cease to generate its present effect of polarisation. If industrial development experiences a real boom, Nigerien labour resources will be called upon to contribute more intensely. The current configuration could, therefore, be profoundly altered. In this regard, it is still too early to assess the impact of the recent devaluation of the CFA franc.

V The Nigero-Chadian confines: a dead end

Even more so than the Upper Senegal River Basin, far eastern Niger appears as an area of demographic depression. Environmental constraints, especially its aridity, which is expressed by the plunge in isohyets at the level of the Chadian Basin, partly explain this depression. However, this argument is not entirely satisfying. It does not account for the abrupt decline which is visible on the population density map immediately to the east of Zinder (Figure 2.2, pp. 42–3). In reality, this limit corresponds to the eastward extension of the Nigerien groundnut basin. It marks the passage between what has long been, and to a certain degree remains, the dividing line between the 'useful' Niger (equipped, administered, supported) and the eastern confines (left on their own, politically as well as economically, and practically isolated from the rest of the nation). Here, the borders neither attract nor help to settle populations. To the south, the large economic and commercial centres of central Nigeria are distant indeed, while on the

Chadian side the political and military instability is, at the very least, unfavourable for the development of sustainable trade. In the end, what characterises this part of Niger is a quasi-symbiotic relationship with its large Nigerian neighbour; the border is almost completely open to people, goods and money.

The combination of these conditions makes the demographic future of this region largely dependent on the evolution of its neighbouring countries, principally the resolution of the Chadian conflict and the position occupied by eastern Nigeria in its nation's political and economic agenda. These are domains in which all current predictions are uncertain.

VI A Sahel in the process of abandonment

Over the whole expanse of the zone covered by this study, the demographic retreat observable in the 'strictly' Sahelian region represents a constant. After an earlier movement north by agricultural populations, this relative depopulation is incontestably the result of the past two decades' droughts. None the less, it carries an undeniable political dimension which is the expression of a decline in the nomadic way of life, inscribed in a framework of conflicting relations between these mobile populations (clinging to their independence) and central powers that, since the colonial period, have increasingly tried to control them. More recently there has been an attempt to 'rationalise' their mode of environmental exploitation, which has most notably resulted in the development of the politics of sedentarisation, which are more or less openly coercive. The violent conflicts between the Malian and Nigerien governments and their Tuareg minorities are incontestably the results of this history. Here again, even if the climatic constraint remains a determining factor, the demographic future of this zone is inseparable from political evolutions that are difficult to predict, namely pursuit or appeasement of tensions between Saharan populations and central powers, and relations between the Sahelian states and Algeria and Libya. Given the present situation it is, nevertheless, difficult to envision a reversal of the tendencies observed today.[14]

CONCLUSION

It is appropriate to emphasise the diversity of demographic issues in the countries of the Sahel. Indeed, beginning with a statement about trends, we can consider that the accelerated growth in population constitutes a major phenomenon engaging the future of this entire zone, as is the case elsewhere on the African continent. However, if we remain at this level of generalisation we cannot uncover the sometimes very contrasting realities encountered on the ground, and, more important, we tend to confuse, in such a singular generalisation, explanatory factors of very different origins whose salience sometimes rests specifically at the local level.

A map of the demographic realities clearly shows that the spatial distribution of population is nowhere uniform and that demographic growth is not everywhere vigorous. The differences that are observable from one area to the next would remain incomprehensible if we isolated them from their context and limited ourselves to seeing only the sum total of behaviours and events which are put into play by individuals. Certainly, births, deaths and migrations are the three parameters that ultimately combine to determine the net rate of growth. When reasoning in terms of the potential impact of demographic variables on the environment, however, the critical element in the end resides in the pressure which, within a defined space, is exercised on its natural resources. In this regard, the unequal distribution of population, the degree of stability of people residing at their place of birth, the destination of their migrations and their temporary or permanent character are all major variables. Yet, as we have tried to show, these factors are themselves dependent on larger and more stable determinants, namely the constraints and incentives of the physical environment, the legacy of history, both ancient or modern, and the balance in relations with respect to economic and political forces across the whole of West Africa. Depending on the time and place, these fundamental determinants combine in varying ways. It is these combinations that we tried to identify and it is from these that we have proposed a division of the region into large territorial units, each of which is identified by its dominant geo-demography. The approach which we have adopted permits us to bypass overly general statements and to grasp better the diversity of concrete demographic situations encountered in the western Sahel–Sudan.

From here, much work remains before a link can be established between the facts of configuration and the risks to which the environment might be exposed. Without reference to the economic, technical and social conditions of resource exploitation, no meaningful relation between population and environment can be established.

NOTES

1 To convince oneself it is necessary only to read H. Barth's remarks as he travelled, at the beginning of the 1850s, through the zone where Damagaram's armies and those of the Fulani of Katsina constantly clashed, which resulted in its inhabitants abandoning the area (Barth, 1965: 449–50).

2 Thornton (1992) argues that slavery was an institutional component of many African economic and social systems that minimised the impact of the transatlantic trade. Slavery certainly was not introduced by transatlantic commerce, but its prior existence, on the contrary, did enable the expansion of this trade. It is, nevertheless, hard to imagine that following the opening of new channels

the intense demand did not have political, social and economic effects on local societies.

3 Here also, H. Barth is an important witness and the events that marked his long voyage through these regions in the middle of the nineteenth century are quite revealing in this regard (Barth, 1965).

4 The first observers obviously overstated the population density by estimating between 30 and 50 inhabitants per square kilometre, whereas it was perhaps about 20 inhabitants per square kilometre, which was already high for the time (Marchal, 1983: 405).

5 It was not until 1947 that forced labour was finally abolished by a law initiated by the African leader Houphouet-Boigny, who was later to become President of Ivory Coast.

6 Henceforth money became necessary in practically all material and social aspects of everyday life (Raynaut, 1977a).

7 Which did not halt (as was so often seen at borders of two colonial empires) the deliberate strategies by each power aimed at attracting its rival subjects (Painter, 1985: 292).

8 The franc of the 'Communauté Financière Africaine': the currency linked to the French franc and used in the French-speaking countries of West Africa (100 CFA = 2 FF prior to January 1994; 100 CFA = 1 FF since January 1994).

9 One must be very prudent concerning the accuracy of the evaluations proposed in this domain. Therefore, according to the sources used here, it is estimated that in 1975–6 the number of Malian and Voltaic migrants living outside their countries were for the former, 450,000 or 420,000 and for the latter, 330,000 or 1 million (Zachariah et al., 1977). Whichever hypothesis is chosen, it is certain that the phenomenon is massive and that its effects reverberate throughout the demographic structure of the countries of origin as well as the countries of destination.

10 The detailed studies undertaken in southern Burkina illustrate how migration for commercial purposes towards the markets located along the route to the Ivory Coast played the role of bridgehead in the Mossi rural colonisation movement (Collectif, 1975).

11 This dependency on the Nigerian economy is such that, in this region of Niger, the everyday use of Nigerian currency strongly competes with that of the CFA franc.

12 A detailed study in 1984 of the medium-sized city of Maradi (85,000 inhabitants at the time) showed that with a growth rate of 8 per cent per year, the annual migratory influx was about 3,500 people, which only represents a drop in the ocean compared with the million inhabitants who comprise the department of which this city is the capital (Herry, 1990).

13 Even if identical populations live on either side of the border, from a social and historical viewpoint this line is not completely arbitrary. It generally corresponds to the dividing line between territories controlled by the Hausa states, which resisted Fulani conquest, and those placed under the authority of the Sokoto empire.

14 For analytical factors concerning the cultural and political foundations of the crisis of Tuareg societies, see Bourgeot (1990).

4

MAJOR SAHELIAN TRADE NETWORKS: PAST AND PRESENT

Emmanuel Grégoire

Among the many elements influencing the way Sahelian societies manage their environment, economic factors are as important as, for example, demographic ones. In fact, the history of the Sahel reveals that economic determinants have had a direct impact on the functional mechanisms of agrarian systems and therefore on the way in which rural communities use space. The introduction and spread of commercial agriculture (groundnuts and cotton) during the colonial period and then later the emphasis on subsistence crops after the drought of 1973–4, by way of the major rural development projects and hydro-agricultural schemes, brought about specific peasant responses. These responses have illustrated how external economic change can disrupt the relations between local societies and their environment.

We shall consider this aspect in its historical perspective, by showing that these societies have always managed their natural environment, not only in order to reproduce themselves materially and socially, but also to participate in larger trade systems that have extended beyond the African continent for close to a century.

THE SAHEL: AN ANCIENT TRADING REGION

The Sahel has always been a land of long-distance trade. This commerce assured the prosperity of great polities throughout its history. The accounts of the first explorers to penetrate the interior of the continent during the first half of the nineteenth century (Heinrich Barth, René Caillé) reveal a dense traffic of trans-Saharan trade networks, movements towards the coast and large regional networks based on local resource exploitation (Barth, 1965; Caillé, 1980).

Sustained trade has linked the Sahel with North Africa for many centuries. It contributed to the birth of large market cities, such as Kumbi Saleh, Oualata, Jenne, Gao, Tombouctou, Agades, Zinder, Katsina and Kano. It was in these cities, in contact with southern Morocco, Tripoli and sometimes even Egypt, that journeys were broken and goods exchanged. From the

Sahel, camel caravans exported gold, ivory, cotton and ostrich feathers, as well as slaves (until the middle of the eighteenth century). From North Africa or Europe they imported crafts, horses and arms that allowed states to maintain their hegemony and to control these very trade movements. The first of these was most likely the Kingdom of Ghana, which originated in the fourth century as a result of the development of trade between the Sudan and North Africa by way of Western Sahara. Ghana was succeeded by the empires of Mali (thirteenth and fourteenth centuries) and later Gao (fifteenth and sixteenth centuries), the powers of which rested on the gold trade, and then later by the Hausa states, where the merchants were linked to those of the Arab world, and finally by the empire of Bornou for which political and economic expansion originated, in part, from its relations with the Mediterranean (Libya, Egypt) through the Fezzan (Suret-Canale, 1968–72). War and trade were thus often complementary, with the former feeding the latter. Similarly, commerce and Islam were closely linked, with Islam's spread from North Africa following the commercial routes into the Sahel (Meillassoux, 1971).

While trade across the desert was important, other trades were also developing towards the south, including salt, natron, iron, fabrics, leather articles and dried onions. Even dates from the Saharan oases were exchanged for forest products. Thus the Hausa merchants wove dense networks into Gonga and Ashanti (present-day Ghana) where cola nuts were produced (Lovejoy, 1978). Kano was connected to Salaga by a whole network of way-stations and some Hausa traders, such as 'Alhaji' Alhassane dan Tata, the founder of a large merchant dynasty in Kano, even settled in Kumasi for several years. This southward commerce also gave birth to large market towns that acted as contact points for Sahelian and forest states. One of the most important of these was the Salaga (Ghana) market, but Bobo-Dioulasso, Kong, Bouna, Bonduku, Kintampo and others were also notable. The Sahel was therefore bordered to the north by a series of 'Saharan ports' and to the south by a group of contact cities for coastal commerce. Several of these agglomerations (Kano, for example) gave birth to truly urban civilisations with populations reaching several tens of thousands by the beginning of this century.

Concurrent with these long-distance exchanges was a multitude of smaller movements that connected the different Sahelian regions to each other. These internal channels carried a few European goods and mainly local products, such as foodstuffs moving from areas of surplus to areas of shortage, as well as local crafts. Yarse merchants, originally from Yatenga (a major Mossi kingdom), would therefore organise caravans headed for Mali carrying cotton products and returning with salt and also dried fish and mats (Izard, 1971).

Nevertheless, during this period peasant communities generally lived within a subsistence economy based on the exploitation of their immediate

91

natural environment, which met all their needs and assured their material and social reproduction. These populations therefore lived on agriculture and the products of hunting and gathering.

The large movements that crossed the Sahel were founded on complementarity and were orientated predominantly north–south rather than east–west, with various Sahelian zones offering analogous products. As we have just seen, these economies were not directed inward but instead outwards towards the Arab world as well as to Europe, particularly in zones close to the Atlantic Ocean (Senegal). They were to be profoundly disturbed by the colonial penetration that accelerated after 1875. France initially took possession of the entire fringe bordering the desert and then, progressing from west to east, completed its domination by the turn of the century with the conquest of the region bordering Lake Chad.

THE COLONIAL CONQUEST: A PROFOUND CHANGE IN THE SAHELIAN ECONOMY

Colonisation profoundly altered the very bases of the workings of the Sahelian economies, especially for agriculture, which would henceforth be forced to join the world trade networks. It had two immediate and fundamental consequences: the trans-Saharan trade disappeared, and local monies were replaced by the franc.

For several reasons, by 1900 trade across the desert had effectively ceased. First, conquest of the Saharan and northern Sahelian zones was difficult, especially in Aïr where the Tuareg, under the command of Chief Kaocen, had resisted the French in 1917. Later, caravans became increasingly unwilling to travel between sub-Saharan and North Africa for fear of the frequent attacks carried out by looting bands of Arabs and Toubous. Finally, and undoubtedly the principal reason for the decline in trans-Saharan trade, transportation costs became more expensive by desert than by sea, especially after the French and British colonial administrations developed the port and railway infrastructure in the south (in 1912 the railway reached Kano in British Nigeria). Trans-Saharan trade therefore slowly dried up. What remained were the Tuareg salt caravans that provisioned the distant oases of Bilma and Fachi with millet, pelts and fabrics and returned southward with salt, natron and dates.

In addition, shortly after the political conquest, the franc became widely used. Sahelian populations were forced to abandon the practice of bartering as well as the currencies in use at that time, principally cowries (small shells mainly coming from the Maldives Islands in the Indian Ocean) and the Maria-Theresa Thalers, for French coins. This substitution caused enormous problems, because the amount of coins put into circulation was insufficient (especially those in high demand, such as low-value coins like the 50 centime and 1 franc pieces) and encouraged the development of a large black market, which operated at the expense of the population whose traditional monies were considerably devalued. Somewhere around 1920 the franc came into general use throughout the Sahel, completing the process of economic domination which was also reflected in the installation of customs barriers at the newly drawn political borders.

Colonisation also led to new forms of trade organisation. Through the major pan-African trading houses, from Dakar to Lake Chad (French West Africa Company, Commercial Society of West Africa, French Niger Company, Maurel and Prom, Personnaz and Gardin, etc.) European commerce progressively strengthened its domination of the Sahel, where the economy became tightly connected to that of Europe. Nevertheless, African merchants were able to maintain control over some sectors of activity that were considered 'traditional' and in which colonial firms had no direct interest (notably in salt, dried fish, livestock and the cola nut). However, the way trade was now organised deprived them of control over the major part of the local trade and placed them in a position of dependence relative to European commerce. Many African merchants simply became intermediaries between the trading houses and their local producers and consumers. This is the group which went into the fields to collect the products destined for export (groundnut, cotton, gum arabic, pelts) and which was responsible for the retailing of products manufactured by European industries (fabrics, hardware, etc.).

Submitting to a new economic order had direct effects on the exploitation of natural resources. While Sahelian agriculture had until then been essentially subsistence in nature, directed towards local self-sufficiency and a medium-range trade, it now served the needs of European industry and this led to the development of groundnut and cotton crops. Trade, and later production incentives, for these two crops (gum arabic was less important and more or less limited to areas bordering the Senegal River) came to be concentrated in several zones which are favourable from the standpoint of nature and accessibility (see Chapter 3). For the groundnut this largely meant the Sérer and Wolof areas of Senegal as well as some parts of the Maradi and Zinder region in Niger. Cotton was concentrated in the central plateau of Upper Volta (future Burkina Faso) and southern Sudan (the future Mali).

In a sense, these already densely populated regions created basins of activity favourable to the development of trade. This gave rise to new economic zones and continental exchanges based on the export of products of commercial agriculture and the import of trade goods. In the preceding chapter we saw how some of these zones became strong poles of attraction for migrants. Within this framework, the spatial distribution of precolonial commercial and market centres was profoundly altered. The demise of trans-Saharan trade marked the decline of trade centres along its routes (Oualata, Tombouctou and Gao, for instance), while the colonial economy

93

favoured the development of markets that collected the products of commercial agriculture. Hence, we see the progressive expansion of trade centres such as Kaolack, Louga and Diourbel in Senegal, of Kayes, Sikasso and Koutiala in Sudan, of Bobo-Dioulasso in Upper Volta and of Maradi and Zinder in Niger. Following the principles of the trade system, these areas also became centres of diffusion for imported manufactured goods and retailing became exceptionally active. In this way the partition of Sahelian economic space that we have already described was established with a few poles of production and exchange alternating with vast expanses dedicated almost exclusively to the exportation of labour power.

Until the Second World War, French colonial policy was founded on the principle that the colonial administration's livelihood and investments should be obtained from local resources. After the Brazzaville Conference (1945) a clear policy shift was brought about with the establishment of FIDES (Fonds d'Investissement pour le Développement Économique et Social – Investment Fund for Economic and Social Development), the prohibition of forced labour and the end of the separate administration for the natives. From this point on the mother country increased the transfer of financial aid to its colonies yearly, notably in the Sahelian zones, in order to create the economic infrastructure necessary for the development of its territories, by means of road construction, railways, airports, the development of industry to process local products (oil-works, cotton ginning plants) and hydro-agricultural schemes. Thus, these investments, along with a favourable world market economy, made the 1950s a period of sharp growth for those Sahelian economies that had progressively inserted themselves into the world trade system.

INDEPENDENCE: POLITICAL CHANGE, ECONOMIC CONTINUITY

The creation of a colonial administrative apparatus encouraged the development of an African 'elite' whose role was essentially that of (elementary) teacher, administrative employee and shop clerk. These bureaucrats would found the political parties and unions that, after 1946, would claim the largest share of native participation in business management. Later, they would express the demands for independence. Financial support for this goal came from merchants, whose capital accumulation was now primed and who hoped to profit from the departure of the colonisers and their economic agents.

Nevertheless, the achievement of national independence through an easy transition at the start of the 1960s did not bring about a clear break in these economies. The cotton and groundnut trade remained, although the new administrations took steps to limit earlier abuses and to control the trade better in order to create an economic base for their own development.

94

Through the creation of nationalised companies,[1] these states maintained an export monopoly on agricultural products, which constituted their only real financial resource (see Grégoire, 1986).

The new geographic polarisations that were created during the colonial period not only survived but actually intensified. There was unprecedented growth in export agriculture in countries such as Senegal, Niger and Upper Volta, which led to the overexploitation of land and contributed to its degradation (see Chapter 5). A value added tax continued to be imposed on peasant labour through groundnut and cotton exports. However, this exaction no longer came from the colonial authority but from those who exercised power in the new independent states. The creation of this nascent class and the maintenance of a continuously growing bureaucracy was also guaranteed by means of a direct tax that, each year, weighed more heavily on rural communities. For example, in 1963 a peasant in Niger had to sell 40 kg of groundnuts at 24 francs per kilogram to pay the tax for one person,[2] whereas by 1970 the peasant would have had to sell 70 kg at 21 francs per kilogram to meet the same obligation (Raynaut, 1977b).

In parallel with the development of this bureaucracy, West African merchants also profited from their countries' independence. In most cases (with the exception of Mali which opted for a socialist-style regime in which the public sector supplanted colonial and private commerce) the new leaders took measures that would support merchant activities, in order to encourage the creation of a national private sector that could compete with European commercial firms. These merchants thereby benefited from banking opportunities and preferential access to state markets. In so doing they strengthened their ties to the political and bureaucratic class, which they backed in exchange for support in business. This collaboration, however, was not always without tension and conflict. On several occasions the two groups openly opposed each other, for example when the authorities considered that businessmen were making excessive profits at the expense of the peasants by commercialising production or by over-speculating on the price of staple products, leading to spectacular inflation that was severely felt in the urban areas. In several countries (especially Senegal and Niger) a co-operative sector was created to protect producers and consumers from the arbitrariness of private merchants. In any case, agriculturalists were quick to realise that state agents charged with 'support-ing' them indulged in misappropriation and embezzlement. Losses from this behaviour were no less than those incurred through trading with merchants. Thus they turned back to these merchants, with whom they had long-established ties and a shared system of values, specifically allowing them to obtain credit denied them by the rigidity of the bureaucratic system. This alliance between peasant and merchant explains the collapse of the socialist regime in Mali in 1968 (Amselle and Grégoire, 1987).

In spite of several unfavourable climatic episodes (such as the 1967–8 drought), until the 1970s Sahelian economies continued to function along principles almost identical to those of the colonial period. Then, in 1973–4, the Sahelian economies experienced a devastating famine following a terrible drought. By the end of this dramatic period in all these countries, especially in zones of low rainfall (the river valleys of Senegal and Niger, the Sahel of Mali and Niger), the peasantry abandoned commercial agriculture (groundnut and to a lesser degree cotton) for subsistence crops (millet and sorghum) of which there was a severe shortage and which were consequently subject to sharp price increases. To some extent, states encouraged this development in the name of food self-sufficiency, which had become a priority. Yet in this way they lost their principal export resource (Niger, for example, which had produced 260,000 tonnes of groundnuts in 1972, produced only 41,700 tonnes in 1975). The elite and bureaucratic classes could no longer reproduce themselves directly from the profits of commercial agriculture. Drought was perhaps more widely felt by nomadic populations, notably the Berber, Tuareg and Fulani, who lost the majority of their livestock.

Perhaps even more than decolonisation, this new situation marked a decisive turning-point in the economic history of the Sahelian countries. Only Niger managed, temporarily, to hold back its effects through the exploitation of its uranium deposits that would bring considerable financial return for nearly ten years (since then the uranium 'boom' has passed and the country has succumbed to an unprecedented economic crisis).

THE YEARS FOLLOWING THE 1973–4 DROUGHT: PROJECTS AND FOREIGN AID

While the drought sounded the death knell of an important source of revenue for the Sahelian countries, it also marked the appearance of a new, unexpected gift. Western public opinion had discovered the Sahel, and for the first time international aid was mobilised on a massive scale.

By the mid-1970s the Sahel had become inundated by a multitude of development projects, such as the large rural development projects financed through major funding sources like the World Bank, the United Nations Development Programme (UNDP) or the European Fund for Development or by bilateral organisations, along with the more limited efforts undertaken by non-governmental organisations. All had the same principal objectives. These were, on the one hand, to develop subsistence agriculture or herding in order to protect the peasantry as much as possible from the effects of drought and, on the other, to guard against the strongly felt threat of desertification. In many cases it eventually became clear that these projects ended most often in failure, and that they contributed more to the development of the cities in which they were headquartered (by favouring the rapid expansion of a local petty bureaucracy necessary for their administration and by stimulating the activities of the private merchant and industrial sectors) than to the peasantry that was the initial object of their activities. The OACV (Opération Arachide et Cultures Vivrières – Operation Groundnut and Food Crops) in Mali illustrates this point well. Initiated in 1974 with backing from the World Bank and the FAC (Fonds d'Aide et de Coopération – French Fund for Aid and Co-operation), this operation was intended to furnish peasants with inputs and technical support, and to market the groundnuts they had been encouraged to produce. After eight years of operation, an evaluation discovered that the project's beneficiaries had been limited to the Malian government, the organisation's own personnel and a few 'pilot peasants' who, for the most part, were in fact merchants, *marabouts* or well-off peasants with government connections. This case is not exceptional and similar situations have occurred in numerous development projects throughout the Sahel (Amselle and Grégoire, 1987). The project years became a series of disappointments. Yamba Boubakar (Yamba, 1993), for example, described just how disappointing, over the course of 20 years, successive programmes for protecting the environment in Niger had been.

After 1974 the peasant economies, already disorganised by the introduction of export crops during the colonial period, continued to face an intense demand (notably for cereals to feed the urban centres where demographic growth was rapid) without the means necessary to improve production (except in a few particular cases such as Mali's cotton region, for a number of reasons that will be described later in Chapter 5). These economies also continue to be very vulnerable to the climatic uncertainties that constantly threaten to disrupt their food supplies. This was the case in 1984–5 when, in spite of improvements in food aid since 1974, severe famine once again struck the region.

Not only was foreign development assistance incapable of attaining its objectives, it also contributed dramatically to the debt of these Sahelian countries. In most cases, this aid was negotiated in the form of long-term loans, although it is true that a part of this debt has been cancelled by the lenders. Nevertheless, the burden continues to be extremely heavy.

Over the years, international aid reached a high level, especially considering that it was often inspired more by greed on the part of the Sahelian states than by necessity. Indeed, there was a great temptation to claim a food shortage solely in order to mobilise external aid, even when such a claim was unjustified by production figures. With the indispensable complicity of donors, which was at times guided by political objectives (in the case of bilateral aid) and at other times guided by the need to justify their existence (in the case of international organisations), 'desertification' and 'drought' too often became convenient pretexts for attracting the interest of potential 'grantors'. It is largely through this largesse that the elite classes would be able to reproduce themselves from this point on, as in Niger where aid

97

replaced the loss of benefits caused by declining uranium production which itself had substituted for groundnuts (Raynaut and Souleymane Abba, 1990). During the ten years from the mid-1970s to the mid-1980s, at every level of the state and bureaucratic hierarchy aid became a new source of accumulation, captured by politicians and bureaucrats, sometimes in connivance with the national business sector. The recently acquired fortunes of many politicians and administrative agents can be explained by the diversion of funds from development programmes, or by the taking of commissions, or through other corrupt practices.

The granting agencies, themselves subject to the constraints of the world economic crisis, finally realised that their aid was ineffective and that the rush to indebtedness could not continue indefinitely. As a result they modified their policies appreciably and increased their efforts to control how funds were being spent. With this goal in mind, they introduced the notion of 'conditionality'. Henceforth aid would be tied to certain economic stabilisation measures within a framework known as the 'Structural Adjustment Programme' (SAP). These measures were considered imperative for overly indebted Sahelian economies, now deprived of export resources and suffocating from an omnipresent state.

THE 1980S: THE ECONOMIC AND POLITICAL CRISIS OF THE MODERN SECTOR

In effect, international organisations (such as the International Monetary Fund) and donor agencies (primarily the World Bank) demanded that the states put their economies in order and improve the efficiency of the public sector and expand the private sector.

As elsewhere in Africa since the mid-1980s, the economic policies of the Sahelian countries have, therefore, been considerably redirected. The goal was to return to a better balance between public and external finances and, at the very least, growth was to keep pace with population.

In order to wipe out the deficits over the past years, the same remedy was administered everywhere in the form of structural adjustment programmes. These advocate, specifically, the opening of the economy and the scaling of internal pricing on the international price system, disengagement of the state, privatisation, general application of the principle of unsubsidised prices and the priority of market forces.

These SAPs reduced state intervention in the economy and introduced an incontestable liberalisation, which benefited the private sector. These policies were reflected in a series of liberalisation and privatisation measures. For instance, with cereal commercialisation, state offices were stripped of their former monopolies. In Niger, SONARA was dissolved and the Office of Food Products lost its monopoly in the cereal trade. Internal commerce underwent a total liberalisation, both in agriculture (more rapidly for millet, maize and sorghum than for rice) and in manufacturing. The Malian (state) Society for Import and Export (SOMIEX) lost its former monopoly and its retail commercial activities when its stores were sold to private merchants. In foreign trade, wider access to international markets was gained, to a greater or lesser extent depending on the country.

A growing number of countries entered into such adjustment agreements because loans by international organisations were subject to the observance of recommendations that reduced the local governments' margin for autonomy, initiative and negotiation (Durufle, 1988). These governments were obliged to follow directives imposed from the outside, and to apply them to populations who had to endure their negative social consequences (closing of public enterprises, personnel cut-backs, the end of automatic administrative hiring for young graduates, etc.).

During this same period, while undergoing economic change, Sahelian countries were engaged, with varying degrees of success, in a process of democratisation largely initiated by their former colonial rulers, who now made aid conditional on the establishment of democracy (as President Mitterrand stated in a speech to the African Presidents Meeting held in La Baule in June 1990).

While Senegal, a pioneer on the continent, had committed itself to political liberalisation well before other countries (Coulon, 1992), other authoritarian military regimes (with the exception of Burkina Faso) began to fall, one after the other. This occurred through popular action as in Mali, Cape Verde (Cahen, 1991) or, as in the case of Niger, through the death of a head of state (Raynaut and Souleymane Abba, 1990). The convening of national conferences, which assembled all the new political forces, sounded the death-knell of single-party regimes and heralded the beginning of multi-party politics (Mali, Niger). These conferences also defined the rules of the new political game (development of constitutions) and designated transitional governments to oversee presidential, legislative and local elections. This process has been completed in Mali, and is ongoing in Niger, where the people voted on a new constitution (December 1992) before electing a new president and representatives (February–March 1993).

Still, the problems of lasting democracy in these Sahelian countries remain. It is certainly reversible, as nearby Togo demonstrates; here the partisans of General Eyadema seem unlikely to step aside for the elected candidates. From a strictly economic viewpoint, many obstacles are likely to appear on the road to democracy. In fact, according to funding sources such as the World Bank and the International Monetary Fund, structural adjustment policies are indispensable for re-establishing a country's budgetary and financial balance and, in certain cases, must be followed with absolute rigour. However, these policies themselves jeopardise the gains of the middle class by reducing the number of civil servants and by freezing salaries and employment in the public sector. A democratically elected regime is hardly in a position to impose such inherently unpopular measures. There is, therefore, a strong temptation to return to an authoritarian power (such as President Sani Abacha in Nigeria), which is deemed more appropriate to respond to the imperative of restoring economic order. In the end, the key issue is one of compatibility between the restoration of economic order, at least in the way it is imposed, and the effective establishment of democracy (Bayart, 1991).

SAHELIAN MERCHANTS: A PERMANENT PRESENCE

In spite of repeated crises and enormous constraints, the economies of the Sahelian countries survive and possess a real dynamism. Alongside the development of 'formal' economies dominated by the state, 'informal' economies have been maintained and even strengthened, animated by groups of African merchants who have always been the principal economic actors of the region. Both in the precolonial era and during the period of colonial trade, commercial exchanges favoured the emergence of merchants of long-distance trade, of whom some of the better known include the Diola, the Hausa, the Yarse and more recently the Berber.

These large merchant communities were Moslem. In fact, Islam was brought to the Sahel by merchants who were in contact with North African Arab traders through the trans-Saharan trade. On the one hand, Islam provided a communal ideological framework which closely followed long-distance trade routes and secured interregional exchanges, and, on the other, helped forge a unique identity (Grégoire, 1993). From Dakar to N'djamena, in order to be recognised as a merchant one must be Moslem and hold the title of 'el haj', a title attributed to those who have made the pilgrimage to Mecca. The acquisition of this title, in fact, confers a symbolic authority and bears witness, moreover, to a professional success and membership of the entrepreneurial elite.

This closeness between commercial and religious spheres is still current today. Merchants encourage Islam's propagation by taking in *marabouts* and well-known "*karamoko*" (schoolmasters), by building Koranic or francoarabic schools, by constructing mosques and by financing pilgrimages to Mecca for family members and close relatives. These religious investments therefore bring together merchant customs and codes of ethics, and support ties of dependency within merchant networks. By imposing a strict code of conduct and ethics on businessmen, Islam provides a framework and a structure for business. It often acts as a substitute for more modern commercial techniques (contracts, banking regulations, etc.) that are difficult to apply in a milieu where many are still unable to read and write.

Throughout all the different economic changes that the Sahel has experienced, merchant communities not only accumulated wealth through traditional commerce, but also by acting as intermediaries between commercial houses and the peasantry, thereby constituting a transitional sector between the spheres of local production and the metropole.

In spite of the many obstacles placed in their path (the commercial trade during colonialism that limited their field of action, and later the policies of independent states that at times deliberately constrained their activities in favour of the public sector) these merchants have always been able to adapt to the difficulties imposed upon them. The decline of West African trade, inescapable according to the 'dependency' theorists of the 1970s, did not occur (Grégoire and Labazée, 1993). On the contrary, Sahelian merchant networks proved to be extremely dynamic. A compromise was found between the organisation of the exchanges developed through the ancient trade circuits and the realities of the contemporary economy. The creation of political borders and later the birth of independent states did not fundamentally disturb their affairs, since they knew how to manipulate them so as to turn obstacles into trade stimulants. Disparities in economic policies and monetary regimes between states became, for them, trade opportunities (Raynaut, 1989c).

These trade networks operate largely on the periphery of the states, mitigating the shortcomings of their formal economies. They eventually came to constitute a parallel economy that provides work and the means of subsistence for a very large number of people. They have their own organisation, often with a decision-making centre and remote locations determined by the imperatives of business, and with these different centres linked through family, client/patron ties or even religion. Hence, upon independence, one of the most important traders from Bamako installed

> his younger brother by the same father and mother in Thies in Senegal, his maternal nephew at Sikensi, the area surrounding Dabu (Ivory Coast) and also a brother by the same father at Kayes in the Sudan. The fortune owned by these four relatives was indivisible, and was controlled by the head of the family.
>
> (Lambert and Egg, 1992: 12)

From an economic standpoint, the activities of these West African merchants were quite diverse, consisting of wholesale and retail commerce, import/export, transport, real estate and even industry. These merchants also make agricultural investments and recently, with the help of a salaried labour force and the occasional use of modern agricultural materials, opened large tracts of bushland for the cultivation of subsistence crops. Others own orchards or market gardens on the periphery of large cities (see Chapter 9). Some of these merchants are also accumulating livestock, with consequences for natural resource exploitation that are discussed in the next chapter.

In spite of their incursions into the area of production, interest by the

101

merchant class in agriculture and herding is largely focused on commercialisation. Their role in livestock commerce is long-standing (Hill, 1966; Bellot, 1982a). However, the liberalisation measures taken over the last several years have been quite beneficial to them, and their range of trade goods now combines local products (millet, sorghum and maize) with imported cereals (rice, wheat). Their networks extend from the small village market to the supra-national economic sphere that includes both Sahelian and coastal countries, and at times extends to other continents where, for example, cereals are traded with multinationals (Grégoire and Labazée, 1993).

Unlike the formal state, which must affirm its sovereignty by defending the principle of an inviolable national border, these businessmen are used to transgressing political entities; they profit from inequalities in regional development and differences in legislation and money. It is they, in the end, who have created areas of homogeneous trade in the Sahel and who are the principals in the present regional economic integration, in spite of the economic and financial crises that exist throughout these modern states.

CURRENT MAJOR TRADE ZONES

Based on the history of present economic conditions of the Sahel, a structuring of Sahelian space from a trade standpoint has been elaborated, taking many factors into account. From this structure, different zones emerge in which can be heard the echoes of the past – distribution of cultural areas and settlements as well as geographic factors such as the presence of labour pools, metropolitan areas, ports and coastal influences.

The Sahel is not therefore a single, unique and homogeneous space, but is instead divided into several economically distinct sub-spaces, all of which are now open to the world market and affected by the movement of products across them. Within these sub-spaces, the principles of trade are intimately linked to the role of the merchants and are essentially part of the 'informal' economy, even though national policies also participate through interactions among states.

There are thus three major and clearly demarcated trade zones (Figure 4.1), each with its own particularities, as we shall demonstrate.[3]

The western sub-space

The western sub-space covers approximately the Senegal, Gambia and Upper Niger River basins and is somewhat unique from the standpoint of history and religion. Hence, Islam is omnipresent from Nouakchott to Dakar and has grown in the Guinean zone with the expansion of the Moslem merchant networks.

From an agricultural standpoint, these countries are similar, having

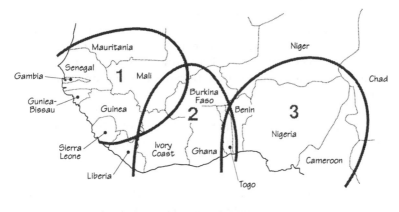

1   western economic sub-space
2   central economic sub-space
3   eastern economic sub-space

original artwork: Charles Cheung
computer graphics: Phil Bradley

Figure 4.1 Sub-regional economies of the western Sahel

experienced the same problem, i.e., extreme variations in production due to a climatic instability that threatens the security of the food supply of their people. For example, in Mali the cereal harvest was half as high in 1985 (the year of the last drought) as in 1989.

A more detailed analysis of the available data enables us to distinguish four large agricultural production zones within this western sub-space:

1 The Senegal valley, which produces mostly paddy rice (the Richard Toll schemes), in addition to millet and sorghum.
2 The zones of Western Senegal (the former groundnut basin) and the Gambia, which produce mostly millet and sorghum, but also maize.
3 The Casamance valley, which produces mostly paddy rice.
4 Southern Mali and the Bend of the Niger, which may be considered the granary for this sub-space (responsible for 50 per cent of millet and sorghum production, 60 per cent of the maize and 20 per cent of the paddy rice).

While millet and sorghum remain the staple foods of the population, especially in rural areas, it is important to note the significant increase of imported products (rice but also wheat) during the last 20 or so years. In fact, rice is preferred by urban consumers (especially in Dakar) both for its price and for the ease with which it can be prepared, compared with traditional grains. Major re-exportation movements have developed from

103

Mauritania, the Gambia and the two Guineas, towards Senegal and Mali, because rice imported from Asia is less expensive than that produced locally. Large merchants have a greater interest in the trading of imported grains than in local cereals. Formerly the indispensable vectors for trade of complementary local products, they became the instruments for a more outward orientation.

Inherited colonial borders and state policies take a heavy toll on trade in this zone. In particular we may cite the existence of different currencies in the Gambia and Mauritania, countries for which only the CFA franc can be used as payment in the international markets.

A phenomenon of widespread extension, which is particularly applicable to the area and type of exchanges of goods just described, should be noted here. Until its recent devaluation (12 January 1994), the long-lasting overvaluation of the CFA franc compromised the competitiveness of countries like Senegal, Burkina Faso and Niger with respect to countries like Gambia, Ghana and especially Nigeria which had weak currencies. The markets of the former were inundated by diverse products imported, often fraudulently, from (and produced by) these latter countries. The result was the strangulation of their own industry (textiles, for example) and a lack of markets for certain of their own agricultural products (notably rice).

Added to the economic situation just discussed is an inadequate communication infrastructure, major demographic disparities between coastal and interior regions and economic policies that are divergent, if not openly contradictory (especially in relation to the commerce in rice).

Finally, although the human geography of this sub-space suggests a long-standing economic and commercial integration, on the contrary the current political and economic map suggests evidence of fragmentation.

The central sub-space

Beginning at the edge of the desert in the north, crossing the Sahel and ending at the limit of the northern savanna of the Ivory Coast, the central sub-space stretches from Mali to Burkina Faso. Settlements are quite diverse and composed of nomadic and sedentary groups.

From the standpoint of land use, this zone is somewhat homogeneous. In the northern sector a pastoral zone is exploited by the Fulani and Tuareg, while further south agriculture begins to dominate, first with the subsistence crops of millet and sorghum on the Mossi plateau and then with cotton in the south-west of Burkina Faso (see Chapter 6). There are also development projects along the waterways: Mali's l'Office du Niger hydro-agricultural schemes in the Kou valleys in Burkina Faso.

As far as trade is concerned, this zone has always been crossed by many movements of various goods, with Sahelian products exchanged for those from coastal countries. These transactions were carried out by Diola merchants, who adapted so well to trade commerce during the colonial period and to the demands of contemporary business that they continue, through highly structured and extended networks, to dominate regional commerce, notably between Bamako and Abidjan. From an economic standpoint, these Diolan expatriates give this sub-space a certain homogeneity. We can distinguish two principal centres of concentrated activity:

- The first is located in the border zones, comprising the cities of Sikasso, Koutiala and Bougouni in Mali, Bobo-Dioulasso and Banfora in Burkina Faso, Korhogo and Odienne in the Ivory Coast and Kankan in Guinea. It was created by a diversity of goods and exchange circuits, by their varying volume (neighbourhood commerce, long-distance movements) and by the diversity of those who take part in the commerce of the frontier (Grégoire and Labazée, 1993).
- A second centre emerges around the Mossi plateau, the Ouagadougou region and south-eastern Mali. It acts as a corridor for Malian products, such as livestock, destined for coastal countries (Ghana, Ivory Coast). Cereals such as millet, sorghum and maize are also traded and exported from Ouahigouya to Mopti by way of the Bwa country.

In this central sub-space, Burkina Faso acts as the gateway between north and south. There also exists a dynamic internal commerce in Burkina, where deficits in cereals in the northern regions are met by surpluses from the south-west.

The eastern sub-space

This sub-space consists of Niger, a Sahelian country forming a hinge between Saharan and black Africa with a long tradition of trade. It is composed of populations that have a long secular history of living together and of commerce (Hausa, Djerma, Fulani, Kanoury, Tuareg and Toubou), although some are focused towards North Africa and others towards Nigeria.

The dependence of Niger's economy on its powerful neighbour to the south is evident and so strong that they appear to have achieved a certain economic integration, founded on their official trade, but also, and above all, on their parallel networks. The dynamism of their trade relations is a response to two sets of factors:

- Structural factors include the dependence of Niger (being landlocked) on the south for its provisioning (manufactured products, energy, etc.) and the agropastoral complementarity of the two countries (traditionally, Niger exports livestock and imports cereals from Nigeria), their difference in size (Niger has less than 8 million inhabitants while Nigeria has more than 100 million) and their economic potential. These two countries also belong to two different currency zones. The CFA franc is very strong while the naira is significantly undervalued, making foreign currency inaccessible.

- Economic factors that stem from disparities in economic policies (customs policies, subsidies, state interventions, protection, etc.) that create trade opportunities and the possibility of manipulating them legally or otherwise (Grégoire, 1986).

Niger is therefore focused and dependent on the south for a large part of its provisioning, notably of food. In turn, it benefits from the dynamism of Hausa merchant networks whose reach extends from the village to large cities, to the markets of Nigeria and, in certain cases, to other continents through connections developed with the multinational grain traders.

CONCLUSION

The colonial, and later the independent states, were largely responsible for subjecting agricultural and pastoral societies to destabilising influences. This was the case early on with trade, whereby certain peasant societies were more or less forced to produce ever-increasing quantities of groundnuts and cotton (the long-standing primary financial resource of both the colonial powers and later the new states). The intensive practices associated with these crops affected both the production systems and the population's food security, notably during the 1973–4 drought. This was also the case, starting in 1975, with the development of projects in the Sahel designed to guard against future famines. The 'technological package' popularised from Senegal to Chad was, therefore, exactly the same throughout; it failed to consider regional differences, be they ecological, social or cultural. These projects could not succeed because the technical themes they promoted did not respond to the peasant's real needs, nor to his capacity for absorbing new technical practices or investment. These factors eventually led the Sahelian states to incur substantial debt. That the major labour reserves were also affected by these massive efforts to diffuse commercial agriculture or the more recent large agropastoral development projects was overlooked. Labour was exported to the centres of Sahelian commercial agriculture and especially the coastal plantations (see Chapter 3).

To these peasants, merchant networks represented an alternative to state intervention. It was therefore preferable to sell their groundnut or cotton harvest to private merchants rather than to the co-operative, since closer economic and social ties were maintained with merchants than with the co-operative agents who often mistreated them. In fact, the trade created through these alternative, merchant-based arrangements gave these rural societies a certain margin of flexibility, enabling them to survive and meet their needs at the margin of the formal economy. Hence, the diffusion of fertiliser in central Niger is primarily the result of merchants, rather than state agencies. Farmers purchase fertiliser brought in fraudulently from Nigeria at reasonable rates instead of the product that is more costly and often not adequately available through official circuits. These exchanges do not always benefit the farmer. Cereals imported at lower cost (notably Asian rice) have long undermined the development of regional production. Rice produced in the river valleys of Niger and Senegal, as well as in schemes developed over the last few years, is more costly than rice imported from Thailand or Vietnam. It is still too early to assess the impact of the recent devaluation of the CFA franc (12 January 1994), although several developments suggest that national rice production may have regained its competitive edge (in March 1994 a bag of Nigerien rice sold for 9,000 CFA francs in Niamey while a bag of imported rice cost 13,000 CFA francs).

Like those of coastal states, the economies of Sahelian countries are, therefore, characterised by a juxtaposition of two systems: a formal economy attempting to order production and to regulate trade within a rigid legal, administrative and financial framework, and an informal economy, following a different principle, animated by merchant networks and the populations that take advantage of these arrangements (if necessary, fraudulently). With the economic recession plaguing West Africa, the dichotomy between formal and informal economies has been accentuated. The crisis in the formal economy is connected to the state (also hit hard by recession) which limits its capacity to function and thereby reduces its revenue and tariffs. States can no longer meet basic expenses, such as recurrent civil servant salary payments, maintaining and developing social infrastructure and the cost of sovereignty which has been on the rise with the advent of democracy and the associated institutions created by it. Although it is better able to absorb its effects, the recession also impacts on the informal economy. Merchant networks have created breathing space in the form of markets outside state control. These markets are based partly on the economic and social organisation of older models of society (including multiple elements, such as kinship, religion and clientilism) and partly on a deeper understanding of market mechanisms, both at the global level and the level of internal continental trade. The strength of these networks, therefore, resides in their ramifications which are primarily economic, but also state-related, social and religious.

In the end, at the margins of the state, these networks order trade by taking advantage of complementarities between production zones, differences in legislation from one country to the next and membership in different monetary zones (for example, Senegal and Gambia, or Niger and Nigeria). To understand a population's economic strategies and the dynamics that order its relationship with the natural environment, in all its diversity, we must consider the informal economy. Trade and production principles can conflict, especially given present circumstances of trade globalisation and the extremely competitive nature of interstate commerce. Within such a context, the Sahel appears more subordinate than ever to the external world and deprived of a voice. These rural populations are exposed to economic constraints beyond their control, yet to which they must respond, that have repercussions for the way they manage the natural environment.

NOTES

1 Office of Agricultural Commercialisation in Senegal, Office of Food Products and the National Company for Groundnut Commercialisation (SONARA) in Niger, Office of Agricultural Products in Mali.
2 This is a lump sum personal tax or 'poll tax'.
3 This analysis is based on work by the research team of IRAM–INRA (Institut de Recherche et d'Application des Méthodes de développement (Institute for Research and Implementation of Development Methods) – Institut de Recherche Agronomique (Institute for Agronomic Research)) and of Benin University (Coste and Egg, 1993; Egg and Igué, 1993).

5

A SHARED LAND: COMPLEMENTARY AND COMPETING USES

Claude Raynaut and Philippe Lavigne Delville

After describing and analysing the variety of natural, demographic and economic situations that divide the Sahel–Sudanian zone, our discussion will now focus on a new axis of differentiation that describes the exploitation of land and its resources by Sahelian societies. Here also, diversity is clearly evident, but beyond merely taking note of this fact, we must try to order the mosaic of individual situations. This can only be accomplished by a progressive approach, beginning with a categorisation that is necessarily static and reductionist, and then slowly introducing nuance and movement.

In earlier times, the exploitation of natural resources in the Sahelian zone largely relied on a combination of various activities, including hunting, fishing and gathering. Hence, historical investigations of settlement patterns reveal that villages established in the last century or at the beginning of this century were guided as much by the search for game as by the quest for new lands (Raynaut *et al.*, 1988). Currently, in many societies in this part of Africa, the uncultivated bush and hunting activities continue to provide an empirical foundation from which to build categories for a representation of the world and of social organisation. Along with many other examples, this can be seen among the Hausa of Niger, with their rites over opening up bush lands (Nicolas, 1975) and the richness of their animal taxonomy (Luxereau, 1972), and among the Sénoufo of Mali, with the major role played by hunting societies (Sanogo, 1989). Despite this, the virtual disappearance of wild fauna means that hunting barely survives as a means of subsistence, even when not officially outlawed or suppressed.

On the other hand, fishing still remains a vital activity, having contributed since the distant past to the concentrations of population around areas where fishing resources were abundant. The prosperity of the ancient civilisations of the Niger River's Central Delta was certainly founded on the trans-Saharan trade, but also owed much to a whole chain of activities linked to fishing: the catch, processing and then the marketing of fish. Today, several population groups that have settled along the banks of the large rivers and lakes of the Sahel live partially or almost exclusively on the products of fishing. Coastal fishing, very much alive in Senegal and

Mauritania, is only mentioned here in passing because it represents an entirely different ecological domain from that being studied here.

While gathering may have declined noticeably with the degradation of self-seeding vegetation, it continues to contribute significantly to the economic and material lives of people in the Sahel. Innumerable tree species are used for fruit, leaves, bark or wood (Bergeret and Ribot, 1990). They provide food as well as primary materials used by artisans. The traditional herbalists also heavily exploit these species as well as other types of vegetation (ACCT, 1987ff.). However, it is the collection of wood, woodfuel or timber (which can be considered a gathering activity) that represents the most widespread and heavy use of natural vegetation.

Clearly all these forms of natural resource exploitation have played a role in the evolution of the Sahelian ecosystem, and continue to do so. We put them aside for now, however, in an effort to categorise and to map the different types of land use. There are several reasons for doing so. In addition to a lack of precise information on gathering, this decision is based on less circumstantial motives. Without proof to the contrary, gathering seems to constitute a generalised practice throughout the zone and appears to be linked mainly to a greater or lesser abundance of available resources. In this region there are no hunter–gathering populations like those found elsewhere in Africa.[1] With respect to fishing and hunting, it is unclear how they may be linked to the degradation of vegetation and soils, the phenomena upon which the whole of our investigation centres.[2] In spite of the substantial impact wood collection can have at the local level (especially in areas where bush land is accessible to large urban agglomerations), at the level of the Sahelian zone its consequence is relatively minor when compared with the clearing of land for agriculture or of trees for pastoralism. While reserving the possibility of ultimately integrating information concerning wood collection into the analysis (a major type of exploitation, the potential impact on the environment of which cannot be neglected) this analysis will focus on livestock and agriculture. These are the production activities upon which the overwhelming majority of the Sahelian populations primarily depend for their survival, populations who are commonly incriminated in the processes of environmental degradation which is incorrectly classified as 'desertification'. On this basis a preliminary typology of the modes of natural resource exploitation can be produced very schematically, starting from two major axes of differentiation regarding livestock on the one side and water on the other. On the basis of these two dividing lines, we can reach a first assessment of diversity.

RELATIONS WITH LIVESTOCK

It is between pastoral nomads and sedentary agriculturalists that we can establish the most obvious opposition between societies in the Sahelian and

Sahelian-Sudanese zones. These two major modes of environmental exploitation create relations which are sometimes complementary and at other times competitive.

Herding and agriculture: complementarity and competition

Complementarity is evident when it translates into the exploitation of very different ecological regions, such as the nomadic pastoralists north of the 300 mm isohyet (the true Sahelian zone), and the agriculturalists south of this limit. Complementarity can also be seen through the sharing of a single territory and the articulation of activities across time and space. Examples are uncultivated lands and fallow areas used as pasture, the residues of the harvest used by the pastoralists' livestock and manuring contracts during the dry season which allow herders to feed their livestock while fertilising peasant lands. Complementarity can, nevertheless, change to competition when the rules of sharing and access to resources are no longer respected. In the cohabitation between agriculture and pastoralism, the movement of livestock over an area is a critical element. Flexibility in this movement is a necessary condition for the optimal exploitation of forage resources by pastoralists, but it must remain harmonious with the more permanent occupation of the land implicit in agricultural activity. An equilibrium must be constantly maintained between the two demands. The solution is generally found in a methodical organisation of the movement of livestock over the course of the year. During the rainy season within the agricultural zone itself, animals are withdrawn into areas that are less favourable for cultivation and, therefore, less densely populated (lateritic plains, zones of compact soils). However, large-scale transhumance movements are also observable which, according to the rhythm of the seasons, alternately move animals and the men that herd them from agricultural regions to northern pastoral zones and back again. This mode of cohabitation between activities spread out over time exists throughout practically the entire Sahelo-Sudanian margin. It can give rise to complex and organised forms of regulation, as is the case in Mali in particular. In the Interior Delta of the Niger, as described by Gallais (1984), the ebb and flow of floodwaters governs the rhythm of access to the flooded pasture for Fulani livestock as well as the agricultural activities of rice cultivators and the Marka, *Rimaibe*, Bambara and Bwa millet cultivators. The norms which preside over livestock movements in this region are largely the heritage of regulation established before the seventeenth century by Fulani rulers and reinforced in the last century in the state of Macina, within the global social and territorial organisation of the Fulani *Dina*. Even if such an extreme level of regulation does not exist elsewhere (in a case such as this, it was the expression of the hegemonic power over space by a pastoral society), oscillating movements of transhumance take place all along the Sahelo-Sudanian zone, driving

111

livestock north in the rainy season and then south at its conclusion. This circulation can sometimes produce complex crossing patterns, as in central Niger where, during the wet period, *Farfaru* Fulani pastoralists from the Nigerian savanna occupy the pastoral zones temporarily abandoned by the *Wodaabe* Fulani, who have left for the salt pastures within the Saharan borders.[3] In addition to these seasonal movements, the passage of animals along the commercial routes that lead Sahelian livestock south towards the African coastal markets illustrates how agricultural lands are subject to an imposed dependence based on a coexistence with pastoral production systems.

Such a coexistence has never been without tension because it demands a conciliation of rival interests. Conflict can erupt when livestock is poorly controlled, and when herds wander on to cultivated fields. This has always had a tendency to occur at critical periods in the annual cycle, particularly during sowing, when herds are late in leaving agricultural lands, and during harvest, if they return too early. Clashes occur when agricultural activities hinder the movement of herds and cut off their access to water sources or pastures. This situation is illustrated in its extreme along the Senegal River with the development of irrigated agriculture (Santoir, 1983).

In spite of inherent tensions in the joint use of the same space, the two modes of exploitation have for centuries been able to coexist without intolerable contradictions. It is the growing scarcity of land and the imbalance between supply and demand of plant resources which has caused competition to become generalised and exacerbated. In this regard, the increase in livestock since the 1940s, through the effects of public animal health policies, has certainly played a significant role (Boutrais, 1992). However, the major phenomenon during the second half of this century is certainly a massive and rapid expansion of cultivation. For example, in Niger the area of cultivated land increased by more than 50 per cent between 1960 and 1980 (Pons, 1988a). This dynamic has primarily resulted in the progressively more complete occupation of lands where agricultural development has long existed. Because movement is severely restricted in such saturated areas, for example, Yatenga Burkinabe (Benoit, 1982; Marchal, 1983), it has become increasingly difficult to reconcile agriculture and pastoralism. At the same time, residues of the harvest have been more frequently appropriated by cultivators for their own livestock, depriving pastoralists of a fodder resource on which, at other times, they could depend (Milleville, 1985). However, the intensification of agricultural settlements within areas already under cultivation was not the only response to the growing demand for land. This demand also provoked a massive clearing of land towards the north, in a movement that pushed pastoralists back towards areas where the productivity of vegetation was increasingly marginal and where the climatic risk grew constantly worse. Figure 2.2

clearly highlights areas where this push exists, as we have already noted (see pp. 42–3, 44–5, 50).

Diversity among pastoralists

While it is convenient for presentational purposes to contrast agricultural-ists and pastoralists, this dichotomy is, in fact, far from being clearly defined. The very category of 'pastoralist' can mask several different realities. To understand this diversity we follow the distinctions proposed by Retaillé (1989). First of all, it is appropriate to avoid the usual amalgamation between pastoralist and nomad. Pastoralism designates a form of pro-duction in which the material existence and the social reproduction of a human group are organised around the appropriation, exploitation and circulation of herds. The term nomadism applies to a mode of residence and occupation of space that is based on mobility. The diversity of pastoral societies can be understood not only by a meeting of these two categories (pastoralists may or may not also be nomads) but it is also necessary to take into account the nuances that are encountered along each of these two axes of classification. There exist many degrees of mobility, with possible gradations from full nomadism, which displaces an entire community and its herds, all the way to the minor movements in which herders simply withdraw animals at certain periods of the year to pastures a short distance from the settlement areas. Moreover, the degree to which a group is mobile is far from being definitively fixed. In Sahelian pastures, the increase in the number of wells has led to the partial sedentarisation of formerly mobile pastoralists. Thus, in the Senegalese Ferlo, permanent settlements have emerged around new water sources where, since their construction, move-ments around them are limited to approximately 10 km (Déramon *et al.*, 1984). The permanent installation of one group of herders in the region has in fact provoked a schism between two Fulani groups: the *Walwalbe* from *walo*, i.e., along the floodplain of the river, and the *Dierdierbe*, from the *dieri* or interior (Santoir, 1977; Niasse and Voncke, 1985). From the point of view of mobility, as in many other regards, opposition between these two categories is not so clear and many subtle variations can exist within the same group. Thus, among the Bundu Fulani of eastern Senegal certain families are transhumant in the dry season and others are not – the deciding variable being the size of the herd (Deshayes, 1989).

Nuances are just as frequent in the case of pastoralism because a population can depend more or less exclusively on livestock for its material and social reproduction. Thus, many traditionally pastoral populations combine herding with other activities, which may play an important role in their cultural organisation and in their way of life. Here, the typology proposed by Retaillé (1989) is appropriate. As opposed to pure pastoralists, illustrated by such groups as the Fulani (specifically the *Wodaabe*), Retaillé distinguishes the 'Bedouin' pastoralists who raise dromedaries (Moor, Tuareg, Toubou) and for whom animal production has had a long association with other forms of activities essential to their survival (such as war and the taking of slaves), and the long-distance trans-Saharan trade or trade between the Saharan saltworks or oases and the commercial cities of the Sudanian zone. In these 'Bedouin' societies, often represented as the embodiment of nomadic pastoralism, the animal (regardless of its symbolic importance (Bourgeot, 1990)) represents but one link in an economic system functioning on a large and diversified base. Here, the reality is made much more complex because specialised activities may appear to be a function of a horizontal division into territorial entities, as well as an expression of vertical stratifications of social status. Thus, in these large Saharan societies, warrior and trade aristocracies are radically distinguishable from their pastoral tributaries and their cultivating dependants or artisans. Sahelian and Sahelo-Sudanian Fulani society is also characterised, along with the *Rimaibe* caste, by a division of labour founded on domination –allegiance relations.

To understand the principle of land occupation and the exploitation of natural resources imposed by the whole society, the entire system of the division of labour and the exchange of goods and services must be considered. This is notably the point Bourgeot makes when he traces the history of human settlement of the *Kel Tamaseq* in the Maradi region, in central Niger:

> Nomadism, the pioneering agent of land use, was followed by villages composed of *Buzu*,[4] who took up agricultural activity (millet, sorghum and beans). In a paradoxical manner, the opening up of new pastures brought about an extension of agriculture and encouraged the diversification and specialisation of production activities, creating or consolidating the technical and social division of labour and re-inforcing the relations of domination.
>
> (Bourgeot, 1977: 11)

The conditions under which pastoral activity itself is carried out are therefore not homogeneous. Their variability contributes to the distinctions among communities and social categories. In this respect, the type of livestock is a particularly important criterion. Thus, within the entire Tuareg society there exists an opposition between aristocrats, *Kel imenas* (those of the dromedary), and dependants, *Kel ulli* (those of the goats) (Bourgeot, 1990). At another level, although the hierarchical connotation is absent, the Fulani divide themselves into distinct groups: those who raise cattle and those who raise sheep. If such differences reinforce cultural and social divides, equally they produce a great variety with respect to herd movements, to their mobility and to the conditions for exploiting grazing resources.

Cultivators who herd, herders who cultivate

With respect to the diversification of production activities, the practice of agriculture and its degree of association with pastoral activities constitutes an essential criterion for differentiation among groups. Only complete material distress, combined with initiatives organised to distribute food aid, could incite the Nigerian *Wodaabe* to attempt, often with little success, to work the land.[5] In other Fulani categories, agricultural practices instead have been part of an ancient patrimony activated or abandoned cyclically throughout history in a pendular rhythm (Dupire, 1962, 1975). Often, therefore, their system of production achieves a close articulation between agriculture and herding, permitting a greater yield per hectare than those of neighbouring agriculturalists (Beauvilain, 1977: 184; Diarra, 1979: 88). In addition, having money at their disposal from livestock, the Fulani can limit the amount of lands devoted to commercial agriculture or avoid having to sell cereals and, consequently, have better food security (Lavigne Delville, 1988). As we have seen, if livestock and caravan trade are highly valued activities within Tuareg society, agriculture is often a component in the economy of certain constituent territorial groups.[6] The high degree of specialisation evident in the forms of land improvement and the use of manure practised by the *Buzu* of central Niger is a frequently cited case (we will return to its major characteristics in Chapter 6, Figure 6.1, p. 143). Concerning differences between modes of exploitation of natural resources, local situations are generally far from being reducible to simple categories. Thus the case of the *Haalpulaar* of the Senegal River's middle valley represents an example of a highly complex combination of differentiated production activities, where, when considered as a whole, the society can be divided into three distinct sub-groups: agriculturalists, herders and fishermen. Throughout the year, each of these groups in turn uses and profits from the same area:

> This rotation is not just a succession on the same spot: it is equally a cyclical process since the result of the activities of one of these social groups creates the necessary conditions for the next group: the residues of flood plain agriculture create a common grazing land . . . while the excrement from animals, along with the vegetal waste, provide nutrition for the scavenging fish. . . . At the end of the food chain, bacteria cause the deterioration of organic waste into energy-poor elements which in turn are assimilated by vegetation, thus completing the material cycle.
>
> (Schmitz, 1986: 69)

With the exception of the particular case of irrigated agriculture in the Saharan oases, it is generally in the Sahelo-Sudanian zone that we encounter pastoralists practising agriculture. They often live among long-established agricultural communities, occupying the interstices between village land-holdings. Haratin, Fulani and *Buzu* are found from the banks of the Senegal to eastern Niger (Nicolas, 1962; Beauvilain, 1977; Hervouët, 1977; Marie, 1977; Santoir, 1977; Diarra, 1979; Benoit, 1977, 1982). Their settlement of this area may be of a relatively long duration (from the beginning of the century or even earlier), but it can also be quite recent, corresponding to a contemporary migration on to new lands. This is notably the case in southern Mali (Raynaut, 1991a) and in the south-west of Burkina (Benoit, 1982). When it is of long duration, cohabitation is generally accompanied by forms of exchange and collaboration which have led to bands of interdependence between agropastoralists and cultivators, such as access to water and consumption of field stubble for the former's livestock and caretaking of animals for the benefit of the latter. This cohabitation may also have led to specific forms of organisation within the region destined to facilitate the coexistence of both modes of land use (see Figure 5.1).

Nevertheless, as previously noted, competitive situations appear today where the effects of heightened pressures on land hinder the herds' mobility, reduce pasturage and force them to fall back on lands of poor agricultural quality, where production of natural vegetation is also ex-tremely poor (Serpantié *et al.*, 1986). Under these conditions extensive livestock production is facing many more constraints and obstacles, leading to profound changes in production activities and in the agropastoralists' way of life. As a result herds are reduced to several milk cows and smaller ruminants; transhumance ceases, after having been gradually reduced to small-scale movements of limited scope; the stalling of livestock on fields becomes generalised; and the relative importance of agricultural activities in the community's economy increases (Diarra, 1979). In such circum-stances, can we still talk of agropastoralism?

In the case of peasant societies, the question should be posed differently. If the term pastoralism is used in a restrictive sense, the qualification of agropastoralism can only be applied to societies whose material existence, like their social and religious values, places an essential importance on livestock. In this sense, this designation is inappropriate for communities whose systems of representation, social structures and, at times, religious practices are primarily organised around agricultural activities. This in no way signifies that the practice of agriculture is exclusive and livestock production is absent. On the contrary, certain peasant communities prac-tise agriculture and herding extensively within semi-intensive systems. In this regard, the example of the Sérer of Senegal has become a classic study (Pélissier, 1966; Lericollais, 1972).[7]

The ownership of large herds is not enough, however, to transform peasants into agropastoralists. In certain cases, the accumulation of animals is a response to the economic objective of capitalisation and has little impact on the functioning of the agricultural production system. From this point

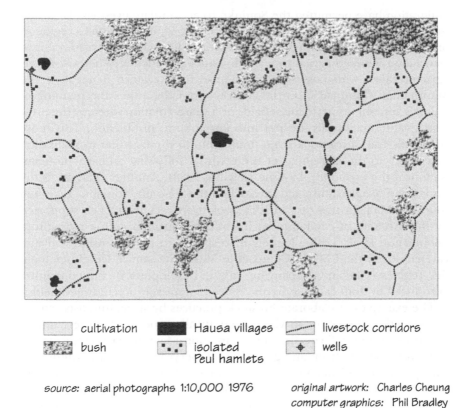

	cultivation		Hausa villages		livestock corridors
	bush		isolated Peul hamlets		wells

source: aerial photographs 1:10,000 1976

original artwork: Charles Cheung
computer graphics: Phil Bradley

Figure 5.1 An example of cohabitation between cultivators and pastoralists

This diagram is an interpretation of aerial photographic negatives covering the southern portion of the Maradi Department in Niger, where Hausa agriculturalists and Fulani agropastoralists now cohabit, since their arrival from Nigeria at the beginning of the century. The hamlets occupied by agropastoralists are often those located on the periphery of village agricultural lands. A network of livestock corridors, at times separated by hedges from the cultivated areas, enables animals to gain access to wells during the rainy season (the right to water animals having been granted virtue of ancient accords with the villagers). These paths also lead to pockets of non-cultivated bushland that serve as winter pastures and help to connect the large axes of transhumance or commercial pathways over which animals belonging to more northern pastoralists are driven. This kind of organisation of space permits a harmonious cohabitation of different modes of resource exploitation. Cohabitation can lead to competition, however, especially in densely populated areas (which is the case here, where there are 50 inhabitants per square kilometre). Since the Fulani are also highly skilled agriculturalists, they benefit most from such a system. The network of livestock corridors is arranged in cells where hamlets and fields are concentrated, thus the 'Fulani crop lands' appear quite isolated in relation to village lands.

of view, the example of the Soninké is very illustrative. At the end of an ancient and complex history, members of this ethnic group currently represent the majority of Sahelian emigrants present in France (Adams, 1977; Bradley *et al.*, 1977; Weigel, 1982). Revenues earned through emigration have permitted the Soninké to accumulate livestock. However, the role attributed to livestock varies according to an individual's region of origin. Migrants originating from the region of Kayes have a relatively minor association between agriculture and livestock production. They are simply agriculturalists who possess herds. Those from eastern Senegal or the Mauritanian Guidimaka, on the contrary, practise either the penning of animals at night on permanent fields or a yearly rotation whereby the prior year's cattle enclosures are put into agricultural production. This more systematic use of livestock is no doubt due to the fact that capitalisation dates back to the beginning of the century, following an early monetarisation of the economy (Lavigne Delville, 1988). Another example is the association of agriculture and livestock in the Sénoufo/Minianka region of southern Mali. In this case, this is a relatively recent practice supported by the dynamic of capitalisation introduced through cotton agriculture (Benhamou *et al.*, 1983). Largely in response to the extension activities carried out by the CMDT (Compagnie Malienne pour le Développement des Textiles – Malian Company for Textile Development), animal manure (produced by draft cattle or herds) is used intensively on cultivated fields.

The examples of elaborate livestock practices by agriculturalists remain the exception, even more so when systematically associated with agricultural practices. However, farms that do not include a few animals (cattle among the most wealthy, sheep and goats almost everywhere) are also quite rare. The purchase and maintenance of animals is probably one of the most widespread methods of saving in the Sahelo-Sudanian zone, especially for women. A new situation is now developing with the adoption of animal-aided traction, which encourages more intensive livestock practices, based on a virtually permanent stalling of animals and the delivery of fodder to them. Where these new techniques have been heavily adopted, draft animals can by themselves make up a sizeable herd. This is particularly the case in the entire cotton zone of southern Mali, which is monitored by the CMDT (Raynaut, 1991a).

The intensification of pastoral practices did not result simply from the diffusion of animal traction. With the increasing scarcity of self-generating fodder resources and in response to high market demands for meat, new strategies create an ever-increasing requirement for agricultural residues, such as millet stova, and residues from groundnut and bean crops. Agriculturalists are pioneers in this evolution. This is understandable to the extent that, in contrast with pastoral societies, modifications introduced to their livestock practices have had few repercussions on social relations.

A framework for spatial distribution

Extreme diversity in local conditions, due to the variable place held by livestock in the material and social lives of Sahelian and Sahelo-Sudanian societies, is thus the conclusion of even the most superficial analysis. However, the difficulty in understanding the situation with the help of simple analytical frameworks does not stop here, because nothing so far described is static; changes are currently at work at all levels.

The pastoral lifestyle may have known many crises in the past, but it has always been limited to temporary shocks (generally linked to climatic or epizootic events) which have not been obstacles to recovery. Today, however, the situation is completely different, given that from now on pastoralism is exposed to the danger of profound disorganisation which could very well lead to its extinction as a way of life.

The multiplication of major well-drilling operations in the Sahel, which has generalised and facilitated access to water without organisation of the social conditions of such access, has disturbed the ancient systems of regulation which, through the control of wells, rationalised the exploitation of pasturelands (Retaillé, 1984). The attraction of these wells has encouraged the sedentarisation of populations and has led to concentrations of animals which are incompatible with the very principles of extensive livestock production.

In the Sahelo-Sudanian and Sudanian zones, the greater and more permanent settlement of agriculturalists has created heavier obstacles for dry season transhumance, which formerly opened southern pasturelands to animals which, for many pastoral groups, were indispensable to the feeding of their livestock. Furthermore, as cultivators search out available lands, the agricultural push towards the north competes with pastoralists in their own territory and reduces the resources on which they can draw.

These factors of structural disorganisation, combined with an increase in the number of livestock over the last 50 years, have subjected the pastoral production system to contradictions and internal tensions that have been dramatically exacerbated by the chronic drought conditions over the last 20 years. In order to portray the situation in its entirety, it is important not to exclude the political dimension of the problem. As Boutrais (1992) has noted, anti-pastoral and anti-nomadic policies have long been in place in the Sahelian countries and efforts to sedentarise pastoralists have been a constantly pursued, if not always acknowledged, objective:

> Nomads have always constituted a 'problem' for states. During the colonial period, the goal was to better control and secure taxes from these elusive populations with unused wealth. Currently, development objectives legitimise sedentarisation: access to clinics, to schools and to an administration all imply the settlement of populations. Victims of drought can take advantage of international aid only if they are settled. Sedentary pastoralists are organised into groups, so that they can manage pastoral lands and evict nomads. Nomads everywhere are experiencing strong pressures and harassment.
>
> (Boutrais, 1992: 122)

Whereas the slow death of Fulani pastoralism is relatively discreet (facilitated perhaps by a more or less latent knowledge of agriculture), this demise is accompanied by a violent crisis in Tuareg societies. The historical circumstances of colonisation and decolonisation, the past internal policies of the Nigerien and Malian states, as well as factors at international level (such as the Spanish Saharan conflict and the interventions of Libya) have all favoured the construction of a Tuareg ethnic identity and the emergence of armed resistance, a process described well by Bourgeot (1990). A bloody repression and the exodus of refugees into the desert has, at times, signalled a real threat of physical extinction for the Tuareg people.

It is clear that if the future of pastoral societies in the Sahel continues along the trajectory set out during this century, the distinctions on which this analysis is based will lose a large part of their foundation because these societies will have ceased to exist as such. Moreover, reciprocal changes are occurring with agriculturalists as a result of the progression (sometimes rapid, more often slow) of a livestock production more and more closely associated with agriculture. The sedentarisation of pastoralists, the development of herding practices among agriculturalists and in all cases the reduction of herd mobility, are all accompanied by new forms of grazing resource management. Thus the already imprecise borders we have attempted to draw become even more blurred, somehow pointing towards a homogenisation of livestock relations throughout the entire region under study. Yet care should be taken not to overstate this last point. What is being observed is a tendency or trend that has not yet reached fruition. Whatever its current status, livestock production remains a side activity for agriculturalists, while for many pastoralists, who have known crises throughout their long history, the fight to retain their way of life continues (some, in particular, are searching much further south than their regular transhumant pastoral zones for the grazing resources they now lack). Even if this represents only a snapshot of a reality in motion, the description given here continues to be a relevant way of accounting for the ways in which the diversity of land use and resources continues to be ordered in the Sahel as a whole. Even if important changes have begun, the partitioning that occurs around livestock production, the combinations, the interdependencies and the competitions established between herding and agriculture today remain essential elements in an understanding of the concrete conditions of natural resource exploitation.

Is it possible to proceed from this description (already oversimplified with regard to a complex and shifting reality) to a cartographic representation that would allow us to locate the types of patterns which have been identified? Such an undertaking is risky, because mapping these complexities solidifies and schematises them even more. It must be attempted, however, for only reference to a common spatial framework can lead to the localisation of these situations. In so doing, the limited value of such an undertaking must be recognised; we must realise that diversified material under analysis is homogenised and simplified. This temporary simplification is justifiable if it ultimately enables us to return to the diversity and complexity of the connections with other analytical criteria. With this caveat, there is no attempt here to translate cartographically all the subtleties, the transitional phenomena and the dynamics so far emphasised in our analysis. We will simply retain several of the large categories that underscore an area's major characteristic from the point of view of the role of livestock and the relation between herding and agriculture. Such a typology is indisputably crude, but it will be ultimately combined with other analytical levels based on different criteria.

Three major zones have been identified, situated in parallel bands which run from west to east, the length of the region (Figure 5.2).

The pastoral zone

This is the most northern zone, where nomadic pastoralism constitutes almost exclusively the mode of natural resource use. It covers approximately the entire Sahelian region, in its strict sense, and its sub-desert margins. Inadequate rainfall prohibits cultivation except in exceptional and very localised ways, i.e., only where special water resources exist. This is the heart of the historical lands of 'Bedouin' pastoralism (Moor, Tuareg and Toubou) which, as we have seen, was once supported by a diverse and largely decentralised economic base. This was principally long-distance commerce and war, but also, in many cases, agriculture that was practised much further south and worked by a slave class[8] but which, nevertheless, played an integral part in the material and social reproduction of the communities who were still based in zones located much further north. Here more than elsewhere, animal mobility was a survival mechanism because the rains at these latitudes are extremely irregular and unevenly distributed, and pasture production is poor, so that regeneration of the area demands that any exploitation be neither too heavy nor too prolonged. The conditions of use of this zone have been profoundly transformed during this century. With the end of warring expeditions, the extinction of the Saharan and trans-Saharan trade and the emancipation, at least in part, of a slave class devoted to agriculture, the economy of 'Bedouin' societies found itself cut off from resources which, in earlier times, assured its overall equilibrium. Communities whose livestock production represented only one component in a variety of complementary activities were thus forced to fall back on this single productive base and, consequently, to increase their vulnerability.

The introduction of new users on to grazing lands in this zone is another significant factor leading to change. Under the protection of a colonial administration that sought to contain a Tuareg population they had found menacing, after rebellious movements marking the beginning of the conquest period, groups of pastoral–nomad Fulani settled in territories

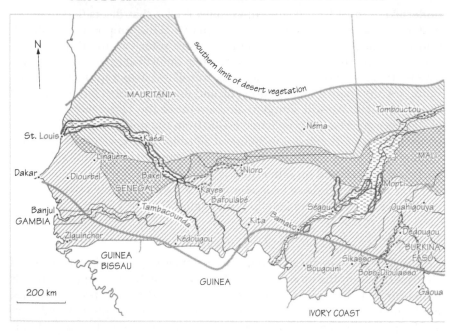

Figure 5.2 Major land use systems of the western Sahel (continued opposite)

which had, until then, been outside their domain. In Niger this push was a continuation of a movement that, after the constitution of the Fulani empire of Sokoto, directed small groups of southern Fulani northward. In the end, other users followed (such as rich cattle-owners who belonged to dominant groups in the Sahelian countries, namely politicians, high-ranking bureaucrats and large merchants), imposing themselves on the Tuareg and Fulani pastoralists and competing for the use of pastures in this zone.

> Cattle-owners with fixed revenues, seeking to invest in livestock are in a better position to speculate than traditional agents. Thus, in only a few years, bureaucrats, merchants, and rich livestock producers, were able to raise substantial herds. . . . These new livestock owners, 'investment herders', who manage their herds through paid intermediary shepherds, thus capture, imperceptibly, the best pastoral sectors.
>
> (IEMVT/CTA, 1988)

Thus, in Mali, between 25 and 60 per cent of livestock belonged to these absentee owners (Pons, 1988a). The movement of these herds, which adheres to a strictly commercial strategy, escapes the social and technical norms through which pastoral communities traditionally regulate the

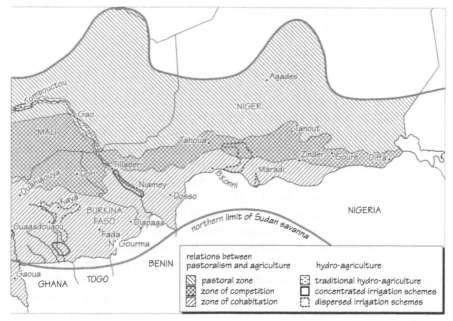

relations between
pastoralism and agriculture | hydro-agriculture

pastoral zone | traditional hydro-agriculture
zone of competition | concentrated irrigation schemes
zone of cohabitation | dispersed irrigation schemes

original artwork: Charles Cheung
computer graphics: Phil Bradley

exploitation of their environment. When combined with a rapid increase in the numbers of livestock and the expansion of drilled wells, which blur the map of social control over the pasture lands, this superimposition of users also disorganises the conditions of resource management and leads to an over-exploitation of land. The constant rainfall deficits of the last 20 years have again aggravated the disequilibrium created by this situation, setting off the catastrophic death of cattle herds in 1973–74 and in 1984–85. In absolute numbers, the reconstitution of herds appears to have been rapid (Boutrais, 1992), but 'investment herders' have profited much more than pastoralists, which has further aggravated the marginalisation of the latter within the global process of resource management.

To locate briefly these facts in relation to the debate that was discussed in the preceding chapter, it is worth noting that what is occurring here is a complex dynamic, of which human population growth is far from being a satisfactory explanation. As Boutrais explains, it is the livestock which is 'the real occupant of space' in pastoral societies (Boutrais, 1992: 110), and their numbers fluctuate according to factors quite distinct from those that drive human demographic growth. In reality, as we have seen, the population in the pastoral zone has experienced only a very slight growth during the last decade. Essentially, then, it is the social, political and economic factors that are responsible for the evolution of the conditions for the use of natural resources in this zone.

The zone of competition

South of the area just described, there extends a second band where rain-fed cultivation of early varieties of millet is possible. Not long ago this area was the exclusive domain of pastoralism, because the climatic risks were too great to attempt agriculture unless absolutely necessary. Recently it has experienced a progressive increase in cultivation as the demand for land, which can no longer be satisfied in the Sahelo-Sudanian zone, has risen. The confrontation between the two modes of resource use generates competition. This is particularly the case in this marginal zone of contact. Even if human population density remains modest in these agricultural frontier lands, the anti-risk strategies which cultivators are forced to apply at this latitude are most often based on maximum extension, where vast expanses are sown each year with cereal crops, while the poverty of the soil reduces the period in which they can be exploited and thereby increases the rate at which land must be cleared (Chapter 6). In addition, because the problem of granivorous birds leads to the felling of trees suspected of providing nesting sites, the advance of agriculture in these regions is accompanied by a virtual disappearance of tree cover.

Besides the quantitative and qualitative reduction in grazing lands caused by this agricultural expansion (a reduction that is not compensated for by crop residues because the climate hazards often drastically lower their volume) the major constraint comes from the presence of fields which limit animal mobility. Therefore, this is a situation full of tensions and contradictions, created by the penetration of peasants into the pastoral domain. The fact that some pastoralists, impelled by the loss of their herds, are currently attempting agriculture only aggravates the situation. The colonial administration and post-colonial powers were aware of this risk, as we have indicated, and endeavoured to check the expansion of agriculture through restrictive regulations. Nevertheless, in the final analysis, this has had only a limited effect and the penetration of agriculture into the pastoral zone has continued throughout the second half of this century. As we saw in the case of the pastoral zone proper, the addition of one more actor on an already crowded stage is, therefore, another factor leading to the present disorganisation.

The cohabitation zone

The Sahelo-Sudanian region is certainly the privileged domain of agriculture, yet this specialisation is far from exclusive. The area has for a long time received transhumant pastoralists from the Sahel and has sheltered more or less sedentary agropastoralists whose lands interdigitate with those of peasant villages. The presence of these agropastoralists is in some areas dense and of long duration, particularly in the south of Niger, in Yatenga Burkinabe, in the Interior Delta of the Niger River and along the floodplain of the Senegal River. Elsewhere, they have only been present for several decades, particularly along the Black Volta and in south-west Burkina (Benoit, 1982). For a long time, trypanosomiasis has kept Zebu away from the southernmost regions of Sudanian climate. Currently, progress towards eradication resulting from chemical prevention programmes has made these areas more hospitable to livestock. In other areas, the clearing of agricultural lands has reduced the breeding grounds of *Glossina* (the disease vector). Forced back by the dangers of drought and pressures on land in the northernmost zones, pastoralists are now establishing themselves here in great numbers. Hence, in the extreme south of Mali, in the Bougouni region, there are 500,000 cattle, of which more than half are owned by herders (FAO, 1991).

Thus, in the whole of the space which was historically their territory, sedentary agriculturalists must coexist to some degree with pastoralists and agropastoralists. Where space is not limited and the cohabitation is organised, a precious complementarity is possible among the various users (contrary to what generally occurs in the transition zone), thereby favouring a sustainable exploitation of the environment. On the other hand, when the pressure on land increases, the inherent needs of the different modes of occupation and land use can become competitive and create a confrontational situation which is detrimental to the coherent natural resource management. A simplistic division between 'pastoral zone' and 'agricultural zone' too often leads those who conceptualise development projects to ignore the fact that cohabitation between sedentary agriculturalists and different categories of pastoralists is one of the major features ordering land occupation and the use of resources. The creation of programmes centred almost exclusively on agricultural improvements can, therefore, only increase tensions and disturb an already fragile equilibrium.

These are then the three major zones that have been mapped. In each one of these the problems associated with animal resources and with the relationship between herding and agriculture are posed in different terms. This division is certainly oversimplified and cannot account for the many nuances raised in our earlier discussion. The limits described are necessarily approximate; they must be considered as margins of transition and not as clear-cut boundaries. In spite of these imperfections this representation does, however, help to identify and to spatialise the dominant situations which will later be refined using other criteria.

RELATIONS WITH WATER

The agricultural use of water other than rainwater is still rare in Sudano-Sahelian Africa, although diverse irrigation systems or other water management programmes do exist on the continent (Marzouk-Schmitz, 1989). In a very localised fashion, and when natural conditions allow it, rice is grown in ponds – often uniquely by women, as is the case with the Soninké (Bradley *et al.*, 1977) and the Hausa (Raynaut, 1989b: 117). Further south, these practices give way to rice agriculture in broader low-lying areas, particularly in the south and east of Burkina Faso.[9] In all cases, these are secondary practices that complete a system of production dominated by rain-fed agriculture. Generally speaking, low-lying areas were little cultivated before the successive droughts of the last 20 years, largely because of heavy soils which are hard to work and the risk of destruction by flood. With decreased rainfall and increased demands on land, the cultivation of bottom lands has increased rapidly and is now the subject of landholding tensions (Berton, 1988).

Dry season market gardening, hand-watered or mechanically irrigated (from an aquifer or surface waters), has developed rapidly over the last 20 years, preferentially but not exclusively on the periphery of urban centres. Traditional market gardening has long existed among the Hausa (Raynaut, 1989b). Here, as elsewhere, this is a complementary activity, individual in nature and oriented towards the market. Although marginal in terms of land use, it is nevertheless an intensive agricultural practice. Innovations (in terms of species cultivated or techniques used) have been more rapid than in rain-fed agriculture. Market garden production follows the evolution of consumer demand, but the organisation of markets and preservation of produce remain the limiting factors for an activity that can quickly lead to overproduction.

While small-scale hydro-agriculture is relatively widespread in the Sahel, there is, on the other hand, no tradition of large-scale irrigation development, even in the valleys of the major rivers where it is theoretically possible. Questioning this, Giri (1983) points out the absence of the two conditions that permitted the creation and generalisation of these schemes in Asia: on the one hand, strong demographic pressure making intensification vital, and, on the other, the existence of an organised state structure over the peasantry. These two arguments merit some discussion.

The first of these arguments operates against the role generally attributed to the demographic variable in the search for constraints on the management of natural resources in the Sahel. Here, it is the availability of a workforce that is put forward, and not excessive demand on primary production. The insufficient exploitation of a resource that brings with it both the possibility of intensification of agricultural practice and of freedom from climatic constraints is attributed to a shortage of manpower. In fact, the causality could just as well be reversed and we could say with equal validity that, in the long term, it was the high levels of production made possible through irrigation that could have stimulated demographic growth and favoured the formation of high concentrations of population. On this point, one can make an analogy with the models of historical explanation relative to the Neolithic revolution – a decisive stage in the evolution of production systems. According to these models, it is clear that demographic growth during the first stage effectively imposed the transition to agriculture and livestock production – a technological change which helped to increase the level of calorie production per surface area (Cohen, 1977). But then, the crossing of this agro-technical threshold favoured, in turn, the pursuit and expansion of this demographic growth (Godelier, 1984).

There is no doubt that the question of relations between population density and the development of water management schemes must be addressed in a similar way, i.e., in terms of a circular causality.[10] In any case, all this is nothing more than conjecture and we should not distance ourselves too much from the current problems in the Sahelian countries. We can, however, take from this reasoning an epistemological lesson: that is that care should be taken with the demographic argument since it can be applied both ways within the domain of relations between society and the environment, as cause or consequence.

Concerning the second explanation, it cannot be applied to the Sahel without qualification because, of all of Africa, these are exactly the regions where the forms of centralised political power have been most highly developed. These structures were nevertheless fragile, not fulfilling the conditions of permanence and of the extended spatial influence needed to carry out vast water management schemes. More profoundly, perhaps, the reason for the lack of interest in developing the major Sahelian rivers is surely due to their commercial orientation which, as we saw in the preceding chapter, drew the necessary means for their material and social reproduction from their geographic position and their control of trade routes with North Africa.

Other determining factors of the physical and natural conditions could equally be researched, but they would not go far to unlock the puzzle. As Lacoste (1984: 195) remarks, although the alluvial deposits transported by the rivers of Africa are less rich than those that have accumulated in the major Asian valleys, Africa's floods are less devastating, and they could, as a compensation, constitute a favourable factor. With respect to the major diseases that are endemic to African valleys (trypanosomiasis, onchocerciasis) and that are supposed to deter settlement, not only are they of little importance in the Sahelian rivers, but, as we have observed, they are just as much the consequence as the cause of lack of development in these valleys.

In spite of an absence of major irrigation schemes, specialised techniques were nevertheless developed by riverine agriculturalists in the major

Sahelian river areas in order to take advantage of the precious water resources that were available within an otherwise arid environment. These techniques all depended on the use of river floods. Thus it is possible to distinguish two major areas devoted to hydro-agriculture.

The Senegal valley

From the Malian border to St Louis there were, in an average year and until the recent construction of dams, 100,000 ha of floodplains put under cultivation the moment the river receded in the dry season (Lericollais and Diallo, 1980). The choice of crops is made in riverine villages by paying close attention to topography. Rain-fed crops are cultivated on high ground (bulrush millet, sorghum, groundnuts) and different floodplain crops are grown in sections left clear by the water's retreat: sorghum in the large sedimentary basins, condiments in the market gardens, and tobacco on the sides of the river bank (Lericollais and Schmitz, 1984).

These production systems combine in time and space different forms of natural resource use, principally wet and dry season cultivation, livestock production and fishing. This diversification in material organisation is itself based on a social diversification which allows cohabitation between different communities whose fields of activity are more or less specialised[11] and which, within each of these groups, guides the technical role and land tenure rights, based on distinctions of status, claim and function.

There is a complex system of resource exploitation at work, through the articulation of work calendars, the division of space, the specialisation of roles and the diversification of tools, within which water and the relations established with it occupy a central place. Clearly, the riverine societies of the Senegal have not created large-scale schemes for the distribution and storage of irrigation water, but does not their investment in the domain of technical and social organisation demand that they be considered as belonging to a civilisation of hydro-agriculture?

Massive changes are occurring today in the form of vast interstate programmes which, with international technical and financial aid, are forging an association among countries bordering the Senegal in order to develop the river and its valley.[12] After several limited attempts which began during the colonial period, efforts were first concentrated in the delta, with the construction in 1964 of a large peripheral dyke on the left bank. This region is vast and flat and is occupied by pastoralists. It has welcomed many major hydro-agricultural schemes, with cultivation being controlled by the colonists under the direction of SAED (Société d'Aménagement et d'Equipement du Delta – Society For the Development and Equipment of the Delta (as well as the Senegal River valley and the Falémé)). Here, irrigation is by pumping, and cultivation uses mechanised techniques (working the soil, harvesting) which are mainly built on the working capital.

From the middle of the 1970s, in response to the drought which extensively disrupted village agricultural systems, the Senegalese and Mauritanian authorities extended the village level perimeters established on the river banks and irrigated by pumping all along the valley. These are rudimentary installations, small in size (20 to 50 ha) and cultivated manually, each being managed by a group of producers. Initially introduced as a response to a climatic crisis, they were also needed in order to initiate the valley people in irrigation techniques, because the ultimate aim was to undertake large-scale schemes throughout the valley and to eliminate flooding altogether. The intention of OMVS (the Organisation pour la Mise en Valeur de la Vallée du Fleuve Senegal – Organisation for the Development of the Senegal River Valley) was to regulate the river flow by constructing two dams: at Diama in the delta, to prevent the backflow of salt water and to raise the level of the river upstream, and at Manantali on a tributary of the Senegal, the Bafing in Mali, for hydroelectric regulating works. The objective is to be able to provide irrigation for 375,000 ha of land in the three countries involved.

These constructions should ultimately eliminate floods altogether and will mark the end of floodplain agriculture, which formed the basis of the farming system in the middle valley.[13] The results will be the conclusion of a process of radical transformation of the relations between man and space which was begun in the valley several decades ago. In a transitory period after the Manantali dam comes into service an artificial flood will nevertheless be created, thus enabling the riverine agriculturalists:

> to continue their floodplain agriculture until irrigated lands can be developed in sufficient quantity. This is therefore a transitory solution, its duration dependent on the rate of development of schemes; but the perspective of a perennial floodplain agriculture on the lower sections of the area, accompanied by a regeneration of the environment, is not totally out of the question at present.
>
> (Seck and Lericollais, 1986: 8)

By the end of the 1980s, about 25,000 ha had been developed in Senegal, half in major schemes in the delta and half in village level perimeters in the middle valley. The two dams are now complete (although the hydroelectric station is not yet in place). Thousands of peasant families now cultivate irrigated rice on plots of a few hectares in the delta and of 0.5 ha in the middle valley. For them, confronted as they are with the disruption of the system that links agriculture, fishing and pastoralism, irrigated rice has become an integral part of the family economy and has a vital role to play in their food security. The rapid deterioration of the irrigation networks, however, has made costly restoration necessary.

The rate at which development is progressing, at all levels, is only half that which was originally planned. Problems with land tenure, the production strategies employed by the peasants and technical difficulties all combine to keep production levels much lower than intended (Crousse *et al.*, 1991). The chains of production encounter major financial problems, associated with the mediocre technical results achieved' on the irrigated areas,[14] and also with the cost of administration and with problems of loan repayments by the cultivators. Thus there is a constant need for new injections of capital, which restricts the overall profitability of the operation: 'Given the current pricing conditions, the more rice produced in the valley perimeters, the more it costs the state' (Seck and Lericollais, 1986: 7).

Paradoxically, whereas it was thought that the development of irrigated rice cultivation would render Senegal self-sufficient in this commodity, it is in fact the taxes on rice imports that fund the chain of production! This being the case, and given the general context of external aid restrictions, the level of investment has slowed considerably. Linked with the plans for structural adjustment (see Chapter 4), state policy is attempting to respond to this situation by withdrawing subsidies and privatising economic services, both higher up and lower down the production chain (especially credit). After rehabilitation, management of the major schemes is transferred to peasant organisations. In short, the aim is to reduce state intervention substantially by rechannelling it into support for producers and irrigation policy co-ordination.

Originally planned to be completed within five years, this disengagement gives rise to a radical transformation in the structure of the production chain as well as in the economic and institutional conditions of irrigation. The process will, in fact, be spread over a much longer time-scale. Although far from complete, this restructuring is already producing some contrasting results. The impact of these increased economic constraints on the producers was to render the already fairly high level of risk to which their activity was subject quite intolerable. This plunged the village perimeters into a crisis situation from which they have, as yet, only partially recovered. A technical overhaul and reorganisation of the production chain were needed in order for those perimeters, which had been transferred to peasant organisations, to achieve net gains in productivity. Entrepreneurs rapidly invested in the provision of mechanised services such as ploughing and harvesting (Démarets, 1991). Moreover, the liberalisation of access to land following the enactment of the law relating to state property and the liberalisation of access to credit led, at the beginning of the 1990s, to a spectacular wave of private schemes. In five years, more than 15,000 ha were developed – more than the SAED had managed in the delta in 20 years! However, as the work had been carried out somewhat hastily, these perimeters soon experienced major problems with productivity and salin-isation of the soil. This led in turn to a spiral of non-repayment of credit, which once again put the entire production chain in jeopardy.

To conclude, even though the Promethean aspiration of developing the entire valley now has to be abandoned because of aid restrictions, the massive investment already injected by the member states and foreign aid has led to a considerable transformation in technical systems and social relations in the delta (the only area in West Africa where agriculture is totally mechanised) and also, to a lesser extent, in the valley. This is a costly form of agriculture, dependent on the market and on the economic risk thus engendered. Constraints on working capital and competition for land give rise to tensions in the management of shared facilities and the use of natural resources (water, land). This is true not only at the local level, but also at the international level, as we have seen in the Mauritania–Senegal conflict. Moreover, the production chain has still not achieved an economic equilibrium, which means that in respect of irrigation and the economy of the entire valley, the future is heavily mortgaged.[15] The fundamental decisions as to whether to maintain or to abandon flooding, decisions which will ultimately determine whether or not the transformation of the society/ nature relationship in this region takes place or not, have not yet been made. The future of the valley hangs in the balance.

The Niger River's Interior Delta

The vast alluvial plain of the Niger River's Interior Delta, its dead delta and, to a lesser degree, its valley which turns back on itself as far as Niamey, forms the second major area of the western Sahel dedicated to hydro-agriculture, traditional as well as modern. This is the oldest centre of rice cultivation in all of Africa: 'It is from the interior delta that floating African rice was propagated, and one species of wild rice, *Oryza perennis barthii*, colonises vast areas at a medium depth. The cultivated varieties are autochthonous rices of the *Oryza glaberrima*' (Gallais, 1984: 95).

As in the Senegal valley, ancient peasant techniques included floodplain agriculture, but rain-fed rice cultivation imposes other constraints and fits into a completely different agricultural calendar to sorghum. Whereas in the basins bordering the Senegal, planting takes place in ground freed by the receding floodwaters (plant growth thus taking place during the dry season), the flooded rice agriculture of the Niger's Interior Delta is pluvio-fluvial, that is operations begin with the arrival of the rains and continue through the flood. This leaves far less latitude for the simultaneous cultivation of crops that are strictly rain-fed, such as sorghum or millet. There is, therefore, much greater specialisation among the rice-cultivating populations of Marka and *Rimaibe* than among the Bambara, who devote themselves to rain-fed millet, sorghum and maize cultivation. According to Gallais (1984), rice cultivators in the delta number 65,000, out of a population of 170,000 which includes Fulani herders. It should also be noted that in the lower delta it is possible to observe forms of floodplain agriculture closer to those found on the banks of the Senegal, such as sorghum, manioc and yams planted after the floods recede.

As on the banks of the Senegal, the existence within the Sahel of abundant water resources has led to projects around the delta of the Niger. There, enthusiasm emerged much earlier and ambition was on an inordinately grander scale. After the more than ten years of study, debate and controversy[16] that followed plans put forward in 1921 by Bélime, the Engineer of Public Works, l'Office du Niger was created in 1932 and was charged with the development of the Niger's dead delta. The goal was to place under irrigation, within 50 years, 960,000 ha of land of which 510,000 ha would be devoted to cotton. Although the construction of the Markala dam and its feeder canals was carried out according to plan, the technical difficulties encountered in the development of agriculture were numerous. In addition, the project's success depended on attracting a massive workforce, because the local population was insufficient. Early on, the forced recruitment of settlers from Upper Volta and southern Mali provided the required labour force, but, at the time, the poor human and sanitary conditions drew strong criticism (Herbart, 1939: 85ff.). The abolition of forced labour ultimately provoked many departures and although voluntary recruitment resumed in the 1950s, it was far from reaching the goals initially set. In the end the financial viability of the scheme depended on cotton cultivation, which proved more difficult and risky than predicted, to the point that it was essentially rice that occupied the greatest land area. The economic basis for the grandiose projects conceived by Bélime failed, and it became necessary to scale down estimates drastically.[17] At present less than 50,000 ha have been put into agricultural production – just 4 per cent of the original goal!

Whereas along the Senegal River, rice is cultivated using pumping techniques and (in the case of the delta) highly mechanised processes, l'Office du Niger benefits from the possibility of gravitational irrigation, which considerably reduces running costs. Moreover, the decision was taken to introduce draft animal power rather than motorised equipment, which is more costly both in investment and in running costs. Last, whereas for the peasants of the middle valley of the Senegal irrigation was simply another factor to be integrated into their already diversified production system, for the settlers with l'Office du Niger, their irrigated plot was their only source of income and could be withdrawn should they be unable to repay their debts. They lived in settlement villages, in homes that were put at their disposal along with their plots. They found themselves, therefore, in the position of quasi-wage-earners – yet with no wages guaranteed! As in the Senegal valley, the aim of l'Office du Niger is to seek financial stability for the system and, to this end, they have embarked on a comprehensive restructuring which has begun to provide more autonomy for the farmers (land security, possibility of pooling resources for threshing organised by the peasants themselves, etc.). Some technical rehabilitation operations (known as ARPON and RETAIL) running in parallel with this restructuring have brought improvements to the irrigation networks and have extended planting, thus producing net gains in productivity and setting in motion a dynamic of accumulation.

Be that as it may, despite a succession of attempts to take the situation in hand, the end result is a very disappointing one more than 60 years after the creation of l'Office du Niger! It is certainly not wise to make comparisons between situations which are not wholly comparable, but one wonders if it would not make sense to reread the history of the development of the Niger's Interior Delta and to draw from it the lessons of modesty and prudence in relation to the Senegal valley.

Irrigation projects in Sahelian countries

While the major Sahelian rivers have been the focus of sometimes Utopian and grandiose projects since the colonial period, irrigation has appeared to be, overall, an acceptable solution for increasing agricultural productivity. Independence reinforced this tendency by putting the focus on food production (most often rice) in the hopes of reaching food self-sufficiency. All the countries have created hydro-agricultural development programmes destined to take advantage of their surface and groundwater resources.

In Burkina Faso, the greater part of the irrigated soil potential is located on the alluvial deposits of the Volta valleys and their tributaries. A few perimeters have already been irrigated, namely Sourou valley, Kou valley, the agro-industrial perimeter of Banfora and the Aménagement des Vallées des Volta. Other, more limited, development schemes in low-lying areas have also been established on the Mossi plateau. There are an estimated 100,000 ha of land suitable for two season irrigation.

In Mali, the scale is completely different and the theoretical potential of hydro-agriculture is estimated at 1 million ha (Afrique-Agriculture, 1990). In addition to l'Office du Niger, a few rice agricultural areas have been constructed along the river in the Interior Delta: *Opérations riz* placed under cultivation 17,000 ha in 1981 (Funel and Laucoin, 1981). In the Upper Valley, different limited programmes for the modernisation of traditional crops have been undertaken, i.e., *Opération Zone lacustre* in Goundam (southeast of Tombouctou) and the action *riz-sorgho de décrue* in the Gao region. Several irrigated village perimeters have also been constructed in the Senegal basin.

In Mauritania, the potential is located along the Senegal valley, consisting of village-level perimeters and larger irrigation schemes developed in the Gorgol, near Kaedi. One estimate places it at 120,000 ha (Funel and Laucoin, 1981).

In Niger, potential irrigation would develop 150,000 ha (after the construction of the Kandadji dam) (Funel and Laucoin, 1981). Several rice agricultural areas have been created over the course of the last two decades along the Niger. Some hill dams have also been constructed in the Ader–Doutchi region (between Birnin Konni and Madaoua), as well as perimeters irrigated by pumping from the underground aquifer of the Maradi valley.

Finally, in Senegal the principal theoretical potential for hydro-agriculture is located in the river valley, as well as in Casamance and eastern Senegal (estimated total 540,000 ha (Afrique-Agriculture, 1990)).

To the extent that these projects secure outputs largely superior to those of rain-fed agriculture, it is generally considered that irrigation must replace traditional production systems that have been judged to be ineffective – suggesting therefore a radical intensification. In fact, the programmes that have been developed have known, for the most part, many set-backs, such as extremely high costs, rapid deterioration of facilities and disappointing technical results (Funel and Laucoin, 1981). Beyond the practical problems associated with the developments (the technocratic conception of the schemes and artificial grouping of producers without responsibilities), this situation is also partly explicable by the importance peasants accord irrigation in their economic strategies. In fact, recourse to such intensification is most often quite localised. Where there already exists a traditional agriculture (rain-fed and/or floodplain agriculture), it is unlikely to be replaced by irrigation, in spite of yields that can be ten times greater. The cost (in labour and money) is such that it can only represent, for the agriculturalists, a complementary activity especially appreciated for the food security that it brings – particularly since the surface areas dedicated to irrigation (often less than 0.5 ha per male worker) are too small for any surpluses to be produced. Thus the irrigated parcel represents for the peasant a supplementary agronomic 'resource', a localised intensification within a production system that maintains elsewhere its strategy of extension. This is particularly clear in the middle valley of the Senegal River, where out-migration generates substantial revenues and competes with irrigation for the use of the labour force and the satisfaction of household economic needs (Lavigne Delville, 1991). Moreover, in spite of the security irrigation is supposed to create, the risk factor has been a more frequent determinant than was first realised and hence the mediocre technical results that have been achieved have strained financial resources. In many cases, water remains the most limiting factor, on the Senegal River (Seck and Lericollais, 1986) as well as at l'Office du Niger. For a long time the supply of necessary inputs from the state suffered many delays. The maintenance of equipment at the desired level has not always been regularly assured. Currently, rehabilitation experiments are being carried out. Coupled with the state's disengagement from supplying inputs and from the maintenance of producer price incentives, these experiments have permitted significant gains in productivity (Cebron (ed.), 1990).

The map illustrating major types of land use (Figure 5.2, pp. 122–3) shows the major zones where hydro-agriculture is practised, distinguishing between traditional forms and more modern schemes. At this scale, it is not possible to highlight more sporadic occurrences, such as cultivation of low-lying areas, women's small-scale rice cultivation and small market gardens, specific examples of which can be found throughout the zone under study. As can be observed, the stated objectives of the major hydro-agricultural development policies, some dating as far back as the colonial era, are still far from being achieved. There is nothing surprising in this, for there is often a great distance between potentialities that can be identified on theoretical grounds and the realisations that can be reasonably envisaged when one takes into account the economic – financial, social and environmental costs of hydro-agriculture. Are these policies a reasonable hope or a costly Utopia? The question remains a subject for debate and polemics, all the more so because the balance sheet of what has already been accomplished is not easily established, as Funel and Laucoin concluded over 15 years ago:

> We think we are getting to the heart of the matter by characterising irrigation as the mastery of water, and the objectives are implicitly contained in this belief: the mastery of water is the control, and therefore the increase of agricultural production. Irrigation is a symbol, one of productivism and success. The results should not – cannot – be seen otherwise. When one examines the results, most development programmes are marked by the paradox of a not entirely convincing success, especially at the social level. Perhaps this explains why development plans are a privileged field of 'discourse': a discourse of justification that seeks to show the advantage of the choices imposed on farmers, the discourse that glosses over the stakes, and finally the social discourse that seeks backing for techniques and their consequences. This paradox explains, in fact, one of the principal contradictions in these development schemes. Are these schemes to be evaluated solely on their technical merit or on a social level, the one being almost always opposed to the other?
>
> (Funel and Laucoin, 1981: foreword)

Be that as it may, contrasting experiences from various irrigation projects show that a radical transformation of production systems at regional level, and of man's relationship with his environment, is a complex and risk-prone operation. State intervention is without doubt a necessary precondition, first, in order to create a favourable technical and economic environment and to provide basic infrastructure, and, then, when the impetus towards private investment begins to emerge, in order to establish and enforce a framework of regulations (quality of irrigation networks, etc.). An ideal intensive irrigation system, with total mastery of the water, requires a perfect technical mastery which is far from being achieved in all cases. Also required, as can be seen from the high production costs of irrigation models developed in the Sahel, is economic security (price ratios, credit system, input provision, outlets). This again is difficult to monitor, especially as it depends on external factors, like the recent reorganisation of monetary parity which has modified the elements of the problem considerably. A successful large-scale intensive irrigation development requires a complex and functional institutional web within which the various protagonists (state, peasant organisations, credit bodies, traders, transformers) can find their place in the context of a coherent, regulated system (ranging from land entitlement to credit guarantees). It is the bringing together of all the component parts of this web that causes problems, even more so than the physical development. Although, for a long time, malfunctions in public intervention structures contributed to the mediocrity of the technical results obtained, mass 'disengagement' by the public sector combined with inadequacies in the regulatory system, can in turn have equally un-favourable consequences – as the upsurge in private irrigation in Senegal has clearly shown.

CONCLUSION

Despite these developments, rain-fed agriculture will long remain the mainstay of survival in the countries of the Sahel. It is the crises that this agriculture faced that created the repeated famines of the last two decades and for the most part it is from the revenues, direct and indirect, from non-irrigated cotton and groundnut crops that the majority of the states in the region secure their financial revenues. If the map depicting the major types of land use does not specifically highlight rain-fed agriculture, it is because it is everywhere present outside the areas of the pastoral Sahel and the flooded valleys, where often, as we have seen, mixed forms (rain and flood) are more common than floodplain agriculture in the strict sense. By considering its different aspects we shall place ourselves at the very heart of the problems of the Sahel.

NOTES

1 This uniformity of gathering is not borne out in the case of wood collection, but it may be assumed that in this case the intensity of the collection is proportional to population density as well as to the distance from urban centres and to their size.
2 To our knowledge, the question of fishing in continental waters has been the subject of very few scientific studies – with the notable exception of the large research programme recently undertaken by ORSTOM in the Niger's Interior Delta (Quensière, 1993; Quensière (ed.), 1994). With respect to gathering and

hunting, we can say that (in the geographical domain being considered) these are areas almost completely ignored by researchers.

3 Raynaut, personal observations.

4 *Buzu*, a Hausa term meaning the *Ighawelen*, former dependants of the *Imageghen* or Tuareg aristocrats.

5 In Niger particularly, the major national programme of 'off-season agriculture' which was started in 1985, linked food aid to the practice of irrigated agriculture. According to Beauvilain (1977), however, certain Bororo/*Wodaabe* groups in the region of Dallol Bosso in Niger may have already experienced a temporary recourse to agriculture at the end of the nineteenth century.

6 An example which can be cited is the *Kel Owey* from the Nigerian Aïr, who have an ancient practice of irrigated agriculture along the valleys of the mountainous massif (Bernus, 1970, 1981, 1989).

7 In their case, the use of the term agropastoralism is justified to the extent that the accumulation of livestock, far from responding exclusively to technical needs, is also a key element in the processes of social reproduction (Gastellu, 1981).

8 Travelling in the mid-nineteenth century through the territories that now form northern Nigeria, H. Barth was astonished by the extent of the lands being exploited by the Tuareg of Aïr (Barth, 1965, I: 48).

9 This may be a relatively recent tradition, as in east Burkina, where rice cultivated in low-lying areas was developed in the 1950s with the introduction of Asian varieties.

10 For a broader discussion of the circular relationship between demographic pressure and productive techniques, see Couty and Hallaire (1980: 41–8).

11 Soninké – agriculturalists; Wolof – agriculturalists and fishermen; *Haalpulaar* – agriculturalists, fishermen or herders; Fulani and Moor – pastoralists and agropastoralists.

12 OMVS brings together in association Mali, Mauritania and Senegal.

13 Although these crops have very much receded due to the drought and the drop in river levels.

14 Especially in the delta's large schemes, the village perimeters having achieved a much higher level of production.

15 Given that a large proportion of production costs are accounted for by imported products, the recent devaluation of the CFA franc may worsen the situation still further.

16 On this subject, a small work published in 1939 illustrates well the polemics which this project created right from its inception, as well as the political battles that ensued (Herbart, 1939).

17 However, one issue that was apparently external to the project itself but in fact intimately linked with it and dear to its defenders paid off. This was the construction, in the absence of the Transaharan, of the Dakar–Niger railroad that was to transport the Office's cotton production (Herbart, 1939: 47, 57, 117).

6

SAHELIAN AGRARIAN SYSTEMS: PRINCIPAL RATIONALES

Philippe Lavigne Delville

While the large majority of Sahelian producers draw their means of subsistence from the surrounding environment through rain-fed agriculture, the specific forms this activity takes are innumerable. Such variability, of course, not only reflects the diversity of the natural environment, but also the multiplicity of agrarian traditions specific to each society. It is also the result of diversification due to recent economic and social developments. If we are to get beyond summary categorisations and misleading generalisations, we must grasp and give order to this variability. We cannot hope to achieve this goal, even partially, except by approaching it from several directions, some of which will be more descriptive while others will attempt to capture how production systems function. We can never understand the full extent of this variability unless we examine it in the light of some of the major questions that dominate relationships between agrarian societies in the area and the natural resources they exploit.

THE QUESTION OF FERTILITY

If there is a single requirement that all societies must meet, it is to maintain the production potential of the ecosystems from which they derive their resources. In order to do this, they implement a series of technical strategies for which the principles and coherence need to be understood.

The 'logic' of a long fallow period

One practice predominates in all Sahelian agricultural systems: that of slash-and-burn followed by leaving the land fallow. The basic principle is well known. Fields are opened up in the bushland, or on land formerly left fallow, cultivated for several years and then abandoned to remain fallow for the time it takes for vegetation to grow sufficiently. The land is then ready to be cultivated once more.

These fallow systems have long been considered as antiquated. Boserup (1970) and Ruthenberg (1980) have shown that, in fact, such systems are widespread in low population density agriculture and not only in Africa.[1] In such contexts they are an efficient means of exploiting the environment, offering the best remuneration for the effort put in. Indeed, after the land has been cleared, crop yields are generally higher and agronomic constraints fewer. After a few years, productivity drops, thus justifying abandonment and the opening up of a new field (when space permits). Moreover, fallow land also has a role to play in relation to property: land entitlement lasts as long as there is a mark of the work put in and, therefore, slash-and-burn is another way of appropriating land (Jean, 1975; infra. Chapter 10). Thus the fact that the fallow system has endured is not an 'anachronism'; there are many agro-ecological, economic and social justifications for its continued use, as long as it remains a relevant and feasible possibility. Even though the practice of long fallow periods is becoming less and less practicable, in the areas we are dealing with here it is nevertheless useful to have a clear understanding of the functions it fulfils.

Agronomic functions

Newly cleared land has the advantage of ideal conditions for cultivation, i.e., the upper soil layers have been enriched by accumulated organic matter over the years, have been broken down by roots, possess good infiltration capacity and the bush and tree cover has restricted the development of herbaceous plants. Moreover, the burning process lays a covering of ash on the surface which represents an immediately available store of minerals for the crops.

The first years' crops therefore benefit from these favourable conditions and produce good yields in return for a minimum of crop maintenance. As the years go by, however, these conditions deteriorate: weeds proliferate and become more and more difficult to keep under control; there is a rapid decline in soil organic matter; and there is a deterioration in soil structure, which becomes compacted, less permeable and more sensitive to erosion. Pest problems increase and there may be a decline in nutrient content as a result of uptake by the crops. These processes are particularly marked in tropical ferruginous soils, which are the type most commonly found in the region studied and which, as we have seen (Chapter 1), are particularly fragile with inherently low mineral and organic matter content and poor structural stability. Even as the work put into weeding increases, yields decline and after five to ten years they collapse completely, stagnating at very low levels. Before this stage is reached, sometimes after a field has been cultivated for three years and just as its productivity begins to decline (often under the effect of the pressure of weeds although before the absence of any clear deterioration of other parameters), cultivators leave it fallow for a period of time.

At this point, the natural dynamics of fallow land evolution come into play, bringing in a succession of plant species which reconstitute the 'mature'[2] ecosystem – annual herbaceous plants are replaced by perennials, then by shrubs and trees. The deep, dense root systems of the herbaceous plants restore the organic matter content and improve soil structure. The regrowth of shrubs eliminates the annual weeds and contributes to the storage of large quantities of minerals in the biomass and, later, to their return to the soil surface layers. Pests decrease. After a number of years, which can vary between 10–15 in Sahelian areas and up to as many as 30 years in a forest region (Piéri, 1989), the dynamics of the ecosystem eventually reconstitute the soil to the point where it is once again capable of a good level of productivity.

The effects of leaving land fallow thus depend on the type of environment and the length of time the land is left, as well as the context in which fallowing occurs. An occasional or short-lived fallow period may be a response to a labour shortage, for example, but will have little impact on the soil. A medium-term fallow period may relieve the pressure of weeds and pests and increase the quantity of fresh organic matter. However, only a long fallow period, which allows hardy plants and trees to emerge, will markedly improve the soil structure, the organic matter content, the level of minerals in the soil and other key properties.

The slash-and-burn form of agriculture thus combines phases of cultivation, during which soil conditions deteriorate more or less rapidly, and phases of reconstitution of varying length, when the physico-chemical and biological characteristics of the soil are restored. This type of agriculture is therefore defined by the relationship between the length of these two phases. For the fallow period to provide maximum benefit, it must be long enough. As the time given to cultivation has to be much shorter than the fallow period (because of the rapid decrease in productivity in the fields) there needs to be a large area of land available, so that new areas of cultivation can continue to be cleared before returning to the initial field. It is estimated that, at any given time, the area of cultivable land available should be three to four times the area actually cultivated (Bosc et al., 1990: 67).

The length of the fallow cycle is not fixed according to any rigid rules, however. It depends not only on overall conditions, including climate and vegetation, but also on a number of other parameters. The first of these is the state of the field when it was abandoned, because the vegetation will recover much faster if the land has not been totally cleared (large trees spared and shrubs simply cut back so they can quickly grow again from surviving root stocks). The presence of a neighbouring fallow field is also important as this helps with seeding. The mode of clearing also has a vital role in the dynamics of the fallow, as does the frequency of bush fires. Last, in any given context, the length of the cultivation phase varies significantly according not only to the availability of land, but also to the cultivation techniques used, which may accelerate or slow down the trend to lower productivity.

The productivity of labour

Boserup (1970) shows that, as long as the land is available, it is the labour force that is the limiting factor in agricultural production. In such a context, extensive systems which invest little labour per unit surface area are the ones which offer the best productivity from this production factor. Ultimately, leaving land to lie fallow is simply letting nature do the work of re-constituting the cultivation potential of the environment. It enables farmers to economise on the labour that would be necessary to increase the cultivation period and maintain the land's productivity with such techniques as weed control, organic matter supplements and anti-erosion work (all these are practical techniques to which we shall return throughout the next two chapters).[3]

This is not to say that permanent cultivation systems cannot exist where the population density is low. We can find such cases, but they tend to coincide with specific conditions such as the existence of certain natural environments (low-lying areas, floodplains), and particular socio-historic circumstances (taking over defensive positions). Conversely, it used to be possible to find forms of farming where mobility was even more pro-nounced than with the long fallow, particularly a semi-itinerant agriculture where local groups abandoned their settlement from time to time to go in search of virgin lands which were not only fertile but also rich in game and other natural products for gathering (see Chapter 5). In combination with slash-and-burn, the long fallow nevertheless represented, not so very long ago, the most widespread practice used to maintain cultivation potential in the area covered by our study. This surely attests to the strong agronomic and economic justification for its use.

Other means of maintaining fertility

The system of slash-and-burn followed by long fallow periods is, therefore, a means of responding to the problem of a decrease in productivity in cultivated land. In addition, by using a cropping/fallow cycle, it is also a way of maintaining the cultivation potential of the environment, using a low rate of land use, yet with a comparatively high rate of labour productivity. At the broader level of the Sahelian agrarian systems as a whole, just as within a single production system, it is also possible to find other means of maintain-ing fertility, with or without the investment of peasant labour.

In certain geographic situations, other natural processes can potentially be substituted for the fallow period. Hence, with clay-rich floodplain soils, fertility maintenance is assured by the floods that deposit fine elements and

141

organic debris resulting from the run-off upstream. At a different scale, the same mechanism governs fertility in the floodplains of the major rivers (the middle valley of the Senegal River, the Interior Delta of the Niger River). During the rainy season, rising waters pour out on to floodplains, each year depositing a layer of silt that makes permanent cultivation possible.[4] Floodplain agriculture, therefore, does not require a fallow period to maintain soil fertility.

The deliberate addition of fertilisers can be a substitute or a complement to the fallow period. For this purpose, many peasant systems add organic matter, i.e., fields closest to dwellings receive domestic waste while those further away benefit from animal manure. The manure can be gathered from livestock pens located within domestic enclosures or in close proximity to villages. Nevertheless, transportation problems remain, especially for the more distant plots. Another method is to leave the livestock, usually cattle, directly on the fields at night. After being fed on bush resources during the day, the animals are tethered in the area to be manured so that nocturnal excreta can accumulate. This process allows the transfer, from the pasture to the cultivated fields, of organic material in a concentrated and digested form. Where this association between agriculture and livestock exists, peasants usually know how many head of cattle and how many nights it takes to manure their fields (Deshayes, 1989). Generally, the location where cattle are kept is moved progressively through the dry season, so that consistency of manuring is maintained. Agropastoralists with substantial herds use their own animals while agriculturalists must often enter into manuring contracts with neighbouring herders or transhumants in order to obtain the necessary quantity of manure. In this case, the transaction frequently takes the form of an in-kind exchange, such as grain in exchange for days of manuring. These contracts, as well as occasional loans of livestock between agriculturalists, aim to balance livestock availability with the surface area to be manured. More elaborate systems have been developed which realise a further association between agriculture and herding. This is particularly true of the *Buzu* (Nicolas, 1962) and the Fulani (Diarra, 1979), in Niger. As Figure 6.1 illustrates, thanks to suitable land organisation in these cases, on-site manuring can take place not only on cultivated fields but on lands lying fallow during the winter season as well. The current field pattern, confirmed by the recollections of several older individuals, indicates that Hausa agriculturalists also formerly practised this technique. In certain cases, as we shall see in the next chapter, the presence of *Acacia albida* directly contributes to fertility maintenance in the zone covered by the grove and indirectly contributes to manure production through fodder for. livestock.

In peasant systems, these two primary modes of fertilisation, fallow and the application of manure, are often complemented by other techniques, such as ploughing between rows (see Chapter 7) as well as rotation and

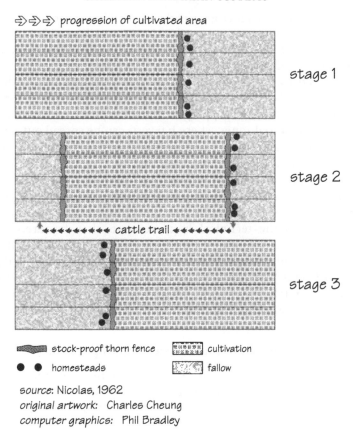

⇒ ⇒ ⇒ progression of cultivated area

stage 1

stage 2

← ← ← ← ← ← ← ← cattle trail → → → → → → → →

stage 3

███ stock-proof thorn fence   ▨▨ cultivation

● ● homesteads   ▨▨ fallow

source: Nicolas, 1962
original artwork: Charles Cheung
computer graphics: Phil Bradley

Figure 6.1 The pattern of crop and fallow organisation in the *Buzu* field system
This figure illustrates schematically how space was traditionally organised by *Buzu* agro-pastoralists in Niger, and how they reconciled and combined the two production activities that were practised. Fields are organised in narrow bands, along which cultivated parcels are opened up as the years go by, at a rate respected by all users in the village. Settlement follows the same pattern, with the straw huts being taken down and reconstructed each year at the edges of the cultivated lands.
　　Cultivated fields are abandoned as fallow, while previously cleared lands that have remained uncultivated for a long time, are reopened. Livestock is stallled at night near the dwellings, on soils that are to be cultivated the following year, thereby taking advantage of the manure. By the end of a cycle that can last several years, once the fields have reached their exploitable limit, the cultivators return to their original starting point and begin anew.
　　Nicolas (1962) was the first to describe this practice with the *Buzu* in the Zinder region. It appears to have long been practised by other groups located in the same geographic area, particularly the Fulani (Diarra, 1979) and even, with certain modifications, by livestock-owning Hausa agriculturalists (Raynaut, 1980a).

intercropping (principally cereals and legumes). Combining different types of crops in the same field produces gains through complementarity due to the exploitation of different layers of soil or air. The different agronomic requirements limit competition and give a higher overall yield. Moreover, by mixing cereals with cover crops, weed control is improved, soils are better protected from erosion and moisture loss is reduced. Diversifying crops also reduces the risk of plant pathogens. Finally, mixing leguminous plants with cereals can favour nitrogenous nutrition.[5] Overall, mixing crops results in greater productivity per unit area. Not all such associations have the same effect, however, and each case should be analysed individually (Dupriez and de Leener, 1987). Mixed cropping is both frequent and varied in the Guinean and Equatorial zones, where harvests can be spread out across the long agricultural season. In the Sahelo-Sudanian zone, however, fewer crop types are mixed and these are often limited to cereals (millet, sorghum, maize) and leguminous plant combinations (groundnut, beans, cow peas). Intercropping, nevertheless, is a very widespread practice among peasants. Agronomists have long tried to discourage this practice, which they found incompatible with the technical improvements they favoured. Others, observing that (on the contrary) intercropping takes maximum advantage of the biological and ecological functioning of the soil and cultivated plants, see in it proof of the 'endogenous agricultural revolution' which contradicts the model of capitalist investment (Richards, 1985). On this point Piéri's findings are more subtle (Piéri, 1989: 423).

The decrease in fertility: a complex problem

There has been reference on many occasions to the 'moderate' reception to 'modern' fertilisation techniques given by the peasants. In sub-Saharan Africa as a whole, fertiliser consumption is derisory – about 8 kg ha^{-1} (Bosc et al., 1990). In fact, its use is mainly limited to the cotton-growing regions and more especially to the irrigated areas, where doses of 150 to 200 kg ha^{-1} are not unusual. Moreover, even in southern Mali, which has been noted for the success with which proposals for technical improvements have been welcomed, soil nutrition is insufficient to compensate for mineral losses due to the exports from cotton/cereal production or to neutralise the acidification induced by nitrate fertiliser application. Van der Pol (1991) thus estimates that the economic losses of soil exhaustion represents about 40 per cent of peasants' gross profit margins in this region. In other words, in today's conditions, 40 per cent of the peasants' income would appear to be accounted for by soil exhaustion. Although these calculations are debatable, they nevertheless indicate the scale of the problem.

Beyond these specific geographical areas, the quantities of fertiliser used are negligible, either because climatic conditions render their use too hazardous (see p. 151), or because price ratios make them unprofitable or, as is most frequently the case, because they are not easy to obtain (not physically available or no credit system). The great majority of peasants in the Sahel–Sudan now understand the benefits fertilisers can bring and would be prepared to use them if conditions for their use become favourable. This is clear from the way the peasants respond so rapidly to any price variations and from the detours they sometimes make in order to obtain fertilisers at an attractive price.[6] However, although various experiments have shown that a moderate dose (about 20 kg of urea per hectare) can be profitable, and of minimal risk, it is unfortunately rare to find cereal crops which have received even this low dose, beyond the specific situations that we have noted.

These different modes for maintaining fertility are hardly mutually exclusive. On the contrary, Sahelo-Sudanian agriculturalists make every effort to combine them optimally in their production system. Household refuse is used on nearby fields and in the *Acacia albida* groves which are grown around the villages (see Chapter 7), manure is transported further afield, on-site manuring is used on more distant fields, rotation with fallow periods is applied on peripheral plots, and intercropping is practised on the vast majority of fields. Fallow periods and active fertilisation are thus complementary practices. Manure is applied selectively on lands that are the most accessible, so that labour can be most efficiently invested. When this spatial and functional complementarity occurs, it often manifests itself in a ring structure, according to the model described and analysed by Sautter (1962).[7] Sometimes, however, when organic matter is in short supply, the manure may be concentrated in those fields which are in poor condition, as is often also the case with fertiliser, where it is used sparingly.

These technical practices are currently undergoing a serious crisis. The practice of long-term fallowing no longer exists except in limited areas. In many places, fallow land has almost completely disappeared, or else the fallow period is much too short to have a significant effect on soil fertility (Piéri, 1989). To facilitate animal traction agriculture or to meet the demand for wood, cultivated land is being completely cleared of trees and shrubs, thus reducing the ecosystem's capacity to reconstitute itself. There are also the effects of drought that reduce biomass production by something in the order of 30 per cent (Bosc et al., 1990). Even the Sérer production system, which illustrates a peasant society's optimal use of environmental resources, has not escaped this development. The numbers of *Acacia albida* have fallen drastically over the last few decades, so that the proportion of cultivated land fertilised by these trees has in some cases declined from 20 to 5 per cent (Garin and Lericollais, 1990; Lericollais, 1990).

The disappearance of fallowing and the cultivation of marginal lands have also led to the loss of grazing land. In extensive herding systems, manure production is linked directly to the number of cattle, which, in turn, is directly dependent on fodder resources. We have had occasion to discuss, above, the competition between agriculture and pastoralism (Chapter 5), but the problem is equally acute within peasant production systems themselves. In the Sine Saloum, for instance, the livestock of Sérer agriculturalists is currently displaced towards the islands which exist in the middle of the marshlands. In the village of Sob, two-thirds of the bovines are now transhumant and the area of manured land declined from 81 ha in 1964 to 10 ha in 1987. In addition, and because wood is in short supply, dung is increasingly used for domestic fuel (Pochtier, 1989). This situation is quite common and is almost identical to that found on the Mossi Burkinabe plateau (Marchal, 1983; Serpantie *et al.*, 1986) and in central Niger (Raynaut, 1975). Piéri summarises the central problem thus:

> The central problem is that of the coexistence of agricultural and herding systems, which are complementary in principle but competitive in practice in an area where the annual production of vegetal biomass is insufficient to meet the energy and subsistence needs of the rural population and the livestock's fodder needs.
>
> (Piéri, 1989: 418)

With the decline in fallowing and in manure contracts, such shifts give rise to much anxiety about the changeable fertility of the land, anxieties that are fuelled by a recurrent problem of decreasing yields, a subject spoken of by peasants and technicians alike. This much debated question merits a detailed examination. We must first note, however, that fertility is not an intrinsic feature of the soil, capable of being measured.

According to Sebillotte, the definition of fertility 'must necessarily refer back to the system of production, as it is this which defines the degree of constraint under which the farming system is placed as a result of such and such a constraint from the environment' (Sebillotte, 1993: 131). He therefore suggests substituting the notion of cultivation capacity: 'a global judgement of the environment from three angles: – potential, for the various crop systems envisaged; – costs involved in producing such crops; – risks, estimated by analysing flexibility and security in the choice and establishment of the crop systems' (Sebillotte, 1993: 131).

Consequently, it is the overall functioning of a biological system (consisting of soil, plants, climate), subjected to technical intervention by the farmer, which is under consideration. We therefore look towards a relative notion, one which relates both to crops and techniques and which depends on means which can be implemented in order to remove constraints imposed by the environment and attain the desired level of production. The physical, chemical and biological characteristics of a soil (which can be measured) define the cultivation capacity, which cannot be assessed except in relation to production objectives. We should, therefore, be extremely cautious when speaking of a 'decline in fertility' because, outside the strict confines of a research station, the various components that define such a notion have rarely all been monitored through time. Techniques change and, as well as there being objective developments to consider (such as the absolute loss of organic matter in the soils), the identification of fertility decline must also relate to more subjective questions. As Sebillotte notes once again: 'it is often because there is a sudden change in practices (over a period of a few decades, or even less) or in their conditions of application, that the subject of safeguarding, of maintaining fertility emerges' (Sebillotte, 1993: 128).

As for yields (which are never the plain and simple expression of the soil's potential because they also depend on climate and techniques), Piéri's (1989) synthesis shows that in the Sahel, over a long-time scale, we note rather a global stagnation in cereal production and a definite increase in cotton (with, nevertheless, some conflicting variations, according to local situations).

Thus, ultimately, we can draw only one carefully qualified conclusion on this subject. Essentially, the fluctuations in inter-annual yields in the zone's principal rain-fed cropping areas can be explained by extreme annual variation in the water conditions. In the long term, however, the evidence suggests that yields may undergo an increasing decline with some loss to the soil's fertility capital (Piéri, 1989: 240). Without making light of the problem, it is necessary to remember that all visions of impending catastrophe are inappropriate, for such extreme views underestimate the capacities of peasant societies to adapt. Agriculturalists are increasingly aware of the serious threat of a decrease in the productive potential of the lands they exploit. This is clearly the case in Burkina (Marchal, 1983), in the Senegalese Groundnut Basin (Lericollais, 1990), in eastern Senegal,[8] as well as in the Tahoua and Maradi Departments of Niger (Freysson (ed.), 1973; Raynaut, 1980b). Faced with the fallow land crisis, they are already deploying a variety of techniques to 'prolong the agricultural cycle' (Floret and Serpantié, 1993; Floret et al., 1993). These enable them, sometimes at the price of a fall in production per hectare harvested, to increase the level of land use, but this is never sufficient to achieve a global equilibrium. Traditional methods of permanent cultivation, which require major inputs of organic matter and much work on the soil, are only partly appropriate in Sahelian Africa. A new equilibrium is likely to involve techniques that slow down the rate of soil deterioration and accelerate its regeneration with short, improved fallow periods. Various techniques exist (organic inputs, working the soil, improved fallowing, chemical fertilisers, control of run-off, etc.). All have a different and complementary impact on the functioning of the climate/soil/plant population system and clearly must be brought together in combination.

A moderate use of fertilisers is certainly indispensable, and has a positive effect when the soil is sufficiently rich in organic matter. The addition of large quantities of well-decomposed manure is most definitely useful, but there are different constraints involved here, namely the availability of capital (for purchasing cattle and carts), and of labour (to prepare and spread the manure). There are also certain limitations when making the transition to intensive cattle-rearing (see the quotation from Piéri,

147

1989: 418 on p. 146). Various experiments are currently underway to improve fallow systems and fodder breaks. These involve accelerating the soil's ability to reconstitute itself (especially the stable humus content and the soil structure) by protecting it from fire, by planting trees at the end of the cycle and even by sowing fodder crops on the fallow land or creating fodder breaks within the cropping cycle.

Although most attractive, these different techniques are still at the experimental stage and their agronomic efficiency remains subject to debate. In any case, if they were to be implemented on a large scale, a massive restructuring of plant and animal production systems would be required, with the use of hedges becoming widespread, common grazing land being discontinued, etc. Some radical changes would therefore be necessary in order to manage the relationship between land and work. Such is the declared aim of certain *gestion des terroirs* projects,[9] many of which are emerging in Burkina Faso and Mali. Apart from a few successes, made possible by a concentration of financial resources, results fall a long way short of ambitions.

The solution must lie in a combination of these different modes of resource management, with peasants making appropriate adjustments where necessary, in accordance with their specific situation and their personal objectives. History has shown that, in this particular area of activity, their capacity to adapt is boundless. There is no reason why they should not continue to show the same talents. They will not succeed, however, unless the overall conditions are favourable to this end. This is particularly necessary in relation to the economic environment which, in a market context, can have a major effect on investment decisions, on the choice of cropping systems and on the allocation of labour. The provision of machinery and equipment and the various forms of input, as well as the acquisition of the know-how to be able to use them appropriately, also has a determining role in the development of technical systems. Lastly, stability in the rules governing access to natural resources is essential, especially the security of land tenure, if practices are to be adopted that require an increased transformation of the environment, and hence capital and/or labour investment. These are not acceptable if there is some uncertainty over the security of access to the benefits (we shall return to land issues in Chapter 10).

Sahelian states do indeed have a heavy responsibility to set in motion these dynamics for change through legislation on natural resources, the setting up of technical and financial mechanisms to support innovation and the creation of a stable and favourable economic environment.[10] Contrary to what is too often asserted today, 'peasant participation' does not represent a preamble to the emergence of a new equilibrium in the 'society and nature' equation in the Sahel. On the contrary, it is acquired beforehand and in accordance with its own choices and its own rhythm, with its own upheavals, its contradictions and its indispensable caution, but only in as much as the obstacles to the readjustments are removed – which is necessary if change is to take place.

THE QUESTION OF RISK

Risk is a constant presence in agriculture and is the determining factor behind many farming practices (Eldin and Milleville, 1989). It asserts itself with particular force in Sahelian and Sahelo-Sudanian Africa. First, risk to cultivation goes hand in hand with the uncertainties of climate, attacks by swarms of locusts and damage by birds, warthogs, etc. Next is economic risk, that is uncertainties over the sale price of produce or the profitability of new techniques (fertilisers, animal traction) and consequences of indebtedness. Moreover, a climatic risk soon becomes an economic risk (Couty, 1989) which can endanger a family's fragile ability for material reproduction. A bad year means that money spent on fertiliser is wasted, and a poor harvest often leads to a spiral of food shortages and debts not honoured.

Clearly, the question of agricultural risk and the peasants' response to it cannot be considered simply in terms of production. In former times, some granaries were devoted to the storage of grain over a period of years, so that supplies were available even when a bad harvest had to be endured. Mutual help between families and networks founded on interpersonal dependence and clienteles played a similar role – as it still does at the present time. Today, the main purpose of the dry season activities (wage-earning employment and market gardening) is often simply to be able to buy in the cereals needed in order to manage until the end of the year. These activities are now an integral part of family strategies for food security. Likewise seasonal migration, even when the net result is a reduction in the number of mouths to feed during the dry season, rather than the production of hypothetical extra income (Lombard, 1993). In short, food security cannot be analysed solely in terms of production. It is none the less true that risk management begins at the production stage and that it can have a profound effect on technical practices in Sahelian societies.

Risk and agricultural production

In the harsh and changeable climate experienced in the Sahel (Chapter 1), anticipating risk has always been a fundamental survival strategy of farmers and, to this day, minimising this risk has been a priority objective on which many technical decisions are based. One anxiety predominates, and that is to provide a regular food supply to cover a family's needs. The repeated famines of the last decades are a stark reminder that, given the prevailing climatic conditions, this can by no means yet be taken for granted. In the Sahel, the length of the wet season barely covers the growth cycle for even the least demanding of crops. In order to minimise the risk, therefore, peasants sow on to dry ground, in other words, even before the first rains or as soon as the rains begin, in an attempt to use the year's rainfall to the full. This represents a considerable gamble, however, as the seed may be lost altogether if the rains are late. The plots then have to be re-seeded, despite the fact that the later the last sowing takes place the greater the risk of the crops not being able to complete their growth cycle before the end of the season. This practice of early sowing is usually accompanied by a complementary strategy consisting of sowing from the outset all the land the peasant has at his disposal, even though he knows he does not have sufficient manpower to maintain the crop should it all survive (Raynaut, 1980b). In doing this, the objective is to anticipate possible spatial variations in rainfall. These variations can be considerable, even on the farms surrounding a single village. The farmer knows that he will subsequently concentrate his efforts on the plots which have received most rainfall and from which the best results can be expected. In fact, risk management strategy is conceived on the basis of covering more than one year at a time. The farmer aims to derive maximum benefit from a year of high rainfall (storing any surplus), while at the same time minimising the risk of zero production if there should be a drought. By sowing over as wide an area as possible, he is achieving these two objectives at one and the same time.

It is easy to understand how, in this context, farmers are very well aware of the degree of risk to which they feel themselves exposed. Their geographical position in relation to the rainfall gradient is obviously a very significant variable in this respect. Taking the region as a whole, the effect is most marked in Niger, in the northern strip of cultivated land and also in the Senegalese Ferlo. The phenomenon is also perceived at a more localised level. Studies carried out in the Maradi region, for example, have shown that the average area cultivated manually per person is greater than 2.5 ha where rainfall averages 400 mm, whereas it drops to 1.4 ha for 600 mm (Raynaut 1980b: 20). In the Sahelo-Sudanian zone, especially the Sudanian part, the rainy season is both longer and less irregular. The practice of dry sowing is not common, with farmers preferring to wait until the rainy season is well established before maximising their chances of success. The unit area worked manually may then drop to below 1 ha, and under these conditions investment in labour and capital per unit of surface area involves much less risk.

The droughts experienced in the 1970s and 1980s had the effect of exaggerating climatic risk factors, forcing peasants to extend their surface areas and to adopt more extensive farming practices. Milleville (1985) shows that in the Oudalan in Burkina Faso, certain farmers cultivate 0.73 ha per worker, whereas they had cultivated only 0.4 ha in 1950.[11]

Crop diversification, the use of a variety of micro-environments which perform differently under the varying levels of rainfall, sowing several varieties which do not have the same growth cycle nor the same water requirements in the same field – these, along with many others, are just some of the practices which can be connected with risk management (in relation both to climate and to plant care).

Risk and innovation

The fear of risk is one of the reasons often put forward for refusing to introduce innovative techniques. Improved seed varieties, which have long been recommended on the basis of their potential in a controlled environment, in fact often perform less well than local varieties when they are transplanted into the production conditions experienced by the peasants. In the Sahel environment, climatic variations are such that the use of fertiliser can represent a major economic risk, particularly as, given that they promote plant development, they render crops more sensitive to moisture stress.

The case of the cotton-growing areas, mainly those in Mali, provides proof of the role played by risk levels in farmers' innovation capacities. Benefiting as they do from more favourable climatic conditions, a secure economic environment and an efficient extension network, these farmers invest in materials, they use fertilisers regularly and correctly and they apply pesticides and even herbicides. Yields have increased spectacularly over the last 30 years. This success clearly confirms the fact that for the peasants, risk is a major factor to take into consideration, and one to set against the possibilities of innovation.

AGRICULTURAL INTENSIFICATION

In the Sahelo-Sudanian zone, the general principle of production systems is most often one of extensivity. As we have seen above, extensivity means here that value placed on work productivity is greater than the value placed on land productivity. Nevertheless, largely because of demographic growth, the aim of increasing yields has, today, become a necessity. At the local level, in the face of land shortage, production gains can no longer be obtained by increasing the area of land under cultivation. At the national level, major surpluses must be produced to feed city dwellers. Given the anticipated rates of urban and general population growth mentioned in Chapter 2, yields would need to increase threefold, and labour productivity fivefold or sixfold merely to maintain the existing food balance! To meet such a challenge demands a good grasp of current agricultural system dynamics. If these dynamics tend towards extensivity, the task is much more difficult than where and when an intensive principle begins to emerge. Once again, the diversity of situations prohibits us from drawing general conclusions.

It is important to define, first of all, the terms 'intensive' and 'extensive', since they are too often used with little precision. Agriculture brings into play three kinds of production factors: land, labour and capital. These are combined by the peasant in varying proportions and he can, to some extent, substitute one for another. The qualification of extensive or intensive is determined by the rate of use of one production factor, labour or capital, as applied to a given area of land. So, for example, American cereal agriculture is labour extensive and capital intensive, i.e., mechanisation is maximised in order that labour may be minimised. Market gardens, on the other hand, are most often labour intensive.

African agriculture is generally labour extensive because in sparsely populated areas labour is the limiting factor in agriculture. As a first approximation, the level of land consumption may be seen as a synthetic indicator of the degree of intensivity of a production system.

Variability of levels of land consumption

In this respect, the first criterion to consider is land use, or the number of years of cultivation set against the number of years of fallow. According to a well-established calculation (Giri, 1983: 85), a family of five cultivating 3 ha for three consecutive years before leaving the land to lie fallow for nine years in fact occupies 12 ha. The land use ratio is therefore 1:4. We have seen that this depends on the environment (soil, natural or secondary vegetation) and cultivation techniques (which may or may not extend the number of years of cultivation). Population density, and hence the availability of land, obviously sets a ceiling on the length of the fallow period. When it is high, population density represents a major constraint. Over the course of the last decades, the land use ratio has increased considerably, although at varying rates.[12]

The second criterion to consider is the cultivated surface per person, which expresses the land/labour ratio in the production process. Although this tells us nothing of quantities of labour or, more particularly, of capital invested per hectare (especially with mechanised cultivation and/or fertilisers), it nevertheless provides a good indication of a certain level of technical capacity. For this reason Figure 6.2, taken directly from the work of Ancey (1977), is useful.

In spite of the fact that this document is out-of-date and that specific estimates and border placements are disputable, it is still quite valuable in that it highlights the variety of man-to-land ratios from one area to the next throughout the zone covered by our study, i.e., from less than 0.4 ha to more than 1 ha per person.[13] If we overlay the map illustrating population distribution, we find no apparent correlation between human densities and land use. For example, cultivated surfaces per person are lower in western Mali, underpopulated though it may be, than in central Niger or western Senegal, where densities are still over 40 inhabitants per square kilometre.

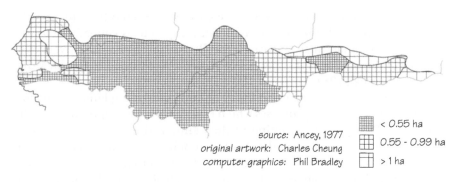

source: Ancey, 1977
original artwork: Charles Cheung
computer graphics: Phil Bradley

▦ < 0.55 ha
▦ 0.55 - 0.99 ha
▦ > 1 ha

Figure 6.2 Cultivated area per capita in the western Sahel

It would appear that what we have here is not simply a mechanistic adaptation to land shortage, but is instead the expression of distinct strategies regarding the use of space.

Several possible factors help to explain this variability. One is the type of agriculture practised, as in the Central Delta of the Niger River, where recession rice agriculture requires investments in labour that are incompatible with the cultivation of extensive areas. In the Upper Senegal region, and probably also in western Mali, the impact of migration on the availability of labour helps to explain why agriculture is confined to a fairly limited area, in spite of the vast stretches of surrounding bush. In the groundnut basins of Senegal and Niger, light soils which are easy to cultivate combine with the excessive demand for land imposed by commercial agriculture, which accounts for the high degree of land use within a context of intense land pressure.

However, population pressure and economic strategies alone cannot account for these differences of degree in the extensivity or intensivity of production systems. Other variables also need to be integrated, in particular the type of relationship which exists between the peasants and the land (see infra, Chapter 10). Put simply, we can say that, historically speaking, the agrarian societies which have always had a mystic relationship with the land and whose economic reproductive base was agriculture and hunting are the ones whose cultivation practices were, in former times, most intensive (and where therefore the lowest rate of cultivated surface per person was found). It is among these societies that we observe practices such as preparing ridges, the creation of groves and a variety of landscape management techniques to which we shall refer again later (see Chapter 9). They often occupied very concentrated village sites, situated in the middle of the bush, mainly as a defensive measure against looting and raids by local warring societies. In the aristocratic, warlike and/or trading societies (see infra, Chapter 9) the economic reproduction base was radically different. The relationship with the land was governed more by political control than by any sacred links. Techniques were more extensive. Such differentiations in the relationship with the land and cultivation practices can still be found today, for example, in the heart of the Burkinabé Yatenga, between the 'people of the land' and the 'people of power' – two categories in the Mossi society (Marchal, 1984).

On the basis of these former differentiations, extending land occupation is the result of different processes which have been superimposed gradually over time. First, at the beginning of the century, there was 'pacification', which released societies from the constraint of insecurity and enabled them to open up fields in the bush. Next, came the need to produce marketable surpluses, without endangering the food security of the family. The dividing up of domestic production structures (see infra, Chapter 11) also played a role in land use by making it more difficult to mobilise large work groups. More recently, the introduction of animal traction has also encouraged expansion (see Chapter 7). Finally, in a context of competition for land (either actual or potential), strategies of anticipation have guided behaviour in relation to land. This is particularly true in areas of recent immigration (the *terres neuves* area of eastern Senegal and western Burkina Faso), where clearing has been carried out on a massive scale with a view to appropriating land reserves in anticipation of future land saturation. This situation can also be seen in zones where occupation was formerly very dense. For example, around Maradi in Niger, it was participation of this kind in the last possible forest clearance a few decades ago that determined the size of farms today (Raynaut, 1988a). Elsewhere, land reform has set the process in motion, by generating the fear of a loss of control of the land (as happened in the Sérer area of Senegal at the time of the law on national property (Gastellu, 1988)). Such phenomena have given rise to some real restructuring in land use, even in agrarian societies, where we have seen former villages completely abandoned, as in Bwa in Burkina Faso (Capron, 1973).

In the context of the global strategy of the slash-and-burn procedure described above, the present diversity in the level of land use results from various combinations of multiple factors. The general trend, nevertheless, is still towards the extension of cultivated land.

A shifting reality

In fact the notions of extensive and intensive are ultimately relative. They express relations between factors and can only be used comparatively. The peasant systems of slash-and-burn and fallow periods are globally extensive only in contrast with permanent agriculture, which aims to maximise the yield per hectare by utilising capital, the norm in the majority of development projects. This norm itself, however, corresponds to specific contexts, namely those world regions where the extension of cultivated areas is no longer possible except on the margins and where a production increase can only be achieved by increased investments per unit area. In an unsaturated area, it would appear to be taken for granted that extensive practices are those which offer the best rate of labour productivity and often the best security – especially when rainfall levels are low and irregular.

It is therefore understandable that peasant systems are for the most part extensive. Nevertheless, this general statement covers a wide variation, both from one region to another and between different crop systems[14] within the same agricultural setting. Indeed, even using manual cultivation, there are a multitude of ways of combining land, labour and, possibly, capital. Thus the circle of fields spread with manure which often surrounds villages may be interpreted as the creation of an intensive farm system. Similarly, in the areas of irrigation development, peasants integrate their irrigated plot, cultivated in the intensive fashion, into a system of rain-fed crops which is largely extensive (Lavigne Delville, 1991). A strategy to extend cultivated areas can, moreover, achieve very successful combinations within the same farm using certain intensification practices recommended by the development services, as has been shown, for example, in the case of some cotton-growers in southern Mali (Djouara *et al.*, 1994).

We are therefore a long way from being able to express the extensive/intensive ratio in the form of a simple categorisation, and even less so in the form of a simple, clear-cut contrast. This expression is subject to different levels of appreciation and to different criteria of judgement. Moreover, it deals with an unstable equilibrium, one that is likely to evolve according to all-encompassing conditions. These conditions may be major trends in agrarian dynamics, or economic developments linked with market opportunities or access to credit. Obviously, any attempt to represent this cartographically would produce a totally arbitrary picture.

It is none the less true that in periods when land becomes scarce, the system of fallowing finds its limit. The classic scenario unfolds. When land is in short supply, the fallow period is shortened and the land no longer has time to reconstitute its production capacities before it is planted once more. When yields decrease, the peasants then tend to compensate by increasing the area of cultivated land. Marginal, fragile areas are then cleared, and suffer erosion. Bush regrowth is no longer sufficient to compensate for the wood that is taken away. A spiral of deterioration is thus set in motion. Although this brief outline is much too generalised to apply to the many and varied realities that we have described without qualification, it nevertheless shows clearly the ambiguity of the relationship between society and nature in the Sahel. There is an apparent paradox (but one that can be explained when we understand the various peasant 'logics' that have just been described) in that, even when the extensive system is in crisis, when land is scarce and the conditions required for the fallow system to operate optimally for lasting results are absent, the peasants can still continue, even

155

increase, their extensive practices. At the individual level, the extension of cultivated land at the expense of investment in labour and in capital per surface unit may, in a transition period, continue to form part of a rational strategy, even though it tends to accelerate the crisis at community level.

CONCLUSION

These crisis situations seem to be characterised by blind leaps into extensification, whereas a swing towards intensification would, from an ecological point of view, be the strategy best adapted to safeguarding the future preservation of natural resources and the production capacity of the ecosystem. This is one of the major contradictions currently facing agrarian systems in many parts of the Sahel. Farmers are confronted with the problem of a limited area of land in particularly difficult conditions, not only in relation to the natural constraints they must endure (soils, climate – see Chapter 1), but also in relation to the historical circumstances in which they find themselves having to effect this radical transformation (rapid demographic growth, trauma of constant droughts, inauspicious economic climate, ambivalence on the part of the state system). Fortunately, mechanisms such as these do not all pull in the same direction. Even as degradation continues, numerous innovations in soil fertility management and in land development (see Chapter 7) show that adaptation is under way and that the peasants are seeking solutions. The crisis in the Sahel is not simply an inevitable fatality, and the example of the Kano region in Nigeria should give cause to reflect on this matter. This region is situated in the Hausa country, in agro-climatic conditions similar to Niger. The population density is over 200 inhabitants per square kilometre (Mortimore, 1989), proof that land saturation beyond the often stated threshold of 40 per square kilometre is thus a relative concept. In another context, but one which may be indicative here, studies in Kenya (Tiffen *et al.*, 1994) show that the environmental crisis, with its tell-tale signs (deforestation, erosion, etc.) may be simply a transitional phase. The Machakos district, which suffered a serious environmental crisis in the 1930s, has now recovered, although its population is much greater than before (see pp. 317ff.). A series of changes, in which government projects have played but little part, have enabled this positive evolution to take place, such as construction of terraces and tree-planting by the peasants, the development of extra-agricultural activities, improvement of market outlets, etc.

Less spectacular, perhaps, is the developing situation in the Maradi region in Niger, which is moving in the same direction. Since the beginning of the 1980s, after a period of clearances which left the soil stripped virtually bare, we now observe in the most densely populated zones the emergence of elaborate practices to protect individual plants which have enabled a significant amount of tree cover to develop (Yamba, 1993). In short, if overall conditions allow peasants to discover new agro-ecological equilibria, then the crisis raging around the extensivity question may be no more than a transitional phase, and it is not impossible that some spectacular dynamics of environmental recovery might be observed in the coming decades, if only in localised areas.

There are many imperatives to which all agrarian societies in the Sahel zone are subject. It is vital to maintain the productive potential of particularly fragile ecosystems; security is crucial in an environment where, to an already high level of natural risks is now being added a no less significant economic risk; and there is need for a constant monitoring of the balance between the value of labour-oriented strategies and land-oriented strategies. It is to these constraints that all agrarian societies in the Sahel are subject and it is within this framework that the dynamics of each must be contained. These important principles should not be ignored if the possible future implications arising from the consequences of, and the limitations on, the diversity of agrarian practices are to be appreciated.

NOTES

1 In Europe too, slash-and-burn (clearance) has long been a very common procedure. It was observed right up to the end of the nineteenth century in northern Europe, in the ancient massifs of central Europe and in the mountains of southern Europe (Sigaut, 1976). Strictly speaking, if taken from a European context, the use of the term 'fallow' is inaccurate when applied to tropical regions. In its European context, the term refers to all the labour judged to be necessary in spring in order to prepare the land for receiving the winter cereals (Sigaut, 1985). It was certainly not a period of rest for the soil or an absence of labour, as is the commonly held contemporary meaning. Because the term is now so widely used in this latter sense, we adopt it here in relation to tropical regions.
2 In ecology, a 'mature' ecosystem (stable, with a large biomass and reduced productivity) is distinguished from a 'juvenile' ecosystem, created by the destruction of vegetation, with a rapidly evolving botanical composition, a high level of productivity but still with little biomass.
3 The problem is expressed in different terms as soon as the use of capital is involved. Animal traction cultivation and the use of fertilisers allow the farmer to increase the number of years of cultivation. Nevertheless, without a massive input of organic matter, he will not be able to avoid a decrease in the productivity of the land.
4 Combinations can be complex at times, like the one mentioned in the Senegal valley (p. 115), where an ecological chain is created linking agriculture, livestock and aquatic fauna.
5 Recent studies have put into context the role of legumes in semi-arid regions. If the water supply is poor, the micro-organisms responsible for symbiotic nitrogen fixation are scarcely able to fix any atmospheric nitrogen. They may even consume nitrogen from the soil, thus making it unavailable to the crops (Piéri, 1989: 357).
6 This is the case in the frontier regions. In Niger, for instance, the peasants have long taken advantage of subsidised prices and favourable exchange rates to buy

in Nigeria supplies that they cannot afford in their own country (Grégoire and Raynaut, 1980).

7 Often this organisation of space creates a gradient of decreasing yields relative to the distance from the village. In one Niger village studied by Grégoire (1980: 68), the following observation was made regarding cereal yields: 477 kg at less than 500 m from the village; 190 kg between 500 and 2,500 m; 130 kg or less beyond 2,500 m.

8 Lavigne Delville, personal observations.

9 We have kept the French terms *terroirs* and *gestion des terroirs*, because an exact equivalent in English is difficult to find. According to Painter *et al.* (1994), the *terroir* was looked at by Africanist scholars, and it was assumed by agrarian communities in Africa, as a significant unit of agrarian social organisation. *Gestion des terroirs* refers to efforts to protect and improve the natural resource base of the *terroir.*

10 In this field, however, the mechanisms involved are far beyond the capacities of the states of this region. They require a commitment from the international community.

11 Marchal noted a similar development between 1930 and 1971 in the Mossi region of Burkina Faso (Marchal, 1983: 319).

12 And, contrary to a commonly held view, the global productivity of the land has therefore increased, despite a stagnation, or even a slight decrease, in yields (i.e., the productivity for a given year of cultivation). Let us consider a piece of land cultivated one year in two with a yield of 0.6 t ha^{-1}, and another cultivated three years in 12 with a yield of 0.8 ha^{-1}. Global production, calculated over a period of 12 years, will be 3.6 t in the first case and 2.4 t in the second!

13 Confronted with the problem of defining worker in the rural context, Ancey rightly prefers to use the criterion of surface area per person.

14 This term designates a group of plots with the same succession of crops, and tended using a similar set of techniques.

7

THE DIVERSITY OF FARMING PRACTICES

Claude Raynaut and Philippe Lavigne Delville

Despite the common background of natural demands and technical options which we have just seen, the range of variation in the Sahel farming systems remains wide. From one place to another, from one society to the next, we find that crops and cultivation techniques can be quite different. We find that different tools are used, space is managed in different ways, and landscapes are shaped accordingly. This diversity is an expression of the distinct approaches used by one group or another to organise its relationship with nature. It also shows the many sorts of pressures and constraints to which each group has been subjected throughout its own specific history.

PRINCIPAL CROP SYSTEMS

This is the criterion selected by Déat and Bockel (1985) to define the agricultural systems of the Sahel. More than merely an objective image and description, the geographic distribution of the principal crops is significant from several perspectives. In the first place, the variety of cultivated plants reflects the effect of climatic constraints. To the north, crop rotation decreases, so that by around the 400 mm isohyet only the intercropping of millet and sorghum, or millet alone, can be found, accompanied by a few 'side' crops.[1] Conversely, where rainfall exceeds 900 mm per annum, diversification intensifies, with maize occupying an increasingly dominant position, along with rice cultivation and yams in low-lying areas.[2]

This variety also indicates diversity on a very local scale. The more diverse the environments that the land exploited by a community cuts across, the wider the range of crops to be found. Even within a limited geographic area (between neighbouring villages, for example) the range of cultivated species can vary widely due entirely to micro-differences in the topography, hydrology or soil types. We will not, therefore, be able to capture this variability at the cartographic scale used here.

Finally, the complexity of crop rotations may indicate a sophistication in agricultural techniques which increases the possibility for rotation or

159

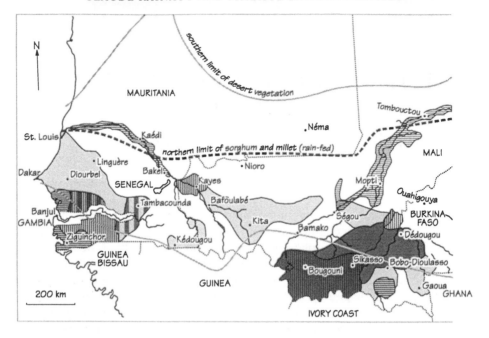

Figure 7.1 Principal cropping zones of the western Sahel (continued opposite)

intercropping. Conversely, in many regions, the trend is towards simplifying production systems due to competition from imported products (with the Soninké, the *Haalpulaar* of Senegal and the Hausa of Niger, perennial cotton and indigo have disappeared), expanding cash crop agriculture (in the cotton or groundnut zones) or perhaps only simplifications that sometimes accompany the colonisation of new lands (Capron, 1973; Benoit, 1975).

Figure 7.1 shows the spatial distribution of the principal crops found in the Sahelo-Sudanian zone. It is based on Déat and Bockel's map (1985) cited by Boulier and Jouve (1988) and has been modified with complementary information drawn from an analysis of additional documentation. Generally, a crop was included if it occupied a certain minimum proportion in the crop rotation (more than 10–15 per cent when the data were available). It is again clear that such a representation oversimplifies and excessively schematises the actual diversity of agricultural systems. On the one hand, it ignores tuber cultivation in the Sudanian regions of Mali and Burkina (principally yam). On the other hand, it does not reflect the many cultivated plants that are occasionally found on a village's *terroirs*: manioc, ground pea, *Cyperus*, sorrel, sesame and many others that are grown along with the more dominant subsistence or commercial varieties. While these crops occupy very small areas, they contribute significantly to resource diversification.

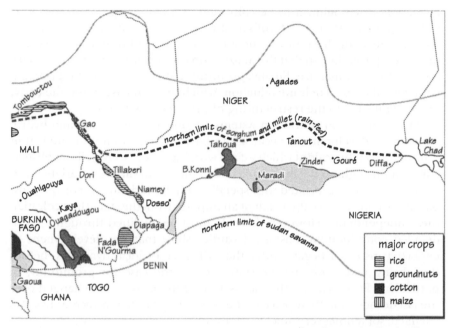

original artwork: Charles Cheung
computer graphics: Phil Bradley

The two main cereal crops in this zone are millet and rain-fed sorghum. They are not specifically shown in Figure 7.1, because they are present wherever rain-fed agriculture is possible. Similarly, the cow pea is not shown, although it is often found in association with cereals throughout the zone, and can play an important role in the local economy – as is the case in central Niger where, for a time, it was more widespread than the groundnut (Grégoire and Raynaut, 1980). It is difficult to know whether the absence of the cow pea in a particular area is real or due to a gap in information, because its contribution to production is most often secondary in terms of volume and because its presence on a zone's agricultural lands is not always apparent in the documents available.[3]

Against a general background of rain-fed cereal agriculture, four major crops are highlighted and localised on the map: rice, maize, groundnut and cotton. This selection was guided by the fact that each of these crops can be considered as the marker of a particular agricultural situation. In a regional context, rice, of course, reflects the presence of traditional or modern hydro-agriculture. Maize is present throughout the Sahel as a 'bridging' crop (which comes to maturity just as food reserves are running out and before the millet or sorghum harvest), although it occupies an important position in crop rotation only when climatic conditions permit, i.e., approaching the Sudanian zone. It is also more common where favourable prices allow it to be cultivated as a cash crop. Maize responds well to ploughing and to fertiliser and, as a result, agencies promoting cotton cultivation have often encouraged maize cultivation after cotton during crop rotation, so that the maize crop can benefit from the secondary effects of cotton manure. Finally, groundnut and cotton are the principal commercial crops. Their introduction, which in many cases was quite early, has generally led to the restructuring of production systems.

Major contrasts in cultivation systems are due to export crops and hydro-agricultural developments because the basic millet/sorghum equilibrium represents, on the scale at which we are working, a relatively well-established constant throughout the area (apart from the use of different varieties and cultural practices associated with specific climatic variations). For the most part these contrasts are the result of an impetus from the state (first colonial, then independent). However, this role must be qualified through a more detailed consideration. Historical studies show that the development of export crops often began under the influence of local dynamics, before then being taken over by the state (this is the case for groundnut, in particular). Furthermore, the localisation of these crops has changed over the last decades and their place in the peasants' production systems has also changed, sometimes very quickly, in response to price variations. The spatial distribution of crops suggested here does indeed reflect the existence of state intervention. However, this distribution of crops must by no means be considered as a fixed and rigid imposition. On the contrary, it should be seen as the result of dynamic interaction between external intervention and peasant adaptation and innovation strategies.

Millet and sorghum, which are rain-fed cereals, are clearly dominant over most of the agricultural zone, their production dictating whether subsistence needs in the Sahelian countries are met. The varietal stock used is, for the most part, local and the result of an age-old selection process (Portères, 1950). The new varieties suggested by agronomic research are still not used very frequently, mainly because they are ill-adapted to peasant conditions,[4] or due to a deficiency in structures for seed production and distribution. However, one must be careful in making such statements, because seeds are circulated widely and the peasants are constantly in the process of adapting the range of varieties at their disposal. If a new variety were brought to a village by a peasant, that is with no direct, official intervention, after a certain time and in the absence of any further investigation, this variety could well be considered to be as 'local'.

Rice cultivation remains necessarily restricted to areas where there is proximity to surface water such as large rivers and low-lying areas. It was recently introduced in the Senegal valley where, as we have seen, recession sorghum continues to dominate. Yet along the Niger River it is an immemorial tradition that developed, and was extended, through an improved control over water brought about by hydro-agricultural projects.

Small-scale peasant rice agriculture can also be found around pools in the Sahelo-Sudanian zone where it is often practised solely by women. It can also be found in the low-lying areas of the Sudanian zone (south-west Senegal, southern Mali, south-west and south-east Burkina). In this instance, although the women now work this crop, it may previously have been a cash crop reserved for the men, who abandoned it 30 years or so ago when the development of cotton provided them with a better income (Lavigne Delville, 1995). Finally, while not illustrated here, rice agriculture is also found in the improved perimeters of the valleys of the Volta rivers, and the Mossi plateau. Maize, for its part, occupies a significant position in Senegal (in Casamance), in two regions of Burkina, and especially in southern Mali where, for a time, its promotion as a commercial crop was the focus of a specific programme by the Compagnie Malienne de Développement des Textiles (CMDT) (Malian Company for the Development of Textiles).

When considering the entire area covered by our map, it is undoubtedly the existence of several, clearly demarcated, major basins, specialising in either one of the two major cash crops (groundnut or cotton) that constitutes the most pronounced factor of heterogeneity. The colonial, and later the independent, state undoubtedly played a crucial role in their geographic distribution through the creation of development structures and commercialisation circuits, as well as in price fixing. As we have already emphasised (Chapter 3), during the colonial period the transport of produce and the global policies for the management of the labour force were particularly important criteria for the location and development of these crops. The current distribution of cash crops gives no hint of their spatial dynamic, which over a period of several decades has been very striking. In Senegal, for example, after following the Dakar–Linguère railroad, the axis of progression of the groundnut basin shifted southward, along the Dakar–Tambacounda line (Lake and Touré el Hadj, 1985). In Mali, the western centres of Kayes and Bafoulabé were abandoned in favour of an extension southward towards Kéniéba, and eastward towards Kolokani, Ségou and San (Amselle *et al.*, 1982). Although the constraint of climate has determined the northern limit of their expansion, the current location of the export crops has been determined more by soil exhaustion and the policies of extension programmes than by the conditions of the surrounding environment. Broadly speaking, the geography of cash crops in the Sahelo-Sudanian zone is organised around three major historic areas.

1 *The Senegalese Groundnut Basin* We briefly described the development of this area when examining its population distribution and mobility problems. This is the site of the oldest hub of agricultural exporting in the Sahelo-Sudanian zone, beginning with the groundnut trade through the coastal trading posts in the last century.[5] Peasant society, as well as production systems and landscapes, bears the mark of a long history of

commercially oriented agriculture. The influx of workers over several decades, some of whom had come great distances, was a powerful factor for change, while the uncontrolled extension of lands devoted to groundnut resulted in the overexploitation of the soil. One consequence of this overexploitation was the progressive abandonment of the crop in the oldest part of the basin. The market demand for groundnut remained strong, so that land surfaces that were lost were compensated by clearing new, peripheral areas (according to the spatial dynamic described above). This mechanism, which has been described many times, continued for several decades. However, following the drought of the early 1970s, and especially because of the deteriorating economic situation (especially producer prices and the cost of inputs), the expansion of groundnut production, which was so evident in the recent past, today seems to have been curtailed.

2 *Central Niger* It is in central Niger that the second major groundnut centre is located. The history of this centre is shorter, dating back to the mid-1930s at the earliest (Péhaut, 1970), yet sufficiently long to have left a profound mark on the local agricultural situation. While this crop has declined enormously since the mid-1970s, nothing in the region today (be it the forms and intensity of land occupation or the evolution of the social structures of production) can be understood without reference not only to groundnut agriculture but also to the economic mechanisms that accompanied the commercialisation of this product and the circulation of money that it created (Raynaut, 1975, Raynaut, 1977a).

3 *The large cotton basin* This basin, extending from southern Mali to western Burkina, is both similar to and different from the situations just described. Beginning in the colonial period, the development of cotton agriculture was considered a major objective for this part of West Africa – indeed, the prospect of an independent supply of cotton to French industry was an important argument in support of the grandiose projects of l'Office du Niger (Herbart, 1939). The set-backs, however, were numerous and cotton production never really took hold until independence, especially in the case of Mali. It actually did take off in the 1970s, when the CMDT took charge of all aspects of its development, from extension to commercialisation. An enclave mainly devoted to cotton was therefore created throughout the entire southern part of the country.[6] In Burkina Faso, the crop was first grown on the Mossi plateau, but has progressively shifted to the newly cultivated valleys to the west and south, where most production is now found.

Outside these large centres of commercial agriculture, groundnut or cotton cultivation is found in many other areas of the zone under study, although their penetration has been more diffuse or has occurred within the framework of recent and more limited development programmes. In

164

Figure 7.1, the groundnut is shown as present in Mali along an almost uninterrupted band extending from the border of Senegal to Burkina. This is misleading, however, since in this vast and often very sparsely populated space the development effort remains diffused, so that the groundnut is far from ever having occupied a position comparable to the one it held in the large Senegalese and Nigerien basins. Cotton can be found in two localised 'pockets'. In one, situated in southern Burkina and corresponding to the operation Aménagement des Vallées des Voltas (AVV) (Operations to Develop the Volta Valleys), it occupies the dominant position. The other, in Niger, reflects earlier intervention in the Ader–Doutchi massif by the CMDT. Yet here, cotton cultivation remains mostly confined to the valley floors and to several small hillside schemes.

If we were to scan the entire zone, the image would be one of pronounced heterogeneity, with a marked contrast between established, clearly identifiable zones of highly specialised agriculture (cash crops, recession agriculture) and a vast background dominated by the rain-fed subsistence crops of millet and sorghum.

Whereas the practice of hydro-agriculture must adapt to the existence of strict natural and physical conditions, the practice of commercial agriculture, by contrast, is subject to an entirely different kind of causality. Economic factors and the intensity of the efforts accorded to their propagation determine their introduction and later development. The monetarisation of peasant economies long imposed by fiscal pressure and by the development of new consumer goods has, almost everywhere, been the initial driving force behind their diffusion (Raynaut, 1977a). Later, and especially since independence, fluctuations in price of both agricultural products and inputs have been the principal determinants of their expansion or contraction. Thus, the decline of cotton in Senegal in 1986–7 is linked to an increase in the price of inputs. The collapse of groundnut agriculture in Niger which started in 1975 can largely be explained by the evolution of the relative price for groundnuts, cow pea and millet (Grégoire and Raynaut, 1980). Finally, in southern Mali between 1983 and 1987, the success of maize as a cash crop and its competitive position relative to cotton was due largely to price supports. Similarly, its recent drop in production is a result of the liberalisation of cereal prices (Berthé et al., 1991).

Even subsistence crops have entered the market and can at times assume the role of commercial crops. Their market status however is complex because these crops possess simultaneously multiple use values within the local economies and only a marginal part of their total production is generally sold (Raynaut, 1973a). To some extent these crops were regularly traded in the past (or exchanged in one way or another), although production specifically for the market is a fairly new, and therefore still limited, phenomenon. It can be seen especially in those areas where agroclimatic risks are limited and where the extension of cultivated land per worker, and/or the existence of outlets in the towns, have made improvements in productivity both possible and profitable. Thus, in the cotton-growing areas, the major cotton producers are also major cereal producers. This remains the exception, however. As there are no price controls and the market is not organised, investing in order to increase cereal productivity is rarely profitable. More often than not, faced with an economic environment that provides little incentive, the production of marketable surpluses is a secondary consideration for the peasants, and as a result the evolution of the surface area given to cereal production follows demographic growth and fertiliser use remains marginal.

The case is quite different for cash crops, which are destined primarily for the export market and are by nature very sensitive to the cost/benefit relationship. For peasant societies this link with an outside market results in a profound change in the way productive strategies are oriented and an increased dependence on demands that are often unrelated not only to the needs of social reproduction, but also to the maintenance of a sustainable relationship with the natural environment (Raynaut, 1977a). In this regard, the land shifts of many commercial production areas (older overexploited zones being abandoned and replaced by an advancing pioneer front) bear witness to the destructuring effect of excessively rapid growth in market production on the relationship between man and the environment. As Rochette describes, this has occurred in the Senegalese Groundnut Basin and also with cotton in Mali and Burkina Faso:

> With the advance of the Senegalese pioneer front, we saw how animal-traction hoes and seeders opened millions of hectares to the groundnut. Fields of cotton, sown in lines in the direction of the slope – for ease of production and so that drainage prevented the cotton from being asphyxiated by excess water – have exposed the thick, rich soils to erosion; like the groundnut, cotton began migrating from north to south – from Ségou to Koutiala and Sikasso or from Ouahigouya-Kongoussi to Houndé and Diébougou – well before the decrease in precipitation.
>
> (Rochette, 1989: 434)

In southern Mali, in sectors where production is concentrated, the effects of a rapid increase in cotton production over the last 30 years (see note 6, p. 182) are currently of central concern to those responsible for rural development. Within the areas where cotton production is concentrated, the current increase in efforts to conserve or restore soils, or to manage natural resources on a better basis, provides additional confirmation of the link that may be established between the position of cash crops in a production system and the toll they take on the environment (Berthé et al., 1991).

The map of major crop distribution not only provides crude information but also shows us the dynamic of production systems. Thus it allows us to distinguish three types of agricultural space:

166

1 Those areas in which peasant production systems have stood aside from organised networks of agricultural development and continue to be based on cereal cultivation. This vast area covers many realities, both ecological and economic; even though the absence of a major cash crop restricts investment opportunities, there is certainly no question of a total absence of change in farming techniques.
2 Those regions that are associated with hydro-agricultural techniques. This often implies a particular type of man/environment relationship, with crop systems relying heavily on the use of inputs (with high production costs), a close interdependence between producers within the collective organisations and, usually, large-scale supervisory structures.
3 Finally, those zones in which the intensity and/or long-term presence of commercial agriculture (even if now abandoned) is accompanied by an environmental impact that has gone beyond the needs of the social and material reproduction of local societies. We will return later (in Chapter 8) to this somewhat broad categorisation since, when cross-checked with other information, it contributes to the identification of environmental situations and helps to assess their degree of vulnerability.

PLOUGHING TOOLS AND PEASANT STRATEGIES

Along with the cultivated plant, the tool is the mediator between people and the land. Indeed, a single tool can have many uses; even minor changes in technical procedures can lead to important modifications in the spatial organisation of land use (Raulin, 1984: 349). The type of tool is, therefore, one indicator of how production systems function. Like other components of farming systems, tools experience rapid evolution that is both cause and consequence of agrarian change. Figure 7.2 shows the geographical distribution of some major categories of ploughing tools within the area covered in our study.

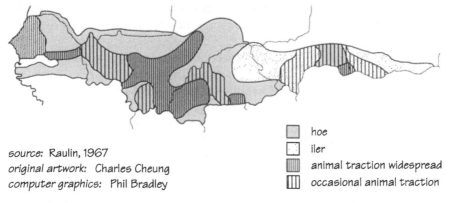

source: Raulin, 1967
original artwork: Charles Cheung
computer graphics: Phil Bradley

☐ hoe
☐ iler
▦ animal traction widespread
▥ occasional animal traction

Figure 7.2 The distribution of agricultural tools in the western Sahel

167

The hoe and iler

Early African agriculture was exclusively manual. One type of hand tool, the hoe, dominates in the Sahelo-Sudanian zone. As a general rule, it is made of an iron blade set at a sharp angle on a short wooden handle, forcing the user to bend over. This generic type in fact covers a wide range of swing-driven tools with many variations which correspond to local traditions, as well as to varied uses within a single farming system (soil preparation, seeding, weeding, ploughing) and different users. Thus, among the Senegalese and Mauritanian Soninké, we encounter seven types of hoe distinguishable by handle length, as well as by the size of the blade, the way in which it is fixed and the angle at which it is set. The choice of tool is determined by the user's gender, the task to be accomplished and the nature of the soil (Chastanet, 1991). There is also a large geographic variability due to cultural differences between ethnic groups (Bernardet, 1984: 380; Marzouk-Schmitz, 1984; Seignebos, 1984).

Among other manual tools is the iler which is a weeding tool prevalent mostly in Senegal and Niger but also found in Mali, Burkina and Chad. It can be described as being:

> made of a crescent-shaped iron blade of about twenty centimetres in width, attached ... to a long shaft ... and ending with a handle. The blade's surface relative to the shaft is slightly tilted (15 to 20 degrees) so that it is pushed ahead of the user, parallel to the ground.
>
> (Raulin, 1984: 342)

The iler works the soil from 2 to 5 cm depth, that is more superficially than the hoe, and cuts off the exterior part of the weeds. It is a tool used on sandy, homogeneous and relatively flat soils. While the variety in styles is less great, there are different types, distinguishable by the size of the blade and the way in which it is used (Raynaut, 1984). In all cases, the iler is a pushed instrument used standing up, whereas the hoe permits, 'through an accumulation of the force communicated to the instrument, more efficient labour than can be done with the iler. Its use necessarily requires more effort' (Raulin, 1984: 113).

The use of either one of these two instruments determines not only different modes of cultivation but two distinct agricultural systems. With less effort the iler allows the rapid weeding of sandy fields that have been sown without soil preparation. The cut grasses remain on the surface to dry in place and at harvest the stalks are often left standing. With the hoe, work takes approximately twice as long as with the iler (Jean, 1975), but the adventitious weeds are buried at the same time as the ground is being prepared. The hoe also allows some control over soil moisture status. In the Sahelo-Sudanian zone, where water recuperation for plants is a prime concern, weeding often includes the digging of furrows around the base of the plants. In contrast, the trend in the Sudanian zone is to earth-up the base to drain the plot. It is from the point of view of fertility maintenance, however, that differences between the iler and the hoe are most significant. With the hoe, weeds are buried in the ridges and concentrate organic material there. Earthing up around the base of plants (as in the Mossi country, for example) favours non-shifting agriculture (the unfertilised furrow is seeded the following year), while ridging between rows, like that practised by the Dogon, allows a 'micro-rotation' and prolongs the period the plot can be used, at times enabling permanent agriculture (Raulin, 1984). The use of the hoe or the use of the iler thus reflects the more or less extensive orientation of an agricultural system. With just the hoe, the various techniques for mounding earth reveal different degrees of agricultural intensification and fertility maintenance.

Ethnic groups that are close geographically and/or culturally may use entirely different agricultural tools, for example, the Manding of Lower Casamance (Marzouk-Schmitz, 1984) or the populations of southern Chad and northern Cameroon (Seignebos, 1984). Even within the same cultural area there can be wide variations. Thus, in the Hausa villages of Niger, which are situated several kilometres apart, some use the hoe and others the iler (Raynaut, 1984). These differences often result from a cultural heritage and, as such, are part of a complex system in which symbolic elements, material techniques and functional solutions are linked.[7] While there have been partial attempts (a collective work by ORSTOM in 1984 (Collectif, 1984) and that of Marzouk-Schmitz in 1989), a comprehensive geography of the agricultural tools of West Africa remains to be produced. Such a study would greatly enrich our understanding of the diversity in technical approaches found among societies confronted with almost identical natural constraints. The only attempt at a comprehensive study remains that of Raulin in his cartographic essay on the distribution of the hoe and the iler from Senegal to Chad (Raulin, 1967). Elements from this document have been used in constructing Figure 7.2, with additional modifications from complementary data, notably for central Niger.

It can be seen that this geographic distribution demonstrates clearly defined breaks in continuity. To some extent this can be explained by differences in soils. The sands of western Senegal and eastern Niger are more suitable to the light and superficial work of the iler. Yet this explanation is not entirely satisfactory, since there are many intermediate situations where either of the two tools could be used. It seems, therefore, that where the compactness of the soil does not prohibit the use of the iler; in the end, technical strategies (towards more or less intensive options) determine the adoption of one over the other.

Certainly, the geographical distribution of tools and agricultural techniques is far from being set in stone. These techniques are not definitively dictated either by environment or tradition. Technical options, along with the instruments that facilitate their application, can evolve to fit the circumstances, such as climate change, the availability of labour and land or alternative methods for managing the workforce. Raulin (1967) has therefore proposed a theory to explain the abandonment of the iler in favour of the hoe (a phenomenon noted in central Niger) as a response to a scarcity of land, making the intensification of agricultural practices necessary. Yet, as we saw in the previous chapter, peasant strategies appear to be more complex than this. In situations of land saturation, Hausa farmers who had previously adopted the hoe are returning to the iler. In doing so they are responding to a need to make the most of their workforce within a context of deteriorating climatic conditions that lower the efficiency of intensive practices (Raynaut, 1984: 510). The Gurmantche of Niger are currently adopting the use of ridging between rows in the first years of cultivation, before returning to earthing up around the base of plants several years before the land is to be left fallow. Elsewhere mounding tends to disappear in zones where new lands are colonised (Raulin, 1984). Many other examples could be cited that illustrate the dynamics of production systems, and it is essential that they be taken into account if we wish to analyse technical innovations or tackle the problem of conserving the ecosystem. In the end, the diversity of traditional agricultural instruments in the Sahelo-Sudanian zone, along with their relatively flexible distribution, suggests that agricultural societies in the region have long had recourse to a large technical reference system, a panoply of know-how and tools from which they can draw, depending on the natural, social and economic circumstances that confront them.

Animal traction cultivation

Although 'tradition' and resistance to change have long been the bywords to explain the 'archaic' nature of the tools used by African peasant societies, all recent studies insist on the flexibility and constant capacity for adaptation and innovation displayed by these societies throughout their history. Consequently, we need to examine the very weak and slow diffusion of animal traction agriculture, a technical item that has the potential to increase greatly the effectiveness of human labour. Giri (1983: 223) correctly reminds us that black Africa has long been in close contact with North African countries where animal traction exists and this technique, like many others, could have been borrowed if the need for use were clearly felt. The widespread practice of long fallow periods, as well as social strategies for organising the labour force (to which we shall return later in Chapter 11), explain to a large extent the limited success this technique has encountered in the past. Figure 7.2 shows that even today, the use of animal traction agriculture is concentrated in limited geographic sectors of the

Sahelo-Sudanian zone. We must therefore explore the principle underlying this heterogeneity. According to Bosc *et al.* (1990: 129):

[Bigot *et al.* (1987)] demonstrate that arguments generally advanced to explain the absence or limited diffusion of the instruments of animal traction agriculture are not valid. Neither the technical knowledge, nor the agricultural or herding specialisations ordinarily linked to an ethnic group, nor the relative lack of draft animals or the influence of trypanosomiasis, in fact deter development projects from adopting animal traction. Instead, whether animal traction is used or not depends on the area's environmental conditions and the economic situation of agricultural exploitations.

At first glance, animal traction agriculture is found in zones targeted by the massive diffusion of commercial agriculture. Most often they are the recipients of specific development projects, but this is not always the case.[8] Although access to materials and the degree of the extension effort may be important, the determining role appears to be the level of financial constraints placed on producers, because unlike traditional tools, animal traction cultivation is a costly innovation that demands an ability to accumulate monetary capital. From this point of view, the existence of credit for equipment has facilitated matters, yet the most important factor has been the terms of exchange between agricultural production and equipment. In this regard, an important factor in the diffusion of agricultural mechanisation is price level and stability (Grégoire and Raynaut, 1980; Bosc *et al.*, 1990). The heavy concentration of animal traction tools in southern Mali's cotton zone reflects the role played by the CMDT and its efforts to guarantee regular outlets and fair remuneration for the cotton produced by peasants (Bosc *et al.*, 1990: 196). Conversely, the absence of mechanisation in many Sahelo-Sudanian regions can be explained not so much by an alleged resistance on the part of agriculturalists as by poor agricultural revenues and structural weakness in supply distribution in sectors that, without commercial agriculture, are of low priority in national development plans. However relevant these economic arguments, agronomic reasons also help to explain the rejection of animal traction cultivation, or at least its 'corruption' by the peasants in accordance with their own strategies. For example, diffusion of the plough, a tool for preparing the soil, has often been given priority; yet, in peasant production systems, weeding is the most labour-intensive task and it is here that hold-ups occur. It is the animal-drawn hoe and not the plough that could respond to this need. Moreover, in the Sahelian zone, when animal traction is used farmers must wait until the soil is sufficiently moist before ploughing, which can delay the date of sowing and cause precious days to be lost, compared with sowing on dry soil by hand. In this context, using animal traction does not inevitably lead to increased production.

In general, the utility of ploughing is not always clear. Thus, working the soil at a deeper level improves rooting and soil moisture status, but it carries the risk that a layer of soil thicker than the humus level is turned over, causing the organic matter to be diluted rather than concentrated. While the positive yield effects of working the soil mechanically can be demonstrated at the agronomic research station, they may not necessarily be the same under conditions of peasant use.

> All the studies done comparing the results of manual and animal traction agricultural exploitations have shown the effects on yields to be minimal. The evidence suggests that to progress from the hoe to the plough increases the yield only if it is accompanied by improvements in the quality of the working of the earth.
>
> (Bigot *et al.*, 1987, cited by Bosc *et al.*, 1990: 130)

Rarely can the agriculturalist manage simultaneously all parameters that would make animal traction agriculture entirely effective, that is the quality of cultivation, respect for the agricultural calendar and the type of tool (Bosc *et al.*, 1990: 130), so that the results obtained often fall below those promised. In addition to reducing constraints which are strictly agronomic, there is the question of the strategy which underlies the production system, especially the relationship between intensive and extensive strategies found in peasant practices. These notions were considered at length in the previous chapter. The choice of draft animal is revealing in this regard. While often imposed by the development project, it can also be a choice made by the agriculturalist – the ox works the soil deeply but slowly, the horse and ass work quickly and superficially. Far from always being a tool of intensification, as agronomists and those who conceptualise development projects would like to think, animal traction has most often been used as a tool for extending the area under cultivation and for reducing the drudgery of the work, as is consistent with the general strategy of production systems. In fact, 'regardless of the zone being considered, the land area exploited per person using animal traction is greater than those exploited manually' (Bigot *et al.*, 1987: 129). When settling colonists from the groundnut basin in the *terres neuves* in Senegal, project technicians conceived models of exploitation making the 'best use' of ox power according to the colonists' own criteria. However, from the first year, the areas cultivated exceeded their forecasts. Economic analysis has since shown that, in spite of lower yields resulting from extended agricultural practices, the peasants' results exceeded estimates, both in total production and especially in remuneration for labour (Benoit-Cattin, 1979; Milleville and Dubois, 1979). In the 'upper basins' region (western Burkina), the area cultivated per person increases from 0.7 ha if worked manually to 1.2 ha when worked with animal traction (Mathieu, 1989). Even in Mali's cotton zone or in the Volta valleys of Burkina Faso, where the technical framework is extremely rigid and respect for extension recommendations is virtually assured, the strategy of extending surface areas has always accompanied animal traction agriculture, sometimes to the detriment of the quality of the work. Everywhere, technical innovation has therefore been 'recast' in terms of the strategies of peasant production systems. Ultimately, the effects of this change are difficult to assess. Animal traction agriculture exerts a positive impact on work productivity by reducing the work time per hectare, but the total quantity of work increases when the area under cultivation is extended. Consequently, 'The effects on net revenues per household per hectare appear positive ... while the effects on net revenues for an individual worker appear to be more variable' (Bosc et al., 1990: 130).

Figure 7.2 clearly highlights the heterogeneous geographic distribution of animal traction agriculture. There often exists a high correlation between the distribution of commercial agriculture and this type of equipment. There are thus heavy concentrations of equipment in areas where cotton and groundnut agriculture have long been established, such as the Senegal Groundnut Basin, Mali's cotton sector and the groundnut region of Niger as well as, to a lesser degree, western Mali. In areas of heavy diffusion the quantity of equipment can be quite high. In the region of Koutiala, the heart of the zone in which the CMDT operates, 80 per cent of production units are equipped either completely (team, plough, combine, seeder and wagon) or partially (Benhamou et al., 1983; Raynaut, 1991a). In the terres neuves in Senegal, 80 per cent of production units are equipped with horse traction and 50 to 60 per cent with ox-drawn equipment (Huguenin, 1989). In the Maradi region of Niger, 20 per cent of all exploitations with some involvement in the local Projet de Développement Rural (Rural Development Project) are equipped, but in those villages closest to the regional administrative centre, the proportion can reach 60 per cent (Raynaut et al., 1988). An identical phenomenon is evident in certain sectors where hydroagricultural schemes are located, such as the interior Niger delta and the Volta Development. By 1960–5, in the Malinke country of upper Niger, 72 per cent of the exploitations located in the rice agricultural zones along the river were thus equipped (Leynaud and Cissé, 1978). On the edge of these areas of concentration a process of dispersion can be observed, in west-central Burkina Faso, south-eastern Mali, and eastern and western Niger. In Yatenga, 10 per cent of production units are currently equipped. In 1975, the total rate of Mossi production units in Bwa country reached only 5 per cent and Bwa exploitations only 1 per cent (Benoit, 1975). Finally, animal traction agriculture can be all but non-existent in large regions of the Sahelo-Sudanian zone: in Mali, between Nioro and the Bend of the Niger, as well as in the Dogon country; in the north and west of Burkina; and in eastern and western Niger. Finally, the non-use of animal traction agriculture can rarely be attributed to a 'rejection' by the peasants. The heterogeneous distribution of this technique in the Sahelo-Sudanian zone ultimately reflects the disparity in local situations, with respect to factors favourable to its adoption. In particular, it shows up an imbalance in state action, in setting up development structures, supplying materials and providing the credit with which to purchase them.

MANAGEMENT OF THE ENVIRONMENT

The human effort to control and to order the exploitation of resources is not made up of agricultural activities alone. An important part of the work and know-how of human beings is specifically invested in practices that aim to shape the nature which surrounds them, according to their plans. These practices, either directly or indirectly, are likely to affect agricultural production.

Groves of trees

The role of trees is one of the most important areas where man's broader relations with nature may be seen, as much in its long-established equilibria as in its present-day dynamics. The tree plays a role in all Sahelian production systems as nutritional supplement or condiment, as nutritional substitute in case of famine, as fodder, medicine, timber or firewood – its uses are extremely varied. The functions and the technical and social conditions for the use of trees with nutritive value have been detailed by Bergeret and Ribot (1990). In this study the tree's larger role, beyond the pure and simple production of wood, is illustrated and its integral position within strategies adopted by agriculturalists for food diversification and in response to climatic uncertainties. This study shows us the importance of considering such trees as a fundamental element in production systems. With regard to herding, it is also important not to forget the essential role of the tree in maintaining the equilibrium of pastoral zones, as Pélissier rightly reminds us, 'above all else it is the tree that allows pastoral specialisation in the Sahelo-Sudan and helps to alleviate the uncertain, unstable and seasonal nature of the herbaceous ground cover' (Pélissier, 1980b: 128). The conditions for tree management can thus reveal the relations between communities and their environment. The most common mode of exploitation is gathering (Chastanet, 1984). Some rural societies have further advanced their relationship with trees by protecting and even planting them. This may be only a defensive strategy, such as in the case of southern Chad (Seignebos, 1980), yet it can also reflect an agropastoralist strategy for the management of land and the renewal of resources.

> Through its conception and the role assigned to it, the tree population in agricultural space gives an indication of the strategy applied by each society to its environment. It is not just a society's needs and techno-

logies that are expressed by the grove, but its very nature and history and, to a certain degree, its structure.

(Pélissier, 1980a: 131)

Thus the close relationship with trees can take on a profound significance with respect to land tenure. According to Pélissier (1980b: 130): 'tree exploitation signifies the right to land exploitation. Similarly, tree appropriation precedes and leads to land appropriation. . . . Did not prohibition against planting trees, long-established in many populations, act primarily to maintain the collective character of land capital?'

This relationship, however, cannot be generalised. On the contrary, in some cases there is a disassociation between the status of land and tree, especially when the tree is only protected and not planted. In the Hausa country of Niger, it is only recently, and after a slow evolution of custom, that landowners have been able to claim exclusive rights to the trees on their property (Yamba, 1993). As long as tree resources remained abundant, the private appropriation of land and the collective use of trees were able to coexist.

Whether the communities that use them consider trees as a common resource or as private property, the tree is an indication, by virtue of its place in land exploitation, of the degree of 'domestication' imposed on nature by human beings. In this regard, groves of trees represent the most complete method of integrating the tree into a system of production. As Raison (1988) notes, not everyone agrees on the definition of a grove. A minimal description would define a grove by 'the regular, systematic and ordered presence of trees on a field' (Sautter, cited by Raison, 1988). However, in dynamic terms, what this type of landscape formation expresses is the existence of peasant techniques for managing vegetation that shape nature in such a way as both to support and to specialise its production. It is thus a form of intensification which, through investment in labour, enhances the productivity of a natural ecosystem. In relation to this subject, there is also great diversity. Groves are not everywhere present in the Sahelo-Sudanian zone, and they can take many different forms. Drawing largely on the typology suggested by Pélissier (1980a), Raison (1988) distinguishes several categories according to their dynamic and the species that comprise them:

• First there are embryonic forms, residual or cleared groves, made up of trees that have escaped the clearing of land for agriculture or that became dominant when clearing created new ecological conditions. These are ephemeral formations that show no true intentionality on the part of agriculturalists and as such do not merit the term grove.
• Next there are selected groves, made up of pre-existing trees that have been protected and maintained by humans. The archetype is the karite grove (*Vitellaria paradoxa*), which is often accompanied by the néré

175

(*Parkia biglobosa*) and the tamarind (*Tamarindus indica*). 'The karite grove is the outcome of a deliberate strategy by people without livestock, that is, without milk, and consequently without animal fat. Contrarily, it is absent from the lands of genuine agropastoral societies' (Pélissier, 1980a: 132). It is also absent from the lands of those who are in regular contact with Fulani pastoralists. The karite bears fruit within 15–20 years and is fully productive in 50–100 years. The presence of such a grove is clearly conceived as enduring through time[9] because current management efforts might only benefit the next generation. It is conceivable only within the context of agriculture that is solidly established in an area and that seeks to assure for the long term the reproduction both of a human community and of its environment. The karite and the species normally associated with it are subject to relatively strict climatic constraints. They do poorly when precipitation declines below 600 mm a $^{-1}$ and are found characteristically along the southern margin of the zone under study, i.e., in southern and south-western Mali and southern Burkina. They are practically non-existent in Senegal, except in the Tambacounda region and the Upper Casamance. In Niger, they are rarely found, except in the Dosso region of Niger and in the extreme south of the Maradi region.

In constructed groves, although humans have not substituted an exotic species for the natural vegetation, they have at least favoured the growth of a species which, without human intervention, would not have developed in the same way. This is particularly the case with the *Acacia albida* groves found in the Sérer country of Senegal, in the *terroir* of western Mossi in Burkina Faso and in the Hausa country of Niger. These groves correspond to a sylvi-agropastoral strategy of intensification, linked to the specific physiology of the *Acacia albida* (Dupriez, 1980). Like all leguminous plants, this species has the ability to fix atmospheric nitrogen and thus to contribute to the soil's enrichment. In particular, its foliage grows in the dry season and falls off in the rainy season so that, at the beginning of cultivation, as the leaves fall they contribute organic material and eliminate competition with the crops. In the dry season it furnishes fresh fodder (in the form of seeds and leaves), contributing, at times decisively, to the reduction in fodder shortages at a critical period and enabling a higher livestock carrying capacity. Increasing the organic and mineral restoration (nitrogen, foliage and especially the excrement of animals who come there to feed and seek shade) improves yields and reduces the fallow period.

The tree is therefore an important link in the agricultural chain. The most advanced example of this system is found on the *terroirs* of the Sérer in Senegal where, for long periods of time, village fields were permanently cultivated while the large fields were organised around a four-year rotation of cereals (sometimes mixed with cow peas), groundnuts and two years of fallow (Pélissier, 1966; Lericollais, 1972). This system permits population densities of between 60 to 80 inhabitants per square kilometre, almost twice that found on more extensive systems. *Acacia albida* tolerates more arid conditions than the karite and thus it is characteristic of the Sahelo-Sudanian band. Nevertheless, it requires a relatively high water-table, making valleys and plains its preferred sites. Its presence in groves is closely linked with areas that have been heavily marked by human use and especially with clustered habitats surrounded by permanent agriculture combined with herding. *Acacia* growth, however, is much faster than that of karite. The fact that a village is surrounded by such a grove may not, therefore, be evidence of its long-standing duration.[10] It signifies, however, the development of an intensification strategy with long-term objectives.

The tree, especially when arranged in a grove, and thereby revealing a plan for the development of land and the organisation of resources, is therefore a witness to the relations between Sahelo-Sudanian peasant communities and their natural environment. Thus it is a witness to history, since many groves are vestigial and reveal the past of a society and its farming system; and a witness also to a contemporary dynamic, since the evolution of a grove reflects in many ways the transformations and adaptations of the communities involved with it.

When exploring the diversity of agrarian situations in the Sahelo-Sudanian zone, as we are doing, a record of the spatial distribution of grove types and their current development would provide a powerful piece of information. Unfortunately, as we have noted, the data allowing us to develop such a map are lacking, but we can nevertheless attempt to establish several relationships between tree management practices and specific local situations.

Following a synthesis of the literature, Raison (1988) notes that karite groves (the species often found in association with néré) indicate stable populations (given the time necessary for a tree to become productive) that have access to sufficient space so that farming can alternate with tree exploitation, and for which herding is secondary. If we add to this the effect of ecological constraints, a relatively well delimited situation begins to emerge. That is the situation of peasant societies at the southern edge of the zone of study, who have long been established in one place but who, at least until recently, did not have a densely concentrated population. Perfect examples of this situation are to be found in the southern zone of Mali with its Minianka and Sénoufo settlements, as well as in the old Bambara country. Fine karite groves are found in all these areas.[11] By contrast, in areas where the population is sparse and subject to a mobile agricultural dynamic or even a pioneering front, and where the additional presence of pastoralists greatly improves access to dietary animal fat, karite grove development is not favoured, even where no ecological constraints exist (this is no doubt the case in eastern Senegal).

However, maintaining a functional grove of karite and néré is rarely compatible with the heavy demands for space and the problems of fertility that accompany high population densities, because when the karite is fully mature, its dense canopy competes with crops for light so that a high density of trees becomes an obstacle to crop growth. Moreover, its presence contributes neither directly nor through livestock to maintaining field fertility. This suggests that karite groves become threatened when production systems are forced towards more intensive land exploitation. The number of trees per hectare, as well as their heights and age distribution (which are also strong indicators of the dynamic of a grove), evolve according to the overall conditions of the farming system. However, this evolution does not always occur in the same way, because the grove may be restricted in order to limit competition with crops and/or to favour the use of animal traction. Nevertheless, we have recently been able to observe the creation of selected groves of néré (especially in the Korhogo region, Ivory Coast) as a result of massive land clearing over the last few years and the good economic value of the néré produce.[12]

Acacia albida groves are also linked with stable populations that are often organised around clustered habitats, but whose historic depth may not be the same as in the preceding case. Because species reproduction requires the passage of the seed through the stomach of a ruminant, the presence of these trees implies a combination of agriculture and herding. When we add to this their greater tolerance to aridity, it comes as no surprise that *Acacia albida* is the preferred tree in the region favoured by agropastoralism, the Sahelo-Sudanian zone. As these trees contribute to soil fertility maintenance by the various mechanisms described above, their presence is of benefit in the event of pressure for space or limitations on fallowing. They are therefore particularly concentrated in areas of high population density, such as the Senegalese Groundnut Basin, the Mossi country and the Hausa country of Niger.

The dynamics of the *Acacia* groves can take contradictory directions. The presence of high population densities within a context of agropastoralism lends itself to their appearance. Yet in the past this density was only relative to the mode of land occupation. Communities were often motivated by defensive needs and were driven to regroup into large agglomerations, even though space remained abundant just beyond the villages. This was the case in the Bwa country of Burkina (Benoit, 1975) and also in the Tessawa region of Niger (Grégoire, 1980; Raynaut *et al.*, 1988). When the conditions of insecurity that necessitated this form of settlement had disappeared, the increase in population caused the village territory to become saturated and agglomerations had to disperse. There was a return to the system of long fallow periods, which made the grove less functional. This situation, associated with the relaxation of systems of collective controls and an increase in land competition by users (native Bwa and Mossi migrants in Burkina, Hausa agriculturalists and Fulani or *Buzu* agropastoralists in Niger) led to a decline in the maintenance of the ancient groves and their degeneration. Such cases create ambiguous landscapes that 'simply reflect political instability, repeated migration, or confrontation between agriculturalist and pastoralist' (Pélissier, 1980a: 133).

Today, many of even the best established groves are in a state of crisis, as a result of excessive land occupation that has overwhelmed the agricultural system's ability to adjust (Thomson, 1988). Demographic growth can therefore play a negative role, especially when aggravated by the demands that accompany commercial agricultural development. Groves in the Sérer country have thus regressed markedly, with a decline of 34 per cent in the number of *Acacia albida* trees between 1965 and 1985 (Lericollais, 1990). Nevertheless, such an outcome is not inevitable. On the contrary, the Dagari from Burkina have planted an *Acacia albida* grove on fields with short fallow periods (Pradeau, 1970). In the Hausa country of Niger, stimulated by private appropriation and following a period of regression, new tree management practices are apparently resulting in at least a partial regeneration of *Acacia* groves (Yamba, 1993).

Within a general framework that counterposes the Sudanian zone (the preferred terrain of the karite) with the Sahelo-Sudanian zone (where the *Acacia albida* flourishes), in the final analysis it is human practices, the strategies used to exploit nature, that have given rise to the groves and guide their form and dynamic up to the present day. These strategies are themselves inscribed within a wider determining framework of which the dominant features in the past appear to have been the duration of settlement, population density and the presence of livestock. Over the past several decades, the most important challenges have probably come from demographic growth and the accelerated development of commercial agriculture. In the context of the recourse to extensive strategies described above, both of these have led to land saturation and the disorganisation of agricultural systems. Although peasant strategies are not linked in some mechanical fashion to the global constraints or stimuli that affect them, they cannot totally escape them, and it is in the compromises they reach that the conditions of tree management provide partial but irreplaceable information.

Management of the land surface

While the presence of trees marks a landscape in the most obvious way, the development of agricultural space takes on other forms which similarly reflect the efforts of peasant societies to establish lasting exploitation by controlling certain natural processes. As data on this subject are unfortunately fragmentary, we are limited here to sketching an outline on the basis of a few examples, without ultimately claiming to produce a complete picture. In several areas of the Sahelo-Sudan varying methods exist for controlling or storing surface water, or for improving its infiltration. These techniques are most frequently found where natural and/or demographic constraints make them indispensable. The best examples are in the Dogon country (Gallais, 1965) where we find terraces, reconstituted soils and micro-irrigation. Here, as in other mountain areas of Africa which have been used in the past as places of refuge (like the Mandara mountains in northern Cameroon (Hallaire, 1992)), populations have had to invent specific land development techniques, which have been extremely labour intensive, in order to be able to live in hostile environments. Micro-dykes to protect the soil from erosion and run-off can be found among societies tied to the land by multiple material and symbolic attachments: the Bwa, Dogon, the Kurumba of Yatenga and the Hausa of Ader in Niger. Recession rice cultivation, practised by women, is generally accompanied by limited management, such as the erection of impoundment dykes or the construction of compartments, allowing just a modicum of control over water flow (Niger, Senegal valley). Particular arrangements, sometimes the same as the preceding examples such as lines of rocks or earth banks, are frequently installed to hold the soil in place and to guard against erosion. In regions with dune soils, peasants often develop simple techniques for fixing or trapping the soil. Examples include barrier grasses (usually *Andropogon*) sown along the edges of fields or branches deposited on areas of hardened soils.[13]

We will not discuss again the land management practices that facilitate cohabitation between agriculture and herding because this has already been presented in the preceding chapter. The development of corridors of passage, hedge planting along the most exposed sections of the route, the digging of deep pools, as well as the demarcation of rainy season pastures, represent investment in work and know-how. They also demonstrate, to just the same extent as do the more purely agricultural practices, the way in which peasant societies exploit and shape the environment in which they live.

Over a long period, peasant societies from the Sahel have experimented with many forms of land management and environmental protection. Faced with the declines in precipitation and soil deterioration, the favourite themes for developers have now become the management of land, water conservation and soil protection. Many techniques have been tried since the colonial period, with varying degrees of success (Marchal, 1986b). After the resounding failure of the technocratic and authoritarian methods employed for soil protection and restoration, a consensus is now emerging as to the use of a range of techniques which are readily accessible to the peasants, such as stone lines, grass strips, quick-set hedges, etc. (Chleq and

Dupriez, 1984; Vlaar, 1992), or even sometimes ideas derived directly from local techniques, such as lines of rocks or zaï.[14] There is also insistence on the need to adopt a 'participatory' approach. There have been many experimental attempts to combat erosion or to put *gestion de terroirs* techniques to use (Rochette, 1989). Even though their technical validity can be demonstrated, especially in the case of Yatenga, their implementation usually needs some external intervention and, although some real dynamics seem to have been initiated, in some places it is rare for peasants to make major investments in such projects. This therefore raises the question of their social validity.

Some points can be raised in response to this question. First, although the principal aim of external intervention is to combat erosion, it must be borne in mind that the objective of the peasants' land management practices has hardly ever been solely environmental protection. Nérés and karites, while marking out landownership, are also a source of extra income, *Acacia albida* provides fodder in the dry season and the strips of *Andropogon* and quick-set hedges are not only land markers (showing field boundaries) but also sources of material for the artisan sector or for construction. Many other examples could be cited. It would not be unreasonable to say that, for the peasants, environmental protection is more a by-product of their economic and land strategies than an end in itself.

More generally, the fact that some techniques were developed and practised by peasants in earlier times does not necessarily mean that they are still relevant, in their eyes, today. A technique's validity derives as much from its efficiency in use as from its coherence with the social and economic principles governing the users. For example, like several other land development techniques, the use of lines of rocks seemed to have been abandoned by the peasants in Yatenga for several decades when development workers decided to reinstate them, in line with current trends. This reintroduction demanded a level of labour investment that was no longer compatible with activities carried out in the dry season nor with the seasonal migrations necessitated by the increased need for cash. Moreover, in former times this technique fitted into the context of a strong symbolic link with the land – a link which has today become very much more tenuous (see Chapter 10). When these techniques are taken up again today, it is not in order to enjoy a return to 'tradition', but rather it is a totally new land investment dynamic, developing in a radically new context, a context which focuses on the monetarisation of the economy, migration and the individualisation of property. The social significance is thus now radically different. The peasants are by no means indifferent to the problems posed by the degradation of the vegetation and the soil. However, they have many other problems to resolve at the same time and often they must respond to more urgent matters. This does not mean that the situation is irreversible. Indeed, some of the examples cited in the preceding chapter show that, under favourable conditions, significant and rapid behavioural changes can occur.

CONCLUSION

The manifestation of agricultural diversity in the Sahelian countries is found in many domains, particularly cultivated plants, tools, the position held by intensive techniques, land management methods and techniques for regulating fertility. This variability is clearly related to the potential of a natural environment and the constraints imposed by it. Yet this variability also reflects the existence of what we might call different 'agrarian cultures', i.e., cultures with distinct social systems and different approaches for maintaining relations with nature (we return to this later, Chapter 9). Finally, the origin of this variability, and this is certainly not the least important explanation, can be found in disparities introduced during the process of market integration, notably by localising 'agricultural development' operations. Regardless of the explanation or the complexity of its origins, this diversity is readily apparent and must be taken into account not just when describing and analysing the current situation but also when searching for appropriate solutions.

NOTES

1 A comparative study of several village fields in the Maradi region of Niger produced the following counts according to their position on a precipitation gradient: 17 species cultivated at 600 mm a^{-1}, 10 species at 500 mm a^{-1}, 4 species at 400 mm a^{-1} (Miranda, 1980).

2 This is by and large beyond the Sahelian and even the Sahelo-Sudanian zone. The diversity of cultivated plants, along with the spread of 'mound' agriculture which indicates a concern with the field's drainage, represents the southern limit of our bibliographic review. Nevertheless, we have had to cross the boundary when a single society straddles different ecological environments (Sénoufo–Minianka in southern Mali, Mossi colonisation in the Bisa and Gourmantche countries).

3 The localisation of niebe agriculture depicted on Déat and Bockel's map does not always coincide with what is found on the ground – especially in the case of Niger.

4 The classic example is that of high potential varieties which give excellent yields in the ideal conditions of the research station but whose productivity plummets in the peasant environment, where they perform less well than the local, more hardy varieties.

5 The groundnut originated in South America, and was introduced into the coastal areas of Africa much earlier by European traders; it then spread spontaneously as a cash crop.

6 Malian cotton production was less than 13,000 t in 1962, rose to 68,000 t in 1972 and to 275,000 t in 1991.

7 Marchal shows the clear differences that exist among the agricultural practices of the Yatenga Mossi: the *Moose* and *Têngbïise* sub-groups, who use different tools,

are also distinguishable by historical origin, political position and ideological function (Marchal, 1984).

8 Thus, in Niger, agricultural supplies have for a long time been acquired mostly in Nigeria, due to the trans-border traffic (Grégoire and Raynaut, 1980). In Burkina's Bisa country, 25 per cent of the production units use bovine traction, exclusive of any development project.

9 Probably in the form of a long cycle (some 20 years) that alternates periods of cultivation with the exploitation of the orchard (Raison, 1988: 20).

10 In the Maradi region, a comparison of aerial photographs taken in 1957 and 1975 highlights the rapid development of trees around villages created during the second half of this century. On the ground, near one village in the Maradi valley, we noted a very thick cover of *Acacia albida*, formed within 25 years (personal observations, Raynaut).

11 According to Raison, zones in Mali with karite densities exceeding 40 per hectare covered more than 36,000 square kilometres in the 1950s.

12 On this point, see the studies co-ordinated by Peltier at the Centre Technique Forestier Tropical (CTFT) (Tropical Forest Technical Centre).

13 Raynaut: personal observations in Niger.

14 *Zaï* is a Mossi technique designed to restore soils that have become too hard to be cultivated. During the dry season, holes (15–30 cm wide, 5–15 cm deep) are dug every 0.5 or 1 metre and are filled with a mix of earth and manure. When the rains come, they are sown with millet or sorghum. After some years of practising this technique, the combined actions of water, of micro-fauna and of the roots of the plants make the soil once again suitable for cultivation (Rochette, 1989).

8

RELATIONS BETWEEN MAN AND HIS ENVIRONMENT: THE MAIN TYPES OF SITUATION IN THE WESTERN SAHEL

Claude Raynaut, Philippe Lavigne Delville and Jean Koechlin

Throughout the preceding chapters, we have examined and described the diversity found in the Sahel, using a number of analytical approaches. From the point of view of nature, demography, the organisation of commercial trade or even the modes of resource exploitation, the Sahel appears fragmented, endowed with a many contrasting situations. Our initial synthesis (see Chapter 3, Figure 3.5) brought together the mosaic of local demographic particularities into a certain number of large geo-demographic entities for which the spatial distribution is shown in Figure 3.6. In order to refine and explore further these situations, we now introduce new factors into this categorisation. Specifically, the diversity of modes of natural resources exploitation will be superimposed on to the sectors that illustrate the uneven population distribution.

By combining two types of data in this way, a template of environmental situations is produced, along with its spatial patterning. This is illustrated in Figure 8.1.

1 *Modes of land use* The data presented in Chapters 5, 6 and 7 have, of course, to be greatly simplified in order to produce a usable map. For agricultural and pastoral relations, both of which are of great significance in determining the conditions for natural resource exploitation, we have retained the same tripartite stratification shown in Figure 5.2: that is the pastoral zone, the highly competitive transitional zone and the zone of cohabitation. With regard to agriculture, areas devoted to rain-fed agriculture are distinguished from those where irrigated agriculture (which here includes recession agriculture) is practised. The first of these two categories has been subdivided into two major types of production systems: those oriented towards subsistence and those biased towards

commercial production. The second category is divided into regions of traditional irrigated agriculture (essentially recession agriculture) and those characterised mainly by modern hydro-agricultural schemes. Although these two categorisations are based on different criteria, the result nevertheless allows us to locate physically the situations we have encountered and described in the preceding pages.

2 *Demographic variables* With these we have had to eliminate much of the detail found in Figure 3.5. Only two classes of population density have been retained: equal to or greater than 20 inhabitants per square kilometre and less than 20 inhabitants per square kilometre. We also contrast the rate of annual growth in a binary fashion using rates equal to or greater than 3 per cent and rates less than 3 per cent. In spite of this simplification, it is still possible to identify the well-defined demographic situations described in Chapter 3: the 'poles of attraction', pioneering zones and areas of demographic 'saturation'. In the pastoral zone, population densities are everywhere well below 20 inhabitants per square kilometre and growth is much less than 3 per cent per annum.

The cartographic representation that results from overlaying these two major dimensions has of course necessitated some adjustment to boundaries, so that the area is not divided up into parcels that are too small. The typology that eventually emerges distinguishes 20 theoretically different situations, of which only 16 are actually illustrated in Figure 8.1. We shall examine these from the standpoint of different modes of land use.

1 In those regions devoted to rain-fed agriculture and subsistence production, demographic pressure generally appears moderate. We calculate that for approximately 60 per cent of the area devoted to this type of agriculture, population densities are less than 20 inhabitants per square kilometre, and that in many cases (44 per cent of the area) the growth rate is less than the Sahelian average. Moreover, we know that the diffusion of agricultural animal traction implements (a determining factor for increasing the area of land under cultivation) is limited. In fact, a substantial part of the workforce has resorted to migration, which provides a major source of income for the population left behind. There are some clear exceptions to this. The first lies on the eastern edge of the Mossi plateau where, despite a low uptake of cash crops, we find significant population concentrations, accompanied on the fringes by a rapid increase in population growth. Another unusual example is in a large part of western Niger (from the Ader massif to the river) where population numbers are high and growth is rapid.

A different demographic dynamic can be observed where agricultural production is more clearly destined for commercial purposes (cotton, groundnut and maize). On 60 per cent of these surfaces, the population density is more than 20 inhabitants per square kilometre, and often very

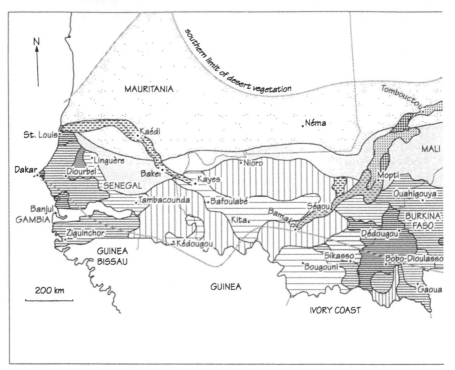

Figure 8.1 The combination of land use and demographic zones of the western Sahel (continued opposite)

much greater. In half these cases, demographic growth remains strong, particularly throughout the Nigerien groundnut basin. Elsewhere by contrast, as in Senegal, saturation occurs, as analysed in detail earlier (Chapter 3). However, the concept of zones of commercial agriculture must always be used with care. There are a number of areas, for example, western Mali, where although groundnuts are cultivated, their popularisation and commercialisation are not the focus of campaigns as intense as those found in the specialised regions of Senegal and Niger. The central region of Niger had been characterised by groundnut agriculture, yet today the crop occupies only a marginal position. Finally, while cotton may be widespread in southern Mali and in south-west Burkinabe, it has not developed in the two cases by the same methods or with the same intensity.

What we encounter in the field, therefore, are combinations of agricultural and demographic situations that vary widely. However, whenever a marked population pressure combines with the presence of a production system strongly focused on the market, we can always be certain that the exploitation of natural resources is particularly intense.

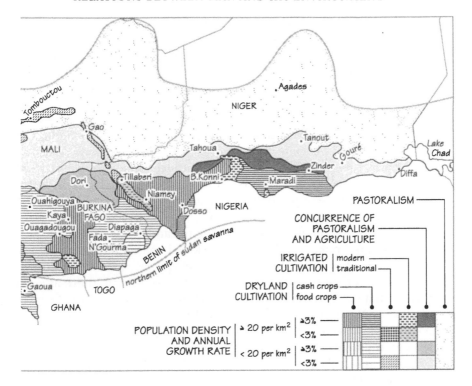

original artwork: Charles Cheung
computer graphics: Phil Bradley

As we have already seen (Chapter 7), high levels of production become necessary with this dual requirement and are realised mostly through an increase in the surface area under cultivation – an increase facilitated by the use of agricultural implements linked to animal traction.

2 Where irrigated production systems are dominant, demographic contrasts between regions are much more sharp. In the valleys devoted to recession agriculture for cereal production, particularly in the large part of the delta and in the upper Bend of the Niger, population densities generally remain low (in spite of strong localised concentrations). Everywhere, demographic growth tends to remain moderate. By contrast, strong population concentrations, accompanied in some instances by a rapid growth, are found wherever hydro-agricultural projects have been developed. It is not our intention here to return to the question of the cause and effect relationship between agricultural and demographic variables. We simply wish to remark on the different combination of situations found on the ground that engender distinct realities in the relationship between people and their environment. Earlier (see Chapter 5) we described the technical and economic implications of

187

hydro-agricultural development. It should be added that the existence of densely populated regions within a context of major technical upheaval is likely to increase the scale of the problems, whether they be the relationship with pastoralists who need access to water or, yet again, the problem of accessing resources not normally present on hydro-agricultural schemes, notably firewood.

3 A weak demographic dynamic dominates where pastoralism represents the primary mode of resource use, i.e., sparse human settlement and moderate growth (in some cases no growth at all). As we have observed for the pastoral zone, this holds little significance when, in actuality, herd size and not population numbers is the significant variable. This is not the case in the zone of competition. Clearly, in the majority of cases, the northward push by agriculturalists has not been accompanied by a concentration of populations. More significantly, since the notion of 'concentration' remains relative, it has slowed considerably over the last few decades as reflected by the weak demographic growth within this zone. However, and most significantly at the local level, there are two regions that run counter to these general trends. First, we note the edge of central Niger from approximately Tahoua to Zinder. Here, high populations densities combine with rapid growth. Second, active population movements continue to occur throughout the northern part of Burkina Faso, in the region of Dori. In both cases, the conditions for an environmental crisis seem to have combined; the extreme competition between modes of natural resource exploitation described in Chapter 5 is occurring within a context of severe demographic pressure.

The map illustrates well the heterogeneity of the environmental situations that are encountered in the Sahel. It is thus a useful instrument for description and analysis and underscores the dangers of simplistic pictures or hasty generalisations. Yet the mosaic that is created brings little understanding of the broad dynamics at work, that structure relations between humans and their environment throughout the region. In order to proceed, the same methodology used for the demographic analysis is applied to a wider synthesis that reduces the numerous specific situations to a more limited number of large spatial complexes and then highlights the dominant dynamics that underpin them. Doing so enables us to go beyond the somewhat reductionist observation to which we are restricted when simply combining demographic situations and modes of land use.

AN ATTEMPT AT SYNTHESIS

Figure 8.2 schematically situates these 'territories' as we examine each of these different complexes in turn. Two observations need mentioning which reveal this document's significance and define its limits. First of all,

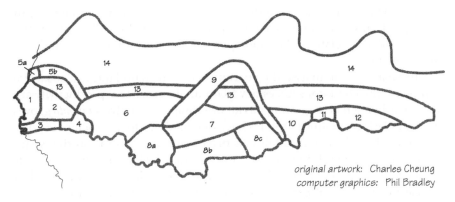

Figure 8.2 The major types of environmental situations in the western Sahel

it should be clear that these delimited sectors are only illustrative. Their borders indicate only approximately the general limits of a particular type of situation surrounding a relatively well-defined local centre. It must also be understood that the characteristics associated with each of these sectors reflect a web of dominant dynamics but capture neither the subtleties nor the particular situations that may be found there. In spite of our desire to depict the reality of the terrain in its diversity, a level of simplification has obviously been necessary, not only because of gaps in our knowledge but also because of the scale at which we are working. Nevertheless, this level of simplification is completely compatible with the objective of reflecting on the dimensions through which environmental variance is expressed. The goal is not to compile an inventory of all the minutiae of this diversity.

The ancient Senegalese Groundnut Basin (sector 1)

Here, the agrarian situations are diverse, resulting as much from differences in the Sérer and Wolof production systems (Ancey, 1977) as from the rainfall gradient from north and south. Nevertheless, there are currently several major characteristics that define a common ground. The intensity of landholding pressure is the principal element, as a result of the combined effects of a very high population density and the increasing demand on agricultural land of the groundnut economy. It is also the result of agricultural practices which consume relatively large areas of land (between 0.55 and 0.69 ha per person, according to Ancey's classification), facilitated by the presence of light soils (sector 1 is partly located in the iler geographic domain), and encouraged by the very widespread use of animal traction (seeders and hoes). An enduring pole of attraction in the centre of a vast labour reserve, more than any other region in Sahelo-Sudanian Africa, this part of Senegal is marked economically, demographically, and socially by

189

its long history of policies intended to develop commercial agriculture. In this regard, this case can be seen as a prime example for reflecting on the dynamic of the relations between peasant production systems and the environment.

A few features summarise the current situation. With land use at saturation point, the conditions necessary to maintain ecosystem productivity are no longer being met. Fallowing has all but disappeared, *Acacia albida* groves are degraded and manure supplies have declined due to insufficient grazing resources. The chronic instability of prices (for agricultural products and inputs) has made itself felt in two different ways: by limiting the possibility for the accumulation of capital necessary for technological change, and by increasing the risks already attributable to climatic hazards, thereby impeding the regular use of mineral fertilisers (Bosc *et al.* 1990: 152). Under such circumstances, the ability to reproduce agricultural production potential is seriously threatened. This obstruction is not recent. Peasant societies have responded less by exploring internal adjustments than by a demographic decompression – a reaction that explains the weakness in growth rates throughout this sector. In its first stage, this decompression was achieved through an exodus towards the margins of the groundnut basin (see Figure 2.3 pp. 48–9). Later, population movements were at least partly redirected towards the urban areas, that is to local towns, but primarily towards the Dakar metropolis which, as we have seen (Figure 3.5, pp. 80–1), exerts a broad influence. In the future, however, the dysfunctioning of a metropolis saturated by uncontrolled population growth is likely to reduce the rate of migration and thereby aggravate tensions at the level of peasant production systems.

The pioneering margin of the groundnut basin (sector 2)

The settlement advance just described first occurred in the northern part of the sector, along the Dakar–Linguère railroad, and on light soils favourable to groundnut cultivation. Subsequently, the constraints imposed by harsh natural conditions and by the encounter with the pastoral economy already present in this area have reoriented the movement southwards towards Tambacounda. As in all frontier zones, production systems are characterised by their extensiveness (of the order of 0.95 ha per person, according to Ancey (1977)). Nevertheless, population densities remain well below those found at the heart of the groundnut basin and available space remains sufficient to permit a long fallow period (Déat and Bockel, 1985: 60). In this band of contact between agriculture and pastoralism, livestock is ever-present. Renewal of soil fertility does not yet pose a problem, but the threat would appear if current population growth rates should accelerate. With the end of the groundnut boom and the partial reorientation of the migratory movements towards the city, demographic growth appears in all cases to be at a standstill in this part of Senegal.

Casamance (sector 3)

From the strict point of view of the analysis of ecological conditions and agricultural production systems, it undeniably goes against reality on the ground to group together into a single unit situations as different as those of the Diola rice-growers of Lower Casamance and the sedentarised Fulani of Fouladougou, both located in the same territory. Yet, bearing in mind the criteria used to define the area covered by this study (see Introduction, note 1, p. 10), we did not specifically want to exclude Lower Casamance from consideration because it forms part of a Sahelian country and is subject to political, economic and demographic dynamics which can be understood only when considering the territory as a whole.

Throughout this sector, moderate population densities and relatively intensive agricultural practices generally limit the level of man-induced pressure on the environment.[1] Space remains abundant, in spite of limits imposed by its insalubrious valleys (onchocerciasis, trypanosomiasis). Beyond the coastal zone settled by the Diola, essentially devoted to flooded rice cultivation, this sector features the cultivation of various cereal crops, combining millet and sorghum with maize and rice. Cotton has recently been introduced as a commercial crop, but its presence has not profoundly altered the production systems or landscapes. Although to a lesser degree, this geographic sector is also under the influence of the attraction of the Dakar megalopolis.

Far eastern Senegal (sector 4)

Here again, there are diverse local situations which are sometimes not suitable to grouping, such as the semi-mountainous agriculture practised by the Bassari. Nevertheless, several dominant features justify their inclusion into one geographic unit. First is the isolation of a region, which even today is scarcely integrated into the national structure, whether from the point of view of the collective infrastructure or from that of development programmes. However, although cotton and especially groundnuts are present, their position remains modest within an essentially subsistence-based agriculture. A second complementary factor is the very low population density, combined with moderate demographic growth. This is the beginning of the vast area of low demographic pressure which extends east to the Bend of the Niger. Despite this relative isolation, the attraction exercised by the Dakar agglomeration is also felt here. Finally, the role of livestock, particularly among the Fulani populations of Boundou, bears witness to a herding/agriculture bipolarity which combines with the weak penetration

191

of commercial agriculture to maintain the low agricultural demand placed on the land. The practice of long fallow periods remains practicable throughout the sector. In the future, however, not all of the space that is apparently available can be valorised by traditional production systems. Thus the unhealthy state of the valleys, which represent one-third of the sector's cultivable land, limits the potential for their agricultural exploitation.

The Senegal valley (sector 5 a and b)

This sector, which has seen a great deal of irrigation development (see Chapter 5), consists of two clearly differentiated sub-sectors (from both the agro-environmental and socio-economic points of view): the middle valley (5a) and the river delta (5b).

In the middle valley, the problem of maintaining soil productivity is experienced in entirely different terms from the rest of the Sahel in general, because here the flood is the basis of organic renewal and fertilisation. A second growing season is made possible by the presence of flood basins. There are rain-fed agricultural lands located outside the floodplain, but wherever the presence of floodplain soils is sufficient, these represent only complementary crops (essentially outside the Upper Senegal valley, which we have not included in this sector).

Until now, the height of the flood and the availability of labour have been the two major limiting factors on agricultural activities. Thus, the existence of high population densities in this sector does not assume the same importance as in rain-fed production systems. The long-established pattern of labour migration which characterises the valley, which is not the result of saturated landholding but reflects a reaction to economic marginalisation, has led to a depression in demographic growth rates.[2] For a long time young men have been leaving, first to the groundnut basin and later to Senegalese cities or abroad, searching for monetary earnings that agriculture has not been able to provide. Irrigated production, as it is currently practised on schemes of limited size, becomes integrated into diversified economic systems where it combines with floodplain and rain-fed agriculture as well as income from migratory labour. In this wider context, developed perimeters represent a specific mode of land use, with a specialised objective that can be simply summarised – money earned through migration is recycled as a way of securing subsistence for family members who remain in the village. The intensification of work input is considerable but the intensification of capital remains moderate; water and input supplies are kept to the minimum level necessary for food security.

The nature of the problem is likely to change radically with the completion of the hydroelectric station at Manantali, which may eliminate flooding. In this case, the maintenance of the fertility of land under

192

continuous cultivation will require systematic applications of organic and mineral fertilisers. If, in the future, the possibility of practising recession agriculture disappears, peasant strategies for land use intensification will need to be revised in order that the productive potential of the valley's soil might be sustained. From this standpoint, one factor which will impact heavily on how decisions are made is whether migration revenues stay constant or decline. This depends on factors beyond local control.

In the delta (sector 5b), rain-fed crops are virtually non-existent and most agricultural production derives from irrigated agriculture. The results and ambiguities of this system have already been described (Chapter 5). This area is unique in the Sahel, both because of the degree to which the environment has been made artificial and because agriculture in the area is entirely mechanised. Here, the question of the environment is linked both with the economic equilibrium of the chain of production, which is a determining factor for the future of irrigation, and also with the resolution of salinity problems. This phenomenon has already led to the sterilisation of some schemes[3] and threatens the entire region. As well as the Diama dam, which prevents the salt water flowing back up the river, a regional drainage network is also under consideration. It will almost certainly be necessary to modify cultivation systems.[4]

The Upper Senegal valley and western Mali (sector 6)

This vast expanse of low demographic pressure, which extends from Upper Senegal to the Bend of the Niger, is characterised by sparse settlement, a weak migratory attraction and only a moderately intense level of agricultural land use (less than 0.55 ha per person according to Ancey (1977)). Overall, land occupation remains sparse and there exists a great potential for opening up new agricultural land. However, there are settlements concentrated along the valleys, especially that of the Senegal, and around localised water resources (as the distance from the river increases) which could lead locally to an increase in landholding pressure and shorten the fallow period. The central part of this region, along the course of the railroad, has in the past been the object of efforts to spread groundnut cultivation. However, the hectarage devoted to this has remained modest (in the order of 25 per cent of cultivated land). More distant territories, especially in the north of the sector, have been the object of agricultural subsistence development programmes.

In both cases, these efforts were accompanied by the diffusion of animal traction inputs at reduced cost (Ancey, 1977: 10) so that in some areas uptake reached relatively high levels (50 per cent of the farms are equipped, sometimes more). Livestock remains important, whether via transhumant cattle, agropastoralism or livestock capitalisation, as is the case with the Soninké population. Where climatic conditions permit, closer to the

Sudanian region (between, approximately, Bafoulabe and Kita) there are beautiful karite groves. In this vast region, human pressures on the environment remain moderate and the problem of reproducing the production systems is not critical.

Of all the areas of Sahelo-Sudanian Africa where the diagnosis of a catastrophe of oversettlement and overexploitation of natural resources does not apply, this is a particularly illustrative example. Even more than an equilibrium between rural communities and their environment, we see the effect of a lethargy in the production systems. This is a situation that is itself the result of an old policy that created a labour exportation reserve out of this region and neglected to build essential infrastructure, particularly roads (see Figure 3.3, p. 71). If this entire pattern were to change and if agricultural development were to benefit from real incentives (market outlets, transport facilities, price security), the current situation could quickly evolve and approach the level of saturation seen elsewhere – all the more so because the presence of the sandstone massifs (Kaarta) and the hardpan plateau greatly reduce the area of potential agriculture.

The high density zone of Mali-Burkina (sector 7)

This is one of the most densely populated areas of the Sahelo-Sudanian zone. In some areas population density can be as high as 40 inhabitants per square kilometre, notably south of Ségou, in Dogon country and on the Mossi plateau, and never falls below 20 inhabitants per square kilometre. This phenomenon is not the result of particularly attractive natural conditions but of an historical heritage (see Chapter 3). In spite of moderate land use intensity (less than 0.55 ha per person according to Ancey (1977)), in many areas this situation leads to land use saturation, which not only prohibits the practice of long fallow periods but also reduces the amount of uncleared land, thereby creating a serious handicap for those who pursue herding activities, such as Bambara farmers in Mali, Mossi farmers in Burkina and, especially, for various groups of Fulani agropastoralists. The most extreme illustration of this crisis situation is found in the Mossi country, although in varying degrees the picture is identical throughout the zone, with soil erosion and a loss of fertility, degradation of naturally regenerating vegetation (the *Acacia albida* groves are heavily threatened) and falling crop yields. The preservation of potential productivity is all the more uncertain because the modest level of commercial agriculture yields a poor financial return, which only permits limited recourse to mineral fertilisation.

Of course, the vast geographic area shown in Figure 8.2 includes a wide diversity of local situations. The sub-region of Ségou, with its proximity to the Niger valley, benefits from the potential for diversity that is unknown elsewhere, including recession agriculture, floodplain rice cultivation and fishing. For its part, the old Dogon country on the Bandiagara plateau bears

194

witness to time-tested traditional practices of managing high population density, particularly those founded on a methodical management of space and micro-irrigation. In Yatenga there is the combination of harsh natural conditions, long-established population concentrations and repercussions from short-lived cotton production development policies. These local differences, significant though they may be, today tend to fade in the face of burdens imposed by land use saturation. Seriously aggravated by the repeated droughts over the last two decades, this major phenomenon makes it impossible for rural communities to secure their material and social needs in an environment whose reproductive mechanisms are profoundly disrupted.

This environmental situation parallels that of the Senegalese Groundnut Basin. Yet in many respects the underlying dynamics are different. In this case, the overexploitation of natural resources is essentially the consequence of population concentrations, itself the heritage of a secular history. The repercussions of the development of commercial agriculture cannot be blamed, for such policies have always been timid, both during the colonial period when the better part of the efforts and financial means were focused on l'Office du Niger's Project, and after independence when resources were mobilised for the cotton zone (in the case of Mali) and for the management of the Volta rivers (in the case of Burkina).

The trend to demographic depression that is occurring (see Figure 2.3, pp. 48–9, the map of growth rates) can only be partly interpreted as a response to land use saturation. It is a part of the continuation of older dynamics that transformed this region (especially, but not exclusively, the Mossi country) into a labour reserve for the rest of West Africa. What results is a demographic situation of high population densities coupled with the flight of the workforce. Under the effects of a pressure on land, population movements have now diversified. On the one hand, we witness the development of urban migration, and, on the other, we also see the colonisation of sparsely occupied lands for agriculture (see pp. 48–9). Furthermore, the propensity to migrate has diffused, so that even the Dogon, who did not have the same traditions of mobility as the Mossi or Bambara, are affected. In these regions in crisis, agriculture can no longer secure basic food needs with any regularity (Bosc et al., 1990: 161, 232), much less monetary needs. Increasingly, agricultural activities are becoming peripheral to a system of economic and social reproduction that mobilises many other remunerative activities such as migration, commerce and crafts. It is important not to lose sight of the fact that, in the past, the social and material life of the societies that historically dominated this part of Africa rested only partly on agriculture. To a certain extent and in new forms, therefore, current economic practices do not represent a rupture, but a continuation of old strategies of resource diversification (see Chapters 10 and 11).

195

The southern zones of expansion (sector 8a, b and c)

At the southern border of this region of high population pressure there extend areas which, at the level of demographic dynamics, articulate with this region yet are clearly distinct in their history, agriculture and natural environment. Settled by ethnic groups which in the past have maintained loose links, often marked by conflict, with the large regional state formations,[5] they currently receive new migrants from their neighbouring and overpopulated regions to the north. Mean population densities are often greater than 20 inhabitants per square kilometre but occasionally fall well below this figure. Agricultural land use is all the more moderate (less than 0.5 ha per person) because the natural conditions (soils, Sudanian climate) impose constraints on the working of soils and weeding. The availability of arable land remains important overall, yet land development by newcomers has long been undermined by obstacles of a social nature, such as control over lands exerted by older occupants, as well as natural obstacles, such as the disease-ridden valleys. Contrary to what we observe in the eastern extreme of Senegal, where the natural and demographic situation presents features comparable to those that have just been described, these regions are served by a road infrastructure that is advantageous to the large numbers of men and to merchandise and thereby facilitates the influx of migrants. Interstate disease eradication programmes have mitigated health problems in the valleys, so that migrants can now settle on lands as yet barely exploited. Within the general framework just described, three sub-regions can be distinguished.

1. The first is that of southern Mali (sector 8a). This region is dominated by cotton production and has practically no equivalent throughout the Sahelo-Sudanian zone. Within the space of 30 years, a shift has occurred in the production systems of the local Minianka and Sénoufo societies. In earlier times production was based on the subsistence cultivation of cereals, particularly millet, which possessed ritual functions, and tubers. These lands have largely been taken over by cotton (40 per cent of the cultivated area). Over the course of several decades, more or less starting in the 1960s, an agricultural economy based fundamentally on subsistence[6] opened up to a market economy through the development of commercial agriculture, primarily cotton and to a lesser degree maize, but also potatoes, rice, etc.

   The strong penetration of the market economy into a peasant society is not an unusual phenomenon in Sahelo-Sudanian Africa, especially in the other commercial agricultural regions such as western Senegal and east central Niger. What is unique in this instance is that this development occurred within the framework of a profound transformation of technical systems, characterised by a major capitalisation of agricultural practices. The use of selected seeds, of treatment products, of fertiliser and of organic manure became widespread (at least on commercial crops), while mechanised cultivation was widely adopted (motorisation could even be found in several cases). The success of the liaison between the results of agronomic research and development which, while admittedly partial, appears exceptional in comparison to other situations in the Sahel, provides a unique insight into the elements likely to favour or block progress towards intensification in Sahelo-Sudanian peasant societies. To sum up, the critical point to bear in mind is that the factors that mark success have less to do with the 'attitudes' of the peasants to innovation than with the favourable characteristics of the overall economic environment and the existence of a cohesion between extension proposals and the economic and social strategies of the agriculturalists.

In the first place, the change was made possible by the existence of a secure environment comprising:

• security in the natural domain, thanks to a level of rainfall that at this latitude allows the effects of technical improvement to be optimised and regularised;
• security in the economic domain, due to the effectiveness with which the commercialisation of production was undertaken and the long-standing and favourable ratio between the price of inputs and that of cotton; and
• security in the technical domain, resulting from the installation of management and a maintenance infrastructure that helped to cope with the unexpected.

A second reason for this success resides in the ability of peasant systems to reconcile the relative intensification of capital with their major objective, which remains the optimisation of labour productivity (see Chapter 6). In this regard, two elements played an essential part. First, the existence of a significant reserve of land that, especially in the case of cereals, allowed the use of draft animals in animal traction agriculture as a means of increasing cultivated area, and, second, the rainy season is of such length that agricultural operations can be spread out, reducing the impact of labour bottlenecks created through the increase in cultivation area.

Thus, what we observe in southern Mali is that rare occurrence of the spectacular evolution of peasant systems towards greater intensification. This does not mean, however, that the establishment of lasting relations between local communities and their environment encounters no problems. As a result of insufficient organic manure, the use of nitrogen-based fertilisers has led to an acidification of the soil, which has in turn resulted in reduced yields[7] in the old cotton basin. The amounts of manure now spread by the farmers still fall well below the level needed to compensate for this phenomenon. Intensive manure production requires major investment in labour and significant changes in herding techniques (semi-permanent paddocking). Moreover, recent studies (Bosma *et al.*, 1993) show that even if these techniques were more widespread, there would not be sufficient livestock to meet the demand. For the intensification of cropping to be sustainable, there must be a significant increase in livestock numbers, but since overgrazing problems already occur, this would require a radical change in herding techniques.

Because of the demands created by the extension of cultivated land, in many areas, land reserves are almost exhausted. This has occurred all the more rapidly because arable land represents only half of the area apparently available. Some areas are beginning to show signs of overuse; signs of soil and vegetation degradation can be seen, particularly in the central part of the cotton basin. Over the course of the last several decades, the influx of migrants from regions in crisis has contributed to increased land pressure in those areas already occupied. With the eradication of onchocerciasis, these movements have since extended beyond Bougouni to areas where the human presence remains quite sparse. A two-way movement, whereby arriving migrants are partially cancelled out by the exodus of local populations towards the Ivory Coast, attenuates the impact of these population fluxes on the growth rate. If public development efforts in regions liberated from onchocerciasis increase, the dynamics of land clearing could accelerate substantially and problems similar to those which have already occurred could appear in the old cotton zone.

2. It is essentially in Burkina Faso (sector 8b) that we find the second sub-region that can be distinguished within the great southern geographic area into which the overpopulated region to the north is expanding. One fact indicates its marked difference from the area previously described. This is that the land development dynamic depends much less on the internal evolution of local production systems than in Mali. It cannot be disassociated from the massive settlement migration originating largely out of the Mossi plateau. The original occupants of the land (Bwa, Bobo, Lobi, Gurumsi, Gurmantche) adopt a defensive strategy and try to block the uncontrolled influx of newcomers. However, unlike their Sénoufo and Minianka neighbours on the other side of the border, they have not received the same means to renew their production systems and to strengthen their hold on the land. The latent confrontation between autochthonous communities and recently settled populations is an important factor in the problem of natural resource management.

While the influx of Mossi settlement is ancient, especially towards the west, it has also been reinforced and partly reoriented by the intervention of the public authority following the Aménagement des Vallées des Volta (AVV) which itself was made possible by the earlier eradication of onchocerciasis. At the periphery of the planned settlement in the developed territories, uncontrolled and undirected settlements have multiplied, thereby compounding the effect. In the western part of this sector, in the extension of southern Mali, cotton cultivation represents the driving force in agricultural development, although the conditions for securing its production and which assured its success across the border are not always present to the same degree. In addition, the landholding situation is radically different in that, in many cases, propositions to change productive practices are addressed not to communities of long standing in their territory (who have woven tight social and symbolic links to the land) but instead to pioneer agriculturalists with extensive practices firmly in mind.[8] In the perimeters developed by AVV, the problem is accentuated by tenurial insecurity created by the appropriation of lands by the state. Such conditions do not encourage the adoption of the long-term perspective necessary to inspire concern for the preservation of soils and the environment. On the contrary, the dominant labour-extensive strategies adopted by the agriculturalists are short term. In a context of rapid demographic growth, they constitute a real threat to the preservation of the environment.

3. On the eastern margin of this major southern expansion zone (sector 8c), the particular situation found east of Mossi country (centred around the Fada N'Gourma and extending beyond the Niger border) is especially notable. The easy availability of land makes this a potential area of relief for neighbouring overpopulated areas, and the high demographic growth rate that we see bears witness to such a tendency. However, the lack of communication channels (from which until very recently these areas suffered, especially on the Burkinabe side) combines with the presence of strong natural constraints to limit settlement, because soils are poorly adaptable to agriculture in the north and water supplies are problematic in the south. Land occupation is not yet very dense and certain Gurmantche communities continue to use their ancient methods of long fallow periods and crop rotation. Nevertheless, the pressure on land is increasing rapidly, especially in Niger, where there is a land settlement policy and where, in addition, the uncontrolled exploitation of firewood for Niamey has led to widespread damage to the woody cover (Yamba, 1993).

The Interior Delta and the Bend of the Niger (sector 9)

As in the case of the Senegal valley, here is an area where the floods partly assure the continued soil fertility. The situation is particularly complex, however, in that the specialisation of modes of resource exploitation along ethnic lines is quite marked. The Marka and the *Rimaibe* of the Central Delta specialise in rice cultivation, recession agriculture is practised by the Songhay in the Bend, rain-fed millet or sorghum by the Bambara, the Bozo and the Somono fish and the Fulani are pastoralists. Population densities are sometimes high (over 40 inhabitants per square kilometre) but here the problem is not so much one of simple availability of agricultural land as one of an adequate articulation between the different production systems. Old forms of social control over resources have weakened while the increase of human and livestock pressure, combined with the impact of repeated droughts, accentuates the competition among the various communities. This checks the functioning of production systems and imposes additional pressures on ever more fragile ecosystems.

In this geographic sector are found different irrigated perimeters, some of which are already old and were developed by l'Office du Niger, while others are more recent and are spaced out more or less evenly from Bamako to Niamey. Many are defective and far from reaching their assigned goals, and they are currently the object of repair programmes. From a global and long-term viewpoint, the issues are not comparable to those that exist in the Senegal valley. Indeed, whatever the possibilities for irrigation development, it is not expected that irrigation will entirely replace traditional recession agriculture.[9] Furthermore, even if there is a major exodus out of the zone, the revenues earned are not comparable to the earnings sent back by migrant workers of the Senegal River. The survival of the inhabitants of the delta interior and of the banks of the Niger therefore depends greatly on the improvement of their traditional modes of natural resource exploitation and, in this regard, little has yet been accomplished. The movement of demographic decompression observable from Ségou to Tombouctou reflects the state of crisis in which agriculture and the economy find themselves in a region that is, nevertheless, favoured by the presence of water.

Western Niger (sector 10)

East of the Niger River, the landscape is marked by the presence of large fossil valleys between lateritic plateaus covered with *brousse tigrée*. Villages are concentrated in the more fertile depressions where water is easily accessible, so that, in spite of a moderate population density throughout the sector as a whole, population concentrations in the valleys can reach more than 50 inhabitants per square kilometre. Moreover, when compared with territories where the hoe predominates, the iler reappears with a net increase in per capita area of cultivation (approximately 1 ha). Given this double pressure, demographic and technical, the demand for agricultural land is intense and, in the valleys, fallow periods are only rarely employed and then only for very short periods. Added to the problems of declining soil fertility due to overuse, we find erosion phenomena on slopes and plateaus. Despite their poor suitability for agriculture, farmers see these slopes and plateaux as the last available land resource. They are, therefore, also cleared and cultivated, at the expense of the surface soil horizon, which is quickly stripped away.

To understand the developments and constraints that are currently

200

manifest in this region, it is important to emphasise that it has always remained outside the development of Niger's commercial agriculture, because cotton and especially groundnuts are concentrated almost exclusively in the central and eastern part of the country. Agriculture is essentially subsistence-based (millet, sorghum, cow pea) complemented by a traditional form of irrigated market gardening. Since the colonial period, this area has been subjected to major migratory movements towards the demographic 'magnates' located along the coast and in Nigeria. As in all areas of sub-Saharan Africa that have long played the role of exporters of labour, the local economy is bipolar in character, associating agricultural subsistence with migration that provides the better part of monetary revenues and towards which the living strength of the population is mobilised.

In many ways, this situation is similar to that which prevails in the densely populated regions in Mali and Burkina (sector 7). It largely explains the poor success of past programmes launched to improve subsistence agriculture. As always in similar cases, most of the proposed interventions call for both an intensification of work and capital. This intensification is incompatible with the local economy's existing orientations. Moreover, unlike the situation we have observed in southern Mali's cotton zone, the preconditions for successful intensification (secure markets, appropriate climatic conditions and the availability of land) are not present.

Contrary to other Sahelian countries, Niger possesses few sparsely populated areas with suitable natural characteristics for receiving migrants from neighbouring overpopulated zones (with the exception of the eastern border areas in sector 8c). Furthermore, while demographic growth in Niamey is very rapid, even in the rural areas that surround it, it cannot absorb a sufficient part of the population increase to moderate the pressures on land. In both this zone and all the densely populated areas of the country, this lack of a 'safety valve' is reflected in the absence of waves of demographic decompression analogous to those observed in similar situations elsewhere. The future of such a situation is particularly threatened, from the viewpoint of maintaining soil fertility that has already declined, and also from the perspective of conserving a fragile physical and natural environment.

The Ader massif (sector 11)

In this sandstone massif with a pronounced relief cut by wide valleys, we find in an even more extreme form a spatial patterning similar to the one just described, where villages are concentrated in low areas and along alluvial plains. In a similar way, population densities can reach what for these latitudes are extremely high levels (80 inhabitants per square kilometre or more). Several important features are nevertheless unique to this region. First is the long-time cohabitation of Hausa and Hausaphone agricultural populations, and ethnic pastoral societies, namely the Fulani, Tuareg and their dependants. Herding is thus very prominent in this sector. The existence of precious water resources, accumulating where the topography creates significant surface run-off to feed floodplain zones and semi-permanent pools, is especially important. Since precolonial times, the existence of these water resources has given rise to a small and traditional form of irrigated agriculture that has taken on a new dimension with the opening of new market outlets abroad (onions produced in the Galmi region are primarily exported to Nigeria and the Ivory Coast). The water resources subsequently facilitated the diffusion of cotton cultivation and, more recently, these environments have been the object of intensive development efforts, particularly the construction of hillside and water retention schemes for the expansion of irrigated agriculture.

This relative abundance of water in a semi-arid environment has favoured the intensification of agricultural production systems by means of animal traction equipment, the application of technical improvements in some of the permanent cotton areas, irrigated agriculture in developed perimeters, specialised onion cultivation in traditional gardens and the addition of other commercial crops such as tobacco. Speculation in cash crop production has been accompanied by an early integration into the market economy and furnishes a potential base for capital accumulation. This is only the case for villages located close to a valley floodplain or a hydro-agricultural scheme. In less favoured areas, migration represents once again the most important source of monetary revenue.

Even in the sub-areas that benefit from the most advantageous natural conditions, intensification is far from reaching the level observable in, for example, the cotton region of southern Mali. This is largely due to much less predictable rainfall, the absence of 'spare' land and the less strictly organised structure of extension programmes and commercialisation. As to the developed perimeters, steady progress is hindered by too many technical defects and inadequate socio-economic organisation to allow them to reach their full potential. Finally, the proximity of the Nigerian border to the south (which offers many more or less lawful commercial possibilities) discourages active individuals from investing substantially in agriculture.

We observe in this geographic sector an average level of agricultural land use comparable to that found in Mali or Burkina with, in particular, the intrusion of the hoe where the iler was previously the norm. The mean per capita cultivated area has been estimated by Ancey (1977) to be less than 0.55 ha. However, a high population density, combined with the incentive of the market, takes its toll. The local exploitation of land resources is leading to land use saturation. In an area of such dramatic slopes, and plateaux covered with a lateritic hardpan, problems of erosion and the degradation of vegetation are obviously acute, causing arable soils to be stripped away while depressions are filled with eroded materials. Wind erosion also occurs, especially in valleys orientated in the direction of the dominant north-east/south-west winds (Rochette, 1989: 199). In low-lying areas, maintaining soil fertility is a less pressing issue due to the nature of the soils and the contribution made by floods. However, it is likely to be of concern when and where overexploitation becomes severe.

The Nigerien groundnut basin (sector 12)

Several unique features characterise this densely populated region, covering the east-central portion of the country. First, the zone is located in a vast area of sandy soils. These are tropical iron-rich soils on ancient dune systems, or sandy pockets on rock outcrops that extend almost continuously from the Ader massif to beyond Lake Chad. Second, and from a historical viewpoint, this region corresponds to the zone where Hausa and Hausaphone populations have been concentrated since the nineteenth century, following the fall of the great southern city-states conquered by Ousman dan Fodio's Fulani troops. Finally, this sector is located in the heart of the old Nigerien groundnut basin and signs of this past specialisation remain to this day, in spite of a sharp decline in production since 1974.

Some of the highest population densities in all of Sahelo-Sudanian Africa can be found here (for example over 100 inhabitants per square kilometre in the Maradi valley), amidst agricultural practices which are extremely land-consuming (frequent use of the iler, with per capita cultivated land often exceeding 1 ha (Ancey 1977; Raynaut, 1980b)). This results in a very intense occupation of the land. Almost all land that can be cultivated with traditional techniques is in use and only a few types of less favourable land (sectors of more compact soils, dry valleys) show a less thorough agricultural development (Koechlin, 1980; Raynaut et al., 1988). The major portion of the land that can be cultivated either manually or by animal traction tools is now placed under permanent cultivation. Land saturation is illustrated by a number of signs, similar to those described in other overexploited sectors, namely deforestation, the ageing of the Acacia albida groves, fallow periods reduced to occasional intervals of short duration, shortage of grazing land and fodder and insufficient manuring of soils. Even taking into account decreased production due to low precipitation, the decline in soil fertility remains a major concern for farmers in the region. To these problems specific to agriculture are added those posed by the sharing of space between agropastoralists and transhumant herders. The little remaining grazing land is falling under the plough and as a result the corridors used to move livestock are cut off. Recently, latent conflicts that had simmered for a long time have erupted violently.

As always, it is somewhat artificial to group such a large variety of local situations within a single descriptive model, which can at best be only partially applicable. In particular, villages located in the southern part of the sector, close to the valleys, seasonal lakes and low-lying areas that make possible the practice of irrigated agriculture, possess advantages that do not exist further north. Nevertheless, limiting our observations to the generally prevailing situation, we see that it is marked by a crisis that not only affects mechanisms for maintaining soil fertility but, more broadly, brings about a profound disruption in the relations between rural communities and their environment. Due to the nature of the soils and the topography, wind erosion plays a major role in this environment with numerous consequences, including the smothering of croplands, uprooted trees and ancient dunes laid bare.

In certain ways, this diagnosis is reminiscent of the Senegalese Groundnut Basin. Yet the differences are remarkable in several important ways. While in both cases the diffusion of commercial cultivation stimulated agricultural production and consequently increased demand for land, development intervention has been much more limited in Niger than in Senegal. The active promotion of groundnut cultivation came about much later in Niger, beginning in the 1950s, and developed more through the organising of commercialisation (private markets and later co-operatives) than by the improvement of conditions of production.[10] This, combined with the negative force exercised by the very low official prices paid for groundnuts, has prevented peasant production systems from intensifying. In spite of an increase in agricultural prices and improvements in technical supports since the mid-1970s, in most of the sector the level of equipment has remained low and the use of fertilisers, after a slow start, collapsed when subsidies were discontinued.

It is essentially through a search for internal solutions that agriculturalists are seeking to respond to the challenges that confront them; solutions such as systematic use of agricultural residues for feeding livestock, technical modifications (drawing on the heritage of traditional tools), the systematic collection of manure, increased controls over woody biomass resources and private land acquisition. These initiatives often have only a limited success, but they bear witness to a fierce will to find, on the ground, a future for the region's resource exploitation. A similar evolution may be observed, of course, in other overpopulated and overexploited regions of the Sahel, but they are more critical in this case where few possible external solutions to local problems present themselves.

Temporary migrations clearly represent one solution within a range of options, although this is a recent tradition that does not entail economic bipolarisation (contrary to what we observe, for example, in the Mossi country). Moreover, and this is one of the factors that contributes to the stabilisation of the active population, the proximity of the Nigerian border gives rise to multiple small trades and a trafficking that constitutes numerable sources of non-agricultural revenues without the need to go into exile in order to find money.

In terms of population dynamics, here as in the other densely populated zones of Niger, the major factor is undoubtedly the absence of an area within the national territory that might provide permanent accommodation for agriculturalists lacking space in their homeland. There certainly exist sparsely populated areas over the border in Nigeria, yet in spite of its reputation for openness this border thus far seems to have effectively blocked population movements. The only expansion of cultivated land has been towards the north, into zones of high climatic risk, the same zones from which earlier settlers are now beginning to withdraw.

Under the influence of these different factors, and in spite of some of the highest population densities found anywhere in the Sahel (comparable to those found in Yatenga Burkinabe or in the Senegalese Groundnut Basin), the growth in population shows no signs of abating, remaining at over 3 per cent per year. The depth of the sandy soil cover in most of the sector makes the consequences of overexploitation less severe than in other regions where, after a thin layer of surface soil is stripped away by erosion, lateritic hardpans are completely exposed. Agricultural exploitation thus remains possible, even if at a minimum level. In the absence of effective measures for maintaining soil fertility, yields progressively diminish to the point that they must rely almost exclusively on rainfall. Rural populations are thus confronted with increasingly severe and uncertain survival conditions and yet, generally speaking, they make progressive adjustments to their production systems, take advantage of the more or less immediate fluctuations in the border traffic and manage to remain on the land. Land use saturation, accompanied by neither demographic stabilisation nor the marginalisation of agriculture with respect to other activities, but which is nevertheless not corrected by a significant take-up of modern fertilisation techniques, creates a unique situation. What internal adjustments do peasant production systems use to respond to these challenges? To what degree will these adjustments push back the limits beyond which they risk setting off an ecological catastrophe? There is no doubt that it is in these terms that the problem needs to be stated. We shall come back to these fundamental questions in Chapter 12, the conclusion, making comparisons with situations observed in other parts of Africa.

The margin of competition between agriculture and pastoralism (sector 13)

This margin represents the large belt that crosses the Sahel from east to west. It marks the zone where agriculture and pastoralism meet, at times contentiously. We will not dwell on this dynamic since it was examined in Chapter 5. In the harsh climatic conditions of this zone, agricultural practices tend to be quite extensive (greater than 1 ha per person). The almost systematic cutting of trees in favour of cultivation combines with severe pruning and lopping by pastoralists, leading their own flocks or those of large urban stock raisers, to initiate a deterioration of the vegetation cover that can only be reversed with difficulty. Wind erosion is very active here. Accelerated by the effect of frequent low precipitation in the last two decades, it can give rise to general or localised movements of sand and to ecological phenomena often associated with an advancing desert. The conditions of life for the inhabitants of this zone are obviously precarious. Pastoralists and agriculturalists alike were hit hard by the 1974 and 1984 droughts. The loss of livestock for the former and the absence of a harvest for the latter triggered a withdrawal to the south. From among these people come most of the refugees who arrive at the edges of cities seeking a means of survival.

Whether we consider the future of these people or the threats to the environment, there is no doubt that it is in this part of the Sahel that the most dramatic situations are encountered and that the most urgent search for solutions is needed.

The pastoral zone (sector 14)

It is in this zone that it is most likely that the demographic explanation for the Sahelian crisis is the least satisfactory. The strictly Sahelian region and the sub-desert margin are not only sparsely settled, but overall are experiencing a relative decrease in population, and still the environmental situation remains critical, with degradation of the tree cover and the ensuing development of wind erosion. In part, this is certainly the result of the worsening climatic conditions of the last several decades, but it is also caused by livestock, almost the sole occupants of the region. Shifting the blame from people to animals may seem plausible. At first glance, the argument appears convincing. However, as we have seen, in the final analysis herd size does not progress solely as an automatic consequence of the success of vaccination programmes. It also results from social, political and economic developments. Public policies for digging wells and the investment strategies of city dwellers have certainly had a greater impact on the state of the environment in the pastoral zone than has the improved health of herds belonging to 'traditional' pastoralists.

In spite of this, from a strictly environmental perspective, the crisis is perhaps less severe than in the preceding sector, where the pastoral economy encounters the same problems, but is aggravated by the demand for land created by agriculturalists returning from the south. Studies have shown that with adequate protection pasturage can rapidly regenerate. The real problem is undoubtedly the progressive extinction of pastoral societies and the collapse of the social means of regulating natural resources that have been applied throughout the centuries.

DYNAMICS AT WORK

This stratification of the Sahelian region into major type-situations is clearly cursory and too often depends on generalisations that mask a variety of local situations. It is quickly evident to both observer and development interventionist that we are confronted with problems on the ground which can vary considerably from one micro region, even one village, to the next. In reality, the diversity is thus infinitely greater than our analysis attempts to reveal. The major agropastoral situations which we have identified, described and crudely mapped, represent only those that are dominant. They represent the combination of major influences to which the exploitation of natural resources is subject.

Retaining this level of generalisation does have its advantages, however given that the aim here is not just to record diversity but also to discern the dynamics that govern its appearance. While nuances may become blurred, stepping back enables us to highlight the heterogeneous geographic distribution of certain major factors and to formulate hypotheses of the way these factors affect a variety of agricultural situations. The major variables that have been selected are those that figure in the model on which we have organised our data collection and treatment. Consideration of the entirety of situations identified here shows how, when combined, a limited number of variables can account for the transition from one situation to another.

Agro-ecological constraints

The influence of agro-ecological variables is, of course, very relative. Their effect on population distribution is limited, as we have seen, and the geographic distribution of crops has as much to do with major political and economic choices as with environmental conditions. Yet we have also observed how the physical environment can favour or hinder the intensification of rain-fed agriculture, depending on the degree of risk that it brings. The success of the CMDT in southern Mali can be partly explained by a climatic regime that makes possible the optimisation and regularisation of the results of new technologies. Conversely, low and irregular rainfall is of major importance throughout the Sahelo-Sudan (particularly in the Sahelian part), limiting capital accumulation and the economic effectiveness of expensive technical changes. This is particularly the case in the Senegalese and Nigerien groundnut basins, in the region of high population density in Mali and Burkina, as well as in the Nigerien cotton belt of Ader.

Furthermore, soils and climates are primary factors in the strategies elaborated by peasants. It is partly the need to respond to the uncertainties of climate that guides their preference for extensive techniques; it was the increased rainfall of the 1940s and 1950s that made possible the northern pioneering front and it was the subsequent decrease in rainfall that forced the recent withdrawal. In the Nigerien groundnut basin (sector 12), the nature of the soils and, in particular, the thickness of the sandy layer attenuates the immediate effects of overexploitation and favours the maintenance of high population growth rates. In a larger context, and whatever questions might be raised on the epidemiological history of the Sudanian valleys, the eradication of onchocerciasis (and hence the elimination of a natural constraint) contributes to the relief of overpopulated Sahelo-Sudanian regions in Mali and Burkina.

Demography

While never the sole factor at issue, human pressure on the environment (with the resulting land saturation) represents a major factor in the disruption of production systems and in disorganising natural resource exploitation. This observation is certainly not new. It is evident, for example, in the Senegalese Groundnut Basin, in the densely peopled parts of the Mali–Burkina region, and in the Sahelian area. On the other hand, what this study underscores, and what has not always been sufficiently stressed, is the variable role played by the demographic factor throughout the region. The image of an overpopulated Sahel was created by a crude generalisation of a geographic phenomenon that only partly illuminates the reality found in these countries. There exist vast empty areas where space remains abundant, such as in the Upper Senegal valley or western Mali. The question to explore, therefore, is the origin of this variability. We have demonstrated that the existence here of population concentrations alongside regions that are practically empty, as well as the degree and direction of migration flows, are a manifestation of the historical and geo-economic context. While the localisation of major areas of settlement closely corresponds to the political centres of the nineteenth century, migratory movements (outlets from overpopulated areas and sources of essential economic revenue for blocked production systems) depend on modern national and international politics.

Population density is certainly a variable that greatly affects the relationships between communities and their environments. However, it is itself the product of factors of an entirely different nature and scale than the often cited individual or collective behaviours with which it is generally associated. Moreover, it can only affect the environment through a dual form of mediation: first, through social structures which determine access to the resources of land and labour, and, second, through the technical means to exploit the environment, which, as we know, change in accordance with the amount of space and the number of people. The labour force constitutes a resource and in some conditions its superabundance can allow mutations in the modes of resource use. As we have stressed in the chapters dealing specifically with population (Chapters 2 and 3), all of this should warn us to treat with caution the neo-Malthusian argument that is currently in vogue.

Market constraints and development interventions

In the Sahel, less than a century after the creation of a currency tax, there is not a single local or agricultural society where market exchanges have not penetrated deeply. Of course, the participation of these societies in a commercial economy did not begin with colonisation (see Chapter 4). Nevertheless, the proportion of production put into circulation and the various forms this circulation takes have been dramatically modified during the past century. The growth in the need for money will certainly be seen as one of the driving forces for change in the Sahel, both in respect of social relations and in the methods of exploiting the environment.

As we have shown in Chapter 3, it is according to very distinctly defined modalities, deriving from an interregional allocation of roles throughout the whole of western Africa, that the different local societies have found their place in the broader economic environment to which they now belong. From this standpoint, the main contrast to be drawn is that between zones of export crops, which are able to generate revenue by using the local labour force, and other areas left at the margins of the market production system. These areas play the role of labour reserves and feed the migrations to the regional economic poles.

Beyond this major historical split, the conditions that exist today for a further integration into the market still have a determining effect on the functioning of local agriculture and on the dynamics of the means of exploiting the environment. This is particularly true in the case of prices and market opportunities, which dictate the comparative profitability of different crops and, as a result, the way in which the peasants rank them in their production systems. Responses to fluctuations in these parameters can occur extremely rapidly. The example of southern Mali is a particularly good illustration of this point. When the CMDT wanted to encourage maize cultivation by offering credit and market guarantees at profitable prices, farmers were quick to adopt the technical change that was suggested. In just a few years cultivation areas increased and yields for this crop grew in a spectacular fashion, to the extent that interest in cotton waned. The CMDT then dropped its support with the result that the prices offered to maize producers plummeted. They immediately cut back on fertiliser use, or abandoned it altogether. At the same time they reduced the area devoted to maize, while retaining the improved varieties with which they had been supplied.

More generally, the stability of exchange rates between agricultural products and inputs, at a level which will allow monetary accumulation and minimise the risks associated with debt, is an indispensable condition for the emergence of a process of capitalisation. Until recently, the Mali cotton region combined the most favourable conditions in this respect, but, in contrast, the groundnut regions of Niger and especially of Senegal have suffered from an instability linked with production costs and the price of inputs. In the latter case, this slowed down considerably the distribution of modern equipment and new techniques. In central and western Mali, on the other hand, a sustained policy of subsidies facilitated the distribution of agricultural materials, even though commercial agriculture has only a modest place in the economy. In many respects, the appearance of Sahelian agriculture in all its diverse forms is the result both of economic conditions experienced in the past, of which it still retains the mark, and of those being created today, to which it is constantly adjusting.

To a large extent, these overall conditions are the expression of economic policies that are most visibly manifested in development operations. Given that their aim is to modify peasant technical practices, their very existence constitutes a significant factor in the present variability of agriculture in the Sahelian geographical region. The introduction of new plants, and of technology and equipment, can have a considerable effect on the manner in which agricultural production systems function and on the condition for natural resource management.

While there is hardly a region without its own 'development project', not all initiatives in this domain are accompanied by a similar mobilisation of resources, nor have all known the same success. The large hydro-agricultural schemes represent a case apart, whether in the Senegal valley, the Central Delta, the valley of the Niger or the valleys of the Voltas. These, in fact, emerge from a quite voluntaristic process that imposes radical changes on the material and social conditions of production. Intervention in rain-fed agriculture is much more dependent on the support of the affected population because the financial burden of the purchase of equipment or fertiliser ultimately falls on the producer himself. The degree to which new technical practices have diffused is extremely variable from one region to the next. Between the cotton zone of southern Mali, the Senegalese Groundnut Basin, Yatenga Burkinabe or eastern Niger, for example, the differences are significant. There are two factors that may explain these differences.

1 *The priorities of development policies* Development efforts have been particularly concentrated in the large zones devoted to commercial agriculture. The most illustrative case in this regard is the Senegalese Groundnut Basin, where nothing of what we currently observe can be understood without taking into account the efforts over three-quarters of a century, at times threatened but never abandoned, to promote the cultivation of groundnuts. Conversely, vast areas, such as western Mali and eastern Niger, have been virtually ignored, receiving only scattered and intermittent attention.

2 *Modalities of intervention* We have seen that the technical and economic environment of agriculturalists plays a determining role in their capacity

to modify agricultural practices and their forms of natural resource management. In this respect, differences in the ways in which development interventions are organised can account for the variability of the observed situations. Irregularities in the supply of technical inputs, an uneven level of market security, credit systems whose level of efficiency can be variable and technical support which is not always as attentive as it should be – these are the main factors that explain the varying impact of the countless development projects dotted throughout the Sahel.

Peasant strategies

Whatever the relative importance of these conditioning factors, peasant societies of the Sahel are not content to endure passively the provocations and constraints to which they are subject. Which lands are cultivated, which tools are used, with which techniques the soil is worked, the place of trees on the land, the role of livestock, the means of fertilisation – these are some of the traits that help to make up the identity of peasant production systems. From one society to another, from one geographic area to the next, clear differences emerge which cannot be reduced simply to the effect of external influences; hence the differences, for example, between the Sérer and Wolof systems in the Senegalese Groundnut Basin. In this regard, the degree of extensivity in agricultural practices, and consequently the consumption of available land which is its synthetic indicator, constitutes an extremely significant variable. As we have seen, throughout the zone under study the average area cultivated per person can differ significantly. This can result from efforts to adapt to different natural conditions (soil type, rainfall), but it can also be the expression of cultural choices arising out of a social group's particular style.

The choice, therefore, of the hoe or the iler, or the adoption of other techniques which may be more or less intensive, are not dictated entirely by technical objectives. They may also reflect the choice of distinct options in relation to the management of the labour force. In the following chapters we shall see that these agricultural production strategies are in themselves the expression of a less immediately perceptible reality, namely the principles involved in organising social systems and the way they shape man's relations with nature.

This non-passivity on the part of peasant societies translates into a propensity to reinterpret external provocations and constraints in terms of their own priorities. From this perspective risk minimisation and the maximisation of labour productivity represent the two major priorities on which peasant societies of the Sahelo-Sudan are organised. The acceptability of externally induced changes largely depends on the possibility of reconciling them with these major strategic options. Thus, to the extent that the other conditions discussed above can be met, animal traction

211

agriculture has been adopted where it not only improves the ability to work the soil but can also be used to extend the area under cultivation, notably in the Volta valleys, in southern Mali and in eastern Senegal. The agriculturalists' capacity for initiative may also be seen in the search for their own solutions in the face of new challenges. This dynamic dimension of the relationship between human communities and their natural environment is particularly clear where the complete occupation of available land poses acute problems of resource degradation. This is especially the case when demographic growth remains strong and emigration does not relieve the increased human pressures on the land. A prime example of this type of peasant-initiated adjustment can be found in the ancient Nigerien groundnut basin.

CONCLUSION

Having come this far, we are now in a position to identify several major agricultural situations, each characterised by a particular combination of a range of different elements: natural factors, demography and the technical conditions for exploiting environmental resources. In addition, we can now identify several of the major factors that determine their geographic variability. We have therefore completed the first important step in our effort to understand the diversity of relations between societies and environment in the Sahel. Yet an essential dimension is still missing if we are to get beyond descriptive analysis and begin to comprehend the fundamental forces that underlie the articulation between human action and the dynamics of the exploited environment. At each juncture we have observed that the individual practices of agriculturalist and pastoralist make sense only in the context of a broader, more general explanatory level, where the properties and conditions for transforming the social, political and economic systems in which these practices are inscribed, manifest themselves. What now remains is to comment on this perspective and explore several key questions, located within the societal principles and the recent evolution of Sahelian organisation. In doing so we can understand better what is really happening today in these territories.

NOTES

1 With the exception of the northern part of Lower Casamance, where we encounter a more extensive plateau agriculture in which the groundnut plays an important role.
2 The pocket of high growth observable on the Mauritanian side may be partly due to a withdrawal southward, which affects the pastoral zones situated further north, and to the effect of the border, which blocks population movement.
3 Especially private schemes, which have been created in a rather makeshift fashion without adequate drainage.

4 P. Boivin, ORSTOM, personal observation.
5 Minianka and Sénoufo in Mali; Bobo, Lobi, Gurunsi, Gurmantche in Burkina Faso.
6 Early attempts to introduce money encountered strong resistance, and efforts made since the 1920s to introduce commercial crops such as cotton and groundnuts ended in failure (Diabaté, 1986: 300).
7 Or at least in productivity of the land, if maintaining yields requires increased investment in inputs.
8 Nevertheless, one study shows that a large majority of cotton producers in this region are members of autochthonous populations with, however, much variability among ethnic groups (A. Schwartz, 1991, 1992).
9 Except perhaps in Niger, should the Kandadji dam be completed, although the project's completion is still in the far distant future.
10 The creation of major development programmes designed to improve the productivity of agricultural systems dates back to the mid-1970s. Their impact on rural areas is largely concentrated in a narrow radius around the urban centre from which they originate and along the principal roads (Raynaut et al., 1988).

9

SAHELIAN SOCIAL SYSTEMS: VARIETY AND VARIABILITY

Claude Raynaut

As well as a manifestation of an ecological crisis, the current situation in the Sahel is also an expression of social upheaval. This is now taking place as many of the former strategies for controlling people, their labour and the resources have become obsolete. The appearance of new potential users of land and vegetation (notably urban dwellers), combined with the weight of the market, the emergence of new forms of social stratification in the heart of the rural world and the beginning of an emancipation of youth (due, in particular, to cash cropping and migration) are social transformations which disrupt the regulatory systems that formerly ordered the exploitation of resources. Situations of competition and conflict among individuals and among groups are multiplying. Disorder is introduced into relations between societies and nature. Space is becoming more scarce and climatic conditions are worsening, accentuating and accelerating the transformations that are underway to the point where the sustainable exploitation of ecosystems is now under threat.

The preceding chapters have elaborated numerous examples of the relationships between social phenomena and environmental disruption. These relationships require further elaboration, in order to both account for local diversity and to reveal major underlying features. In order to do this, we begin with two observations.

The first concerns the changes in social conditions that control factors of production (land and natural resources, labour, tools). This is an essential dimension in the evolution of systems of natural resource exploitation. For a century, powerful factors of change have combined to bring about a profound change in the organisation and functioning of Sahelo-Sudanian societies.[1] The most important arenas whereby the forces of change are exercised arise from:

- the political trusteeship inaugurated by the colonial conquest and pursued by the independent states;
- the integration of local communities into a global market economy; and

214

• the introduction of new ideas through education and mass media (especially radio).

These processes of change have not taken the same form or intensity everywhere. Certain enclosed and remote areas have, until now, escaped direct state intervention by being isolated from the major efforts to promote commercialised agriculture and by acting as a labour reserve in the division of labour imposed on the Sahel since the colonial period. Others, on the contrary, have been the target of change, and have experienced the vast increase in development operations and the endeavours to create new collective organisations (co-operatives, political committees) all supported by an expansion in the road network. Certainly, the variations in the nature and weight of external influences represent a major determinant in the uneven degree of transformation in social systems and, therefore, in their current diversification.

Second, however, we must not seek to explain all variation as the product of external domination. No more so than any others, Sahelian societies are not purely and simply the passive toys of constraints imposed upon them. Even before they were subjected to outside influences they were not all organised on the same principles and their points of resistance and their potential weaknesses were not identical. Therefore they have responded in different ways to the new challenges which confront them. In the Sahelo-Sudanian zone, while there is diversity in the functioning of social systems and in their relations with their environment, this diversity is no doubt due as much to the intrinsic characteristics of each as to impositions from without. To pose the question of the diversity of the relations between society and the environment in the Sahel therefore suggests that we investigate first the enduring structures that underpin the heterogeneity of these social systems. This is the goal of this chapter.

In so doing, cartography loses its relevance as a tool because we are dealing here with a complex and subtly conditioned social reality which cannot easily be bounded spatially (a necessary property of a map). Furthermore, the very small number of in-depth studies carried out on Sahelian societies would make any attempt at spatial generalisation quite hazardous. Therefore, we do not attempt here to map the observations which follow nor do we attempt to refine by further subdivision the broad zones that we have so far mapped. Our objective is more modest, to develop the components of a template or framework that can help to capture the many forms of the relationship between Sahelian societies and the resources they exploit.

ELUSIVE ETHNICITIES

Taking account of the variety of forms of social organisation found in the Sahelo-Sudanian zone demands that we must be able to identify the systems

215

to be described and to define their limits. To the extent that ethnic diversity creates an obvious template of reference for social and cultural discontinuities, this may appear relatively easy. Whether it be a way of life, a language or a social structure, there are indisputable differences which enable us to distinguish a Soninké from a Sérer or a Diola, a Bambara from a Dogon or a Sénoufo, a Mossi from a Lobi or a Gurmantche and a Tuareg from a Fulani or a Hausa. Agricultural and pastoral communities which inhabit the Sahel therefore appear to constitute a number of distinct entities, composed of socio-cultural groups that cannot be likened one to another and which are identifiable by certain distinct features that confer specificity and differences to these groups. Starting from this point, a detailed nomenclature of the Sahelian populations was developed during the colonial period. These efforts culminated with the development by the Institut Française d'Afrique Noire (IFAN),[2] after other less methodical endeavours, of an exhaustive map of the ethnic groups in West Africa (Grandidier, 1934, Froelich, n.d.; Richard-Mollard, 1952; IFAN, 1959, 1960, 1963; Poncet, 1973). Until now, this nomenclature has offered a convenient framework both for identification and for referencing objects in museums and for scientific analysis (especially the innumerable monographs which focus on a particular ethnic group).

For many observers, in the Sahel as in the whole of Africa, the category 'ethnic group' appears to provide a meaningful tool for the description of social realities; witness the number of analyses that continue to rely on it in order to account for the political conflicts which are tearing apart the countries within this zone. The reality, however, is far from being as simple as may be suggested by these apparently clear-cut ethnic classifications (which have been taken up, for instance, in the series of atlases produced for individual countries and published by Jeune Afrique (Peron and Zalacain, 1975; Toupet, 1977; Bernus and Hamidou, 1980; Pélissier, 1980c; Traoré, 1981). On the ground, these identified entities lose their clear-cut homogeneity, their borders blur and their scientific validity becomes problematic. Amselle (1990) proposes a convincing critique of the notion of ethnicity, supporting his case with the specific example of Mali. If, under closer scrutiny, these clear-cut ethnic distinctions begin to blur, it is because contrary to the image often projected, far from being engraved in the stone of ancient 'tradition', an ethnic grouping represents in many ways a stage in a continuous process of construction. Thus, as we understand it today, it is in effect the historical result of two orders of combined factors.

The notion of ethnicity bears the mark of the bureaucratic demands of colonial powers which, to be able to administer, had quickly to order and to label a reality of which they had little understanding. To accomplish this, they often fell back on superficial differences and hastily recorded traditions deprived of the critical study necessary for their interpretation. In a large number of cases, even these ethnic denominations corresponded to nothing in the language of those to whom they were supposed to apply. This is the case, for example, with the Bambara, the Mossi, the Fulani, the Sénoufo, the Toucouleur and the Songhai, to cite just a few. At times, the name assigned was a crude phonetic approximation, at other times it reflected the pejorative names used by neighbouring groups, and elsewhere it invoked historical references that no longer make sense today. This state of affairs is illustrative of the indifference of the classifiers with respect to the reality they sought to order.

On a more profound level, inscribed in a longer history, ethnic groups are often also the product of strategies that formerly drove agricultural and pastoral societies of the Sahel to associate with one another, to affirm themselves or to distinguish themselves from those who surrounded them, either to better resist or to dominate. How many among them, treated by the administrator and also the geographer or anthropologist as so many identifiable social and cultural entities, are in reality composites, assemblages of populations of varied origin united only by a common language (in fact, often at the cost of important dialectal variants)?[3] This is particularly the case with the vast Hausa group, a melting-pot (around large, federated political poles, namely city-states) in which until very recently populations originating from every possible direction (the Aïr, the banks of Lake Chad, the southern regions) were combined. This is equally the case with the Mossi political formation, consisting of a mosaic of societies of diverse origins with the old background of autochthonous Dogon and Kurumba populations, on which were superimposed conquering groups coming from the south, laying the groundwork for a common structure to which other smaller groups of Fulani, Soninké and Songhai came ultimately to be associated (Izard, 1985; Marchal, 1983). With respect to the Bambara and Fulani, Amselle (1990) has shown all the vagueness that surrounded the definition of their ethnic grouping. Attempting to define the Djerma–Songhai group, Olivier de Sardan (1984: 20) talks of the superimposition of 'immigrant strata more or less ancient and disparate'. Concerning the Sénoufo–Minianka complex, its composite character is such that the descent system differs from one group to the next. Strongly underscoring this heterogeneity, Sanogo states:

> The Sénoufo have continually assimilated foreign elements. Certain neighbours were clearly influential such as the Bambara, the Dioula, the Akan etc. . . . The existence of several types of marital organization, of several initiation systems, and of several sub-groups as well as the evolution of their respective socio-political systems all support this analysis.
>
> (Sanogo, 1989: 15)

One last example, that of the Sérer, also supports the same observation of heterogeneity. Until recently, immigrant strata have been superimposed and remain visible in the current organisation, while among local groups, clear differences exist in the domain of political organisation (Gastellu, 1981).

In the final analysis, ethnic designations which delineate a homogeneous social and cultural reality are the exception. Very often, within a single generic term, they represent somewhat forced amalgamations of populations whose differences are at least as important as the similarities that are used to homogenise them. By contrast, there sometimes exist important similarities among ethnic groups nevertheless considered to be distinct. Thus certain clan names like Culibali, Turé, Sissoko are the same among the Bambara, the Sénoufo, the Malinke and the Soninké, attesting to linkages which have so far been poorly clarified.

In many cases, the establishment of an official ethnic nomenclature has fixed divisions that represented nothing more than a transitory historical state, the result of the complicated dynamics of assimilation, fusion or domination. For southern Mali, Amselle (1990) has clearly shown how a change in ethnic identity was part of strategies for survival or for conquest of lineage groups. These phenomena of fusion and assimilation are sometimes all the more difficult to identify because it is not rare to find conquerors adopting the dominated group's language and customs. In this fashion, at the beginning of the nineteenth century, Fulani conquerors adopted the culture and politics of the large Hausa states they had seized.

However, whatever the importance of such a critical analysis of the notion of an ethnic group, it would be wrong to conclude that the distinctions created by this notion are totally arbitrary and have little significance for those to whom they are applied. It would be neither pertinent nor useful to merely point out the transience of ethnic categories. To do so would cause us to ignore the active character that a nomenclature affirms, at the very moment at which we speak, in the national political confrontations or more locally in sometimes violent land conflicts. Bourgeot (1990) clearly demonstrates how the Tuareg identity is at once an unavoidable fact in the political landscape of Niger and the result of a contemporary fabrication process of part manipulation and part makeshift cultural fabrication.

The ethnic template often expresses differences confirmed through a 'naive' experience on the ground. It is scarcely disputable (to give several examples that do not depend on the marked contrast between agriculturalist and pastoralist) that the Sénoufo and the Bambara in southern Mali, the Sérer and the Wolof of Senegal, the Mossi and the Gurmantche of Burkina Faso, and the Hausa and the Djerma of Niger all think of themselves in terms of a distinct 'us' and it is evident that their value systems, as well as their forms of organisation, differ in more than one respect. Certainly, these splits often have a political dimension, in the larger sense of the term (i.e., they serve current confrontations over power, influence and interest), and mask exchanges of people, ideas and institutions that may have existed in the past between these ethnic groups. It is none the less true that the groups thus defined think of themselves as distinct, one from another.

The social reality covered by the notion of ethnicity reveals its complexity in the fact that the recognition of identity borders with the outside does not necessarily imply that the communities thus defined see themselves as homogeneous within these borders. The levels of 'otherness' are, in a sense, capable of interlocking. Izard (1985) has done a remarkable job in analysing this dialectic between the 'us' and the 'other', between identity and difference, in the Mossi society of Burkina. In order to progress in our understanding of the complexity of Sahelian social systems, we must surely relinquish the notion of an object's unity constructed along the lines of homogeneity and continuity which originates from our own culture. A different outlook is needed, which sees in 'the all' the combination of a number of elements, at once different and complementary. In the language of symbols in which concepts are often expressed in African cultures, 'the all' might for example be 7, that is the result of combining 3 and 4, or it might be the fertile couple or a result of a fusion between man and woman (Nicolas, 1968). Those who participate in such a system of representation can see themselves as members of a community, without being forced to deny the heterogeneities, at times the antagonisms, on which a communal unity is based.

Faced with the question of social and cultural diversity of agricultural and pastoral communities in the Sahel, we are thus confronted with a paradox. Diversity exists because these communities are claiming distinct identities, often they are governed along different principles of organisation, and the relations they maintain with nature give proof of a variability most evident in the breadth and scope of their production systems, their technical practices and their tools. Nevertheless, as soon as we try to grasp the substance of an ethnic group, it most often dissolves to the point that it is reduced to the lowest common denominator, such as a common language. How can we resolve this paradox?

SEVERAL AXES OF DIFFERENTIATION IN SOCIAL SYSTEMS

Following Amselle's (1990) analysis, it seems to us that ethnic divisions can be considered not as a kind of classification table in which each element occupies in a definitive manner the place assigned to it by virtue of a certain number of stable attributes, but rather as a reference point within a system of transformation that authorises shifts, regrouping and schisms. Like all systems, there are phases of disruption and recomposition, succeeded by periods of equilibrium and stability. It is during the latter that collective bodies impose themselves within a group, like an established tradition. Considered in this way, the terms of the problem become inverted. The objective is no longer to draw from a repertoire of ethnic groups in order to isolate that by which each defines itself as against others; it consists instead of locating several major axes of differentiation by which the properties of social systems can be ordered, and of sketching, on the basis of these axes, a frame of reference within which ethnic groups exist as concrete illustrations of particular 'co-ordinates' within the conceptual space defined by these axes.

We do not claim that the axes that we are about to use here are the only ones that permit a valid differentiation of Sahelian social systems, nor even that they necessarily highlight their most fundamental properties. We simply propose to combine three dimensions which can help us to grasp what is liable to differ in the manner in which social groups organise relations among their members, and to underscore those aspects that may subsequently shed light on the variability of relations between these groups and the environment from which they draw the means for their social and material reproduction.

The degree to which power is centralised

Following Fortes and Evans-Pritchard (1963), the opposition between state societies and so-called acephalous societies has become one of the criteria of differentiation currently used by African political anthropology. It is based on the difference that we are able to establish between, on the one hand, systems in which the political function (that is, to simplify, the regulation of conflicts both internal (deviations from norms, conflicts with social relations) and external (war)) is carried out by a specialised central apparatus with the ability to use force, and, on the other hand, systems in which this function is the product of a consensus among segments with identical status, each of which is organised around a community of filiation (clans, lineages). Attractive though this typology may be, it cannot be applied casually on the ground. All Sahelian societies, indeed all African societies, are based on a foundation formed by a juxtaposition of segmentary lineages. Conversely, all stateless societies require institutional forms which enable them to control, to regulate and to arbitrate relations between their constituent segments (if only to permit the allocation of space or the exchange of goods and women). Any opposition, then, will be one of degree rather than of type. It is no less true that the segmentary or centralised component can have quite a different weight depending on the society under study.

Thus the political formations of the Hausa, Mossi or Fulani of Macina took the form of powerful chieftainships, led by a monarch invested with civil, religious and military functions and surrounded by a specialised executive apparatus, charged with transmitting and carrying out his orders (Nicolas, 1975; Gallais, 1984; Izard, 1985). The power of such states enforced respect over large territories for common laws applicable not only to nationals but also to any foreigner who resided or carried out activities there. The existence of regulations that were widely accepted was indispensable to the development of commercial trade, for it required that significant wealth could circulate securely and be exchanged under stable conditions. Even if Sahelian Africa had no great builder states, the subjection of great numbers of individuals to a common authority nevertheless made collective forms of land development and management possible. Certainly, such efforts were most often limited to the erection of defensive structures. Yet, the enactment of common rules for the management of natural resources was not unusual. We can cite the example of the Sultan of Damagaram (Niger) who, in the last century, strictly protected the *Acacia albida*.[4] An even more enlightening example is that of the Dina, a strict regulation imposed by the state of Macina in order to organise the coexistence of agricultural and pastoral activities in the Niger's Interior Delta.

An authority's ability to be strict and effective did not always depend exclusively on the spatial extent of its authority. Many Sahelian societies were organised as small aristocratic chieftainships whose rule extended over several villages at most. This was the case with the Djerma (Olivier de Sardan, 1984), with the Soninké (Pollet and Winter, 1972) and with certain Bambara and Fulani chieftainships (Amselle, 1990). Such divisions were not necessarily synonymous with political powerlessness because, limited although it might be in extent, the organisation of power in such societies was based on a central authority that had the ability to impose its decisions on the entire community. Moreover, the distinction between large and small chieftainships is far from clear-cut as the largest of state structures were themselves often subdivided into more limited political and administrative entities, such as provinces, local chieftainships, or even villages. Each of these would have had a certain margin of autonomy, thus subtracting a measure of homogeneity from the whole. Thanks to the work of Izard (1985), a detailed analysis of this type of organisation exists, showing that it is formed by the succession of interlocking powers, namely supreme chief, peripheral command and village chieftainships. Whether it is manifest in a complex structure subdivided into several levels of power or takes the form of juxtaposed autonomous political entities, power is exercised within a specialised institution which is legitimised by the tacit agreement that links a chief to his subjects at the time of enthronement.

Segmentary lineage societies are organised on a different basis, with notable examples coming from the Sénoufo and the Minianka of Mali, the Lobi or the Goin of Burkina, as well as certain Sérer groups from Senegal(Gastellu, 1981; Diabaté, 1986; Dacher, 1987; Jonckers, 1987; Père, 1988). In these cases, kinship is the dominant institution through which

221

social relations take shape. Authority is not the prerogative of a specialised apparatus, but is instead an inherent attribute of an individual's age and position in the genealogical succession.

Within the local community framework inscribed in a territory, each lineage thus reproduces an identical structure that subjects youths to their elders and that places all members of a single filiation under the tutelage of a patriarch. This patriarch is invested with essentially ritual functions and his power lies more in his role as mediator with the supernatural world than in his exercise of coercive force. His function is to assure the preservation of an inherent order that associates the profane and the sacred, rather than to make and execute decisions. In the case of a transgression, he does not intervene so much to punish as to protect and restore a threatened equilibrium through the appropriate rites. In such systems, power therefore has an entirely different character than in centralised structures. This does not entail an absence of decision-making or organisational authority between lineages. Rather it indicates that this authority can take the form of a hierarchy which designates a *primus inter pares* among the lineages and, as in the case of the Bwa in Burkina Faso (Savonnet, 1979), it can manifest itself as a council of elders or even as a village chief. It remains no less true that the first principle of organisation for interpersonal relations is kinship and that no decision can be imposed on a territorial community unless it is based on a consensus among the constituent lineages (under threat of rupture and schism, as often occurs in these segmentary societies).

It is appropriate to mention one additional criterion of differentiation to which lineage societies are subject, because its implication at other levels will be seen later. This is the distinction between patriliny and matriliny.[5] Patriliny predominates among Sahelian societies, although examples of matrilineal societies are not lacking, for example, the Sérer and the Bassari (Senegal), certain Sénoufo groups (Mali), the Lobi and the Goin (Burkina Faso). In most cases, the important structural element resides in the tension formed by the coexistence of these two descent lines within the same social system, one maternal and the other paternal, whereby each orders a different part of social life and relations among individuals. Hence, with the Sérer, Gastellu (1981) reports the existence of an institutionalised contradiction between the paternal line, which organises daily life (the organisation of work and consumption carried out in the father's domestic sphere), and the maternal line, along which goods and, to a certain extent land, are transferred. This internal contradiction can introduce an additional segmentary factor into the heart of societies already divided. Hence, Dory (1988) stresses the fissions that are periodically produced by the contradictions between matriliny and patriliny in Lobi society. In fact, the tension between paternal and maternal lines is present in all kinship systems. While perhaps less visible in societies where all inheritance follows the paternal line, whether of symbolic attributes or of material goods, it is no less evident in cases of polygamy where the relationship to different mothers draws the lines which separate children of the same father.

Each of the two major models for the management of power relations, centralised and segmentary, could be refined by detailing their many variations. Let it suffice to underscore the fact that a clear distinction between the two is not possible. Hence, within social formations endowed with a centralised political institution, the relationship between the state apparatus and its subjects is often expressed in the form of an opposition between autochthonous populations (which even today are organised on the basis of lineage) and foreigners who, in a more or less distant past, imposed a centralised chieftainship apparatus. This opposition is illustrated in Mossi country, with the *Nakombse*, people of power, and the *Têngbîise*, people of the land (Izard, 1985), and also in Hausa society, with the *Yan Sarauta*, dignitaries of the chieftainship, and the *Anna*, autochthonous lineages practising pre-Islamic religions (Nicolas, 1975). Regardless of the real historic foundations of these divisions, it is no less true that, in this type of social formation, segmentarity and centralisation explicitly appear as principles of organisation at once antagonistic and complementary. Rather than argue in terms of a clear opposition, there is every reason to consider that, in the Sahel, each society realises a particular combination of these two principles, and thus occupies an intermediate position along the conceptual axis that links them.

Stratification and horizontality

Whether organised primarily on the basis of lineage or by a centralised political structure, many societies of the Sahelo-Sudanian zone are characterised by the fact that their members are stratified, by birth, into social categories which we might call orders and which are distinct and highly differentiated on the basis of status, and between which mobility is theoretically impossible (in particular, intermarriage is forbidden). The most frequent divisions place at the summit of this hierarchy the aristocrat/ warrior, followed by the free man, then by the artisan (especially the blacksmith) and finally by the captive. With certain variations, this division by hereditary status can be found in the majority of Sahelian societies, but the rigour with which it is applied, the degree of social distance which it determines and the mark it makes on daily life can vary considerably from one society to the next and even at times between local communities within the same ethnic group.

Hence, for example, in Soninké society, among the Djerma–Songhai, the Tuareg and certain Fulani groups, the place occupied by each individual in the hierarchical order remains, even today, a major determinant in interpersonal behaviour, in participation in the decision-making process, and in the mechanisms of control over factors of production (Chassey, 1972; Pollet and Winter, 1972; Bonte, 1975; Bourgeot, 1975, 1979; Olivier de Sardan, 1984). In the Soninké villages of Guidimaka (Mauritania), for example, the *komo* (captives) still reside in separate quarters and are required to perform menial tasks (water and wood transport) that a *hore* (noble) could not perform without being demeaned (Steinkamp-Ferrier, 1983). Societies organised on rigorous stratification generally have a strict framework of social values and behavioural norms which correspond to the hierarchy. The noble must demonstrate strength of character and courage and he must exhibit, in all circumstances, perfect control over his emotions. These obligations are more or less directly related to the heroic code of values.[6] The artisan and, to a greater extent, the slave may shamelessly speak improperly, beg or run away from danger. This discrimination rests on such deep stereotypes that, at times, confirmation and validation are sought out in differences in physical and morphological attributes of the different orders. Nobles will therefore be described as slender, supple and elegant, as opposed to captives who are stocky, stiff and awkward (Olivier de Sardan, 1984: 37).

In other societies, this notion of division according to a hereditary order, even if not totally absent, is not so fundamental a principle in ordering social relations. Almost always, a particular social position is attributed to certain crafts (blacksmith, potter, weaver) or to social functions (*griots*, or praise singer). Nevertheless, the resulting distinctions are based more on the principle of complementarity among social roles and specialised rituals than on that of hierarchy and social distance. Although the status of slave has existed almost everywhere in the past within the Sahelo-Sudanian region, the population to which it is attributed has sometimes played only a modest role in the functioning of the local economy or, where such is not the case, has not everywhere been the object of such derogatory stereotypes as those mentioned above.

This minimal relevance of stratification by status is fairly characteristic of societies where the segmentary component is strong, whether it be the Sénoufo of Mali, the Sérer of Senegal or the Lobi of Burkina. Differentiation engendered by birthright in the heart of lineage groups is much more important than that ascribed to some scale of hereditary status. Moreover, slavery has never played more than a minor role in the functioning of the production systems of these groups and, currently, no strong social stigma remains attached to this status. However, this relatively poor division into orders is not the exclusive domain of lineage societies. In complex state formations, like those created by the Hausa and Mossi societies, the hierarchy in political leadership, on the one hand, and the opposition between people of power and people of the land (Izard, 1985), on the other, offers an analytical framework for social differentiation that is more pertinent in many ways than that which may be established on the basis of hereditary order. Certainly, in Hausa society, derogatory stereotypes exist with respect to *griots*, blacksmiths and butchers, and it is rare, even today, that families of different rank are willing to exchange women with them, although these barriers are far from being completely impassable. Among the Mossi of Yatenga (Izard, 1985), there exist endogamous professional groups (blacksmiths, craftsmen–tradesmen) although, within the framework of a general submission to the central political authority, their status is much less connected to a concept of hierarchy than to the specificity of their activities and their origins (in fact some are considered as originating from Songhai or Soninké stock). In this same social formation, the notion of captive is most applicable to 'royal captives', who are the prince's court dignitaries or form part of his land control policy (Marchal, 1983: 701–6), and it does not designate a social category of inferior status.[7]

Given the position held by slaves in the functioning of Hausa agriculture in the nineteenth century (Smith, 1954; Hill, 1972, 1976), it is surprising to observe the way in which their status dissolved to the point of no longer constituting an important reference in the current Hausa system of identification. In Hausa communities, unlike their neighbours the Songhai–Djerma and even more so the Tuareg, a person's servile status is hardly ever mentioned in daily conversation. Within the population of a village, only a meticulous genealogical analysis would identify descendants of former slave families. Nothing in their behaviour or duties would suggest their former status today. This lability of slave status in Hausa lands may perhaps be attributed to the place held, since precolonial times, by commercial concerns in the scale of social values. As Hill notes (1972: 205), the slave was considered an instrument of production in the same way that later a team of plough animals was to be. However, contrary to societies dominated by an aristocratic and military ideology in which the affirmation of social distances is indispensable to the reproduction of the system, there was no symbolic benefit in emphasising their devaluation with respect to free men.[8] At the other end of the social scale, chieftains as well as the dignitaries around them are invested with an incontestable prestige. However, independently of the effective exercise of the responsibilities of their power, nothing definitively distinguishes a class of aristocrat/warriors, which would benefit from an uncommon status. If Hausa society is submitted to a highly structured hierarchy, it is due much more to differences in function and in wealth than to privileges arising out of birth.

The greater or lesser place occupied by such stratifications in the social system, as well as the way they govern economic and social relations and impose individual behavioural norms has, in many respects, influenced the way societies have reacted to the new constraints that confront them in the present century. It is for this reason that, following the colonial conquest, the emancipation of slaves had a much greater destabilising effect on the Fulani of southern Mali, where agricultural and artisanal production was almost exclusively the work of slaves, than on their Sénoufo neighbours, where slavery was of only minor importance (Amselle, 1990). Where aristocratic status strongly influences the system of values and is accompanied by the devaluation of manual labour, it is practically impossible for a noble to imagine acquiring wealth by cultivating his own lands, even with commercial agriculture. He must therefore search elsewhere for money, which has become necessary for his material subsistence and for the fulfilment of his social obligations. Migration allows him to go away and carry out activities which, at home, would have been seen as degrading. In the global context of incentives and constraints created by the geography of these vast labour reserves, this avoidance strategy is a likely factor in the attraction of the adventure of exodus for populations of strong aristocratic traditions, such as Soninké, Djerma–Songhai and Fulani (Painter, 1985: 477). It is because of a difference in social perspectives of this nature that Amselle (1990) attributes the success of cotton among Sénoufo cultivators of southern Mali, compared with their Fulani, Bambara and Malinke neighbours who have shown little interest.

The circulation and distribution of goods and wealth

The greater or lesser role of individual accumulation of goods and wealth in the system of exchange, as well as its social significance, are domains in which the diversity among Sahelian societies is also strongly expressed. Gastellu (1981) provides a lengthy analysis of the Sérer society of Senegal, emphasising its 'egalitarian' character. He described, as we have already mentioned, the existence of institutionalised competition between lines of descent: one patrilineal, through which the organisation of daily work life and the distribution of goods operates; the other matrilineal, through which inheritance is transmitted and prestige goods are accumulated (livestock, jewellery, animal traction equipment). According to his analysis, this contradiction results in a limit to individual accumulation of wealth and maintains a periodic redistribution among lineages (Gastellu, 1981). In a case such as this, the circulation of goods is strongly linked to the system of social relations (as it contributes to its reproduction), thus reducing the margin for individual manoeuvring.

> The dissociation of economic units results in the non-equivalence between, on the one hand, the core formed by individuals from the same production/consumption unit, and, on the other hand, those belonging to the same accumulation unit. This dissociation, combined with the constraints of a norm which prohibits the use of collective goods for individual ends, keeps producers from realising an individual accumulation of goods.
>
> (Gastellu, 1981: 761)

It is not always certain, however, that tension between patriliny and matriliny is necessary to restrict an individual's accumulation of wealth within any given social system. Tight control by elders over the factors of production, and over products, can lead to a similar result within a framework of patrilineal structures. In this case, a concentration of goods (and thereby of power) is most definitely in the hands of certain individuals, the oldest sons of the oldest lineage branch, but this role results from criteria pertaining to their birth and over which they have no influence, namely age and position in the line of primogeniture. The control they exercise over wealth, and which allows them, in particular, to regulate matrimonial exchanges, is the expression of a collective function with which they are invested. They act as managers of a common trust, much more than as possessors of goods from which they can draw personal advantages. The determining element here is the primacy that lineage-based kinship relations exercise on the life and destiny of individuals. Strong symbolic barriers (from the expression of a collective reproach to threats associated with the use of sorcery) restrict the emergence of publicly displayed personal wealth that could only appear as a challenge to the established order.[9]

At the opposite end of this spectrum of economic life, other social systems exist in the Sahel, within which the circulation and distribution of wealth follow completely different principles. The constellation of Hausa states offers an appropriate example. Their centralised political structure was formed by overlapping hierarchical levels corresponding to as many intermediary platforms of responsibility and power, from the prince all the way down to the simple subject. This pyramid was nurtured by an ascending current of tribute, both in goods and in labour levies, and by a descending current of rewards and aid. Every key person in this structure (village chief, local dignitary, court member, prince) played at the same time the role of relay in the general upward movement of goods and services and the role of intermediary for accumulation and redistribution. Thus each notable was expected to use his wealth for the benefit of the persons who were dependent on him and who had contributed to the production of that wealth. In particular, he was supposed to help them in times of need. As in the case of the lineage societies mentioned above, when considered globally such an economic system functioned as an expression and renewal of social bonds by means of confirmation of allegiances and renewal of protection. The essential difference here resided in the fact that accumulation operated in the hands of persons whose positions were most often revocable or interchangeable, because they were political. In such a context, control over goods of value (food, money, livestock, etc.) was not simply, as in a lineage-based society, the product of an immutable order imposed by the succession of births and intimately linked to the balance of the sacred and the profane. It expressed something of the personal value of the individual who exercised it. The public manifestation of personal wealth was from then on not only accepted but encouraged, with all that this implies for exhibition and ostentation.

The Hausa world found itself, moreover, at the heart of a trans-continental, interregional and local trade network which, in the eighteenth and nineteenth centuries, covered a large part of West Africa. As we have noted previously, it was from trade and all its related activities that the Hausa world drew the best part of its wealth. In this context, those in power benefited from their privileged access to the means of increasing wealth and consolidating the pre-eminence of their social status.[10] The ambiguity that was thus created between rank and wealth made possible a semantic inversion by which men enriched by trade could accede to the rank of notable. This promotion became tangible through the network of individuals bound by obligation, who themselves benefited from the generosity of their patron. When one belongs to such a social system, the channels of individual access to wealth are legitimate as long as they are accompanied by the practice of redistribution, and even of squandering. This type of strategy for social ascent is excellently analysed by Nicolas, who describes such a person's behaviour as follows:

> He makes a spectacle of himself. He is extremely interested in anything which might potentially grant him the appearance of super-ior status: a clay house-front, costume, dress mount, liberality, etc. . . . In this enterprise, the 'griot' is a driving force, to the extent that he gives praise in proportion to the size of the gifts he receives: he who gives like a prince obtains the praise of a prince.
>
> (Nicolas, 1975: 453)

The two extreme models just described illustrate two contradictory philosophies of the social order, hence the contribution the economy is called to make in a society's reproduction. One follows a principle that seeks to limit individual wealth and, moreover, in the case of the Sérer favours a periodic horizontal redistribution of accumulated goods between sym-metrical lineage units. The second follows a hierarchical principle that legitimises inequalities between individuals while still maintaining a re-distributive dimension through the means of a vertical reciprocity that exchanges allegiance for protection, and services for gifts. In this case, the accumulation process becomes one of the driving forces of a social dynamic; it fulfils a political function while fuelling the competition for power and leaves a substantial space for personal initiative. By contrast, in the first model, social control of wealth, even if it leads to localised phenomena of concentration, responds profoundly to a cyclical circulation scheme which periodically tends to re-establish equality between accumulation units structured along the basis of lineage and leaves little room for individual enrichment.

A third type of situation is possible, whereby economic inequality exists but is combined with a lack of social mobility. In certain respects this was the case with the Mossi society of Yatenga, within which the attribution of authoritative functions was strictly based on lineage and where the practice of trade was reserved for a particular social category (Izard, 1971, 1985). The appropriation of valuable goods followed strict rules. Only members of the chieftainship, for example, had the right to own large livestock and, more generally, any public demonstration of wealth had to be equivalent to that person's recognised social rank. Any transgression from this obligation meant exposing oneself to severe sanctions by those in power who felt defied. The absence of *griots* within Mossi society, who can make or break a reputation depending on the generosity they receive, reveals an absence of the mechanisms of social competition nourished by the circulation of goods, which can be found within the Hausa society. Generally speaking, the more highly a society is stratified and imbued with aristocratic values, the more this social hierarchy is fixed. The enrichment of a person of inferior status will never allow him to cross the symbolic gap that separates him from the socially dominant categories. Even today, regardless of how much he has earned at the end of a long period of work in France, a Soninké 'captive' can never aspire to take a direct part in the affairs of his village.[11]

These different modes of economic organisation reveal different properties and have brought about distinct reactions with respect to contemporary factors of change, particularly with the generalisation of trade relations and the penetration of commercialised agriculture. Well before the colonial conquest, Hausa society had extensive communication with the outside world through trade mechanisms. As a result, the development of a trade economy dominated by the exchange of local groundnuts for imported manufactured goods from Europe encountered few fundamental obstacles.[12] However, this is far from saying that this did not alter social relations and the functioning of the production system (Raynaut, 1977b). In a different historical context, an identical argument can be made for the Wolof society of Senegal.

Within social formations that follow different principles of organisation, the insertion of the local economy into the marketplace has demanded more difficult adjustments. In effect, market relations separate economic exchanges from their social context. Contrary to the gifts and counter-gifts that were the basis of non-market systems of exchange, buying and selling are anonymous and do not require the existence of specific social relations between the partners. Old social mechanisms for regulating the circulation of wealth are thus short-circuited. This represents a potential factor of disorder for those societies where inequalities were previously strictly controlled. Efforts to distance themselves from this threat largely explains the initial resistance of these groups to the penetration of commercialised agriculture and beyond that to the series of changes which inevitably would follow, such as the possibility for those of dependent status to negotiate, at least in part, the product of their labour and submission to foreign economic demands. The Sérer have long been known for their reticence towards commercial development and indebtedness, unlike their Wolof neighbours (Gastellu, 1981). The Sénoufo–Minianka of Mali have been the object of similar observations (Diabaté, 1986). In both cases, the rejection of cash cropping is far from definitive – quite the contrary, one society eventually made a great deal of room for groundnuts and the other for cotton. Nevertheless, this adoption required some internal adjustments on their part ('accommodations' to use Gastellu's term) destined to limit the social impact of the new economic practices associated with the adoption of these crops. These accommodations consisted essentially of laying down the ground rules which ensured that the elders would come to control the new form of production.

With this axis of differentiation of social systems, as with the two that preceded it, we must avoid seeking completely distinct dichotomies. No single concrete case illustrates perfectly the models that we have described and the more complex a social system the more likely that different, and at times antagonistic, principles of organisation will coexist. Hence, in the large state societies of the Sahel, the vertical movement of wealth and the processes of accumulation have always gone hand in hand with other forms of circulation, associating lineages, domestic units and individuals according to principles of exchange based on a dialectic of equality and invested, above all, with functions of social reproduction (Raynaut, 1973b; Nicolas, 1975). Conversely, in societies that aim to limit personal concentrations of wealth, the limitation of inequalities is the product of a constant battle against emerging individual strategies, which have no doubt always existed but have been encouraged by the development of monetary exchanges (Gastellu, 1981).

THREE MAJOR MODELS OF SOCIAL ORGANISATION

There are numerous other features by which social formations in the Sahelo-Sudanian zone might be differentiated. Our choice of axes of differentiation is justified to the extent that they account for two fundamental principles of organisation: the first concerning power relations between people; the second based on the control and circulation of goods. These are the two fields of reality, the importance of which should not be overlooked if one wants to understand the social principles subtending the relations of production and thereby the systems for managing resources. It would obviously be impossible to propose objective criteria for situating each social group precisely on the gradient of the concentration of political power, on the level of social stratification or on the degree of egalitarianism. Yet, nothing prohibits us from roughly qualifying a social formation as a function of the polarisation it exhibits on each of the three axes that have been described.

This is of course somewhat arbitrary, but is it more so than ranking a population, according to a binary principle, in the category of acephalous or state societies, of caste or casteless societies, or of an egalitarian or non-egalitarian system? All these are categories currently in use.

On the basis of this reasoning, it would be tempting to draw up a typology that would account for all the variations along which these three axes of classification combine, thereby drawing profiles intended to characterise real social formations observed on the ground. Even by simplifying this arrangement considerably, there would be too many eventualities to make their interpretation and application useful. It seems more judicious to remain at an intermediate level, identifying three major situational models, illustrated by frequent examples taken from the Sahel. These situations are summarised schematically in Table 9.1. The three variables used to describe the type situations correspond to the three axes just defined. The level at which they are manifested in each case is evaluated according to a summary scale which distinguishes between two contrary poles, i.e., strong (++) and weak (--), while maintaining an indeterminate area (+-). It is clear that such a classification is extremely reductionist. It makes sense only in relation to the more detailed analyses that were carried out above for each of the axes. Its interest is in helping to characterise clearly distinct models of society which can, as a first approximation, serve as reference paradigms for ordering the diversity of concrete social formations observed on the ground. None of these is entirely reducible into one or another of the models. We are generally confronted with social formations in which different principles of organisation have in some sense become superimposed throughout history, namely principles of kinship lineages, centralised chieftainships, stratified status and roles, military activities and trade. In such a comparison what is at play is more a confrontation between hierarchies of characteristics than a binary analysis of their presence or absence.

Table 9.1 Major types of social organisation

Models of social organisation	Concentration of power	Social stratification	Accumulation of wealth
Major trading states	++	+-	++
Warrior aristocracies	+-	++	+-
Lineage-based peasantries	--	--	--

Notes: ++ Strong; +- Intermediate; -- Weak

Major trading state

The Hausa societies of Niger and Nigeria offer a concrete contemporary illustration of a 'major trading state'. Combining what has been said earlier in more detail, the major characteristics of this category include:

- the fact that power over people and things is concentrated within a specialised authority which has at its command the institutional apparatus and the power of coercion necessary to make its decisions respected over an extended area;
- primacy of the political hierarchy over a stratification of immutable categories as a principle of organisation of people; and
- open circulation of goods and wealth through trade, which makes accumulation of wealth a tolerated means of expression of individual value and social standing.

Mossi society is similar in certain respects – the centralisation of power, the unimportance of slavery as a criterion for social stigmatisation and the importance of trade as an economic activity. However, Mossi society differs markedly from Hausa society in the status associated with wealth and the limitations on social mobility. The large state and trade organisations in the Sahel have, throughout the course of history, known many vicissitudes, emerging or disappearing in the course of war and the displacement of major commercial centres. They were extremely fragile structures, to the extent that they were often only held together by the physical and symbolic constraint of a charismatic personality. They continued to disintegrate after the colonial conquest (at least those whose territory came under the French administrative system) as the new power could not tolerate competition from a strong political structure. They therefore subsist today only in residual forms, though none the less still capable of strongly influencing attitudes and behaviour.

Warrior aristocracies

The Songhai–Djerma societies (Niger and Mali), the Soninké (Mali, Mauritania, Senegal) and to a certain degree the Tuareg (Niger and Mali), are all clearly types of 'warrior aristocracies'. They are characterised by the existence of specialised powers that, within the same cultural group, are organised into much smaller entities (often reduced to several villages, or to several groups of nomads) that frequently enter into rivalries and armed conflict. Social stratification is determined principally by birth, creating a particularly strong split between free men and captives. The marking of social distances represents a strong organising principle within this type of society. At times, trade has been an important activity (with the Soninké and the Tuareg, for example), but warring and raiding represented a highly valued method of obtaining wealth (the capture of slaves or livestock). This type of organisation is often created out of the decomposition and break-up of great historical empires (Songhai, for example) and the nearly general state of conflict which the slave wars created throughout the zone in the nineteenth century.

Lineage-based peasantries

The Sérer of Senegal, the Sénoufo of Mali and the Lobi of Burkina Faso are several examples, although there are others, of 'lineage-based peasantries'. These are social formations whose organisation rests primarily on the coexistence of homologous lineage segments (this same structure often exists in the models described above, but it is held in place by strong, transverse, political structures). Adherence to the lineage and, within the kinship structure, the order of generations, primogeniture and the division of the sexes constitute the major principles of organisation which order individual relations. In these societies, consensus among family groups is the dominant mechanism for decision-making and, when norms are transgressed, the restoration of disturbed social and supernatural balances is seen as more important than acts of sanction. The practice of agriculture and a strong attachment to the land, demonstrated by the strength of agrarian religious cults, often characterise societies of this type. In any case, the social organisation of certain nomadic pastoralist groups, like the Fulani *Wodaabe*, best approaches the features of this model. As Amselle (1990) stresses, it would be a mistake purely and simply to contrast these segmentary societies with the large Sahelian political organisations that have existed alongside them for centuries because they often represented peripheral formations within these major systems organised around trade, complementarities and confrontations.

CONCLUSION

We have chosen not to apply a static nomenclature in our analysis of Sahelian social formations. Instead, in exploring the differences between social systems along tentatively defined axes of discrimination, we have used an approach that is better able to draw on observations of the diversity and complexity of social realities in the Sahel. In so doing, we have attempted to create the basic elements of a broader comprehension. Of course, this attempt is only partially complete, but we shall see that such an initial effort can shed light on the analysis of social modalities of productive resource management.

NOTES

1 External influences, of course, existed far earlier than the formal colonial conquest, in particular through commerce and the slave trade. To be convinced, we need only consider the disruptions introduced into West African economies during the nineteenth century by the massive injection of shell money by large European commercial companies (Johnson, 1970).

2 The Institut Française d'Afrique Noire (the French Institute for Black Africa) was responsible for much of the scientific research conducted in western Africa during the colonial period.

3 Thus the Dogons, whom ethnographic description has made an archetype of the peasant culture of the Sahel in its specificity and uniqueness, are divided into several groups speaking dialects sufficiently different to make comprehension sometimes difficult.

4 Sultan *Taniimum*, who ruled from 1851 to 1884, forbade that they should be cut on pain of death (Salifou, 1971: 7).

5 Filiation, or descent, is said to be patrilineal when it is transmitted exclusively through men and matrilineal when it is transmitted exclusively through women. Diverse goods or non-material attributes (name, specific cult, social status, profession, etc.) can circulate from one generation to the next following one of these two lines of descent.

6 The analogy between aristocratic values and courage in battle can be so close that, according to oral tradition, certain Soninké families from Dyahunu (Mali) lost their noble status because of their ancestors' cowardliness in battle (Pollet and Winter, 1972: 234).

7 Neither Izard (1985) nor Marchal (1983) make an issue of this in their analyses of stratification in the Mossi society of Yatenga.

8 On the marking of social distances as an instrument of production and reproduction in aristocratic society, see the analyses of N. Elias (1985) on court society. Although the historical and cultural context to which this analysis is applied (seventeenth-century France) is totally different from the context under study here, the comparison is enlightening. On the notions of distance and hierarchy, see also studies by Louis Dumont (1966).

9 Processes of accumulating wealth are not always absent from the functioning of segmented societies. In particular, they can be found in the Sérer, studied by Gastellu (1981). Héritier (1975) described this process in the Samo in Burkina Faso, where cowrie shells were accumulated as a means of social competition. Nevertheless, it remained a purely internal process within the lineages and was resolved by squandering, which had the role of regulator/arbitrator.

10 Contrary to what is often observed in aristocratic or warrior societies, trade was not prohibited to nobles of the Hausa society.

11 The information on Mossi society contained in this paragraph is a result of personal communications with Michel Izard.

12 J.P. Olivier de Sardan (1984: 208) undertakes a similar analysis when, conversely, he attributes the failure of the Songhai–Zarma society in Niger to spread the practice of export agriculture to the narrowness of their commercial channels.

10

THE TRANSFORMATION OF SOCIAL RELATIONS AND THE MANAGEMENT OF NATURAL RESOURCES

1 The Birth of the Land Question

Claude Raynaut

That the degradation of vegetation and the erosion and overexploitation of soil in the Sahel are the manifestations of complex agro-ecological dynamics is no longer in dispute. The contested but convenient term, 'desertification', has been applied to this phenomenon, which can be linked to the combined effect of a certain number of objective factors: harsh natural conditions, increased demand for plant production (due to the combined effects of demographic pressure and economic constraints) and the introduction of new tools. However, the unequal form and intensity of these ecological changes, across different geographical locations (a diversity illustrated above, see Chapter 8), is inseparable from other realities that are of an immaterial nature and which concern the relations rural societies maintain with the environment they exploit. We can picture these as the collective means of control of a set of resources, as well as the organisation of the procedures and means for their exploitation. These relations of production cannot be separated from the social system of which they are a manifestation. They reflect the fundamental principles that underlie the global relations among people.

Our understanding of the diversity that characterises the relationship between man and nature across the Sahel cannot be advanced by simply producing a list of the objective determinants of this differentiation. It is also necessary to develop an understanding of how Sahelian societies have been the subjects of their own change. Through their internal principles of organisation and their own dynamics of change they have been able to adjust the relations they maintain with nature to interpret the constraints on their capacity to manage productive resources (relations to the land, control of the labour force and control of technical implements).

From this perspective, the land question is recognised today as an essential aspect of the analysis of Sahelian systems of production. Whether territory is used for hunting, fishing, gathering, pasture or agriculture, the use of land always implies the existence of social norms that regulate individual or collective access. Studies in this area are few, and they have long remained more standardised than truly informed by a detailed knowledge of local social and cultural realities. There has, however, been a renewal of interest in land problems (Le Bris *et al.*, 1982; Crousse, *et al.*, 1986; Downs and Reyna,' 1988; Le Bris *et al.*, 1991).

Nevertheless, the dearth of land tenure studies in the past is not simply the result of chance. In fact we would suggest that in the Sahel, outside of very localised geographic areas, the control of land as a medium for productive activities has only very recently become a real issue. Very often we find that the 'customary rights' which are supposed to record 'traditional' controls, are nothing other than the product of an effort on the part of dominant social groups to take advantage of new arrangements introduced by the colonial administration (Olivier de Sardan, 1984: 219–42; Amselle, 1990). It can even be said that, in this precise historical and geographical context, the concept of land as capital had practically no relevance in earlier times. Of course, the concern inherent to all human societies to defend their autonomy, to ensure the physical security of their members and to protect their communication routes has always necessitated territorial policies, a concern of even greater urgency in the troubled periods of the nineteenth century. Nevertheless, for a long time, for systems of resource exploitation which were for the most part superficial and mobile within a context of resource abundance, the exercise of private control over land as a means of production was not a functional necessity.[1] The more critical problem, therefore, was to control humans and their labour power, an issue to which we will return in the next chapter.

Today, on the contrary, the demand on primary ecological production has reached such a level that the 'finite' character of space has made itself felt (Giri, 1983). We are therefore in a transition period in which the affirmation and recognition of individual rights over a land resource that is growing increasingly scarce, and the restrictions of access to it, have become vital necessities for both individuals and groups. Through strategies of conquest, or the defence of acquired gains, through the accumulation of land by the strong at the expense of the more vulnerable, a redistribution of access to land is taking place in the Sahel today which, for a long time to come, will determine the future state of power relations with respect to land. This is a complex phenomenon, for it brings into conflict multiple actors in the form of states and rural populations, pastoralists and agriculturists, rural and urban dwellers, rich peasants and victims of past famines. It generates competition, which at worst may explode into open conflict, but which at the very least always brings disorder to the exploitation of ecosystems. It is clear that the practices of land and natural resource

236

management employed by Sahelian populations cannot be understood by using a purely quantitative approach (human and animal pressure on land) and by ignoring the social dynamics of which these practices are largely a manifestation.

From one Sahelian country to another, certain general tendencies are at work which position the land question around three major axes:

• the role of the state in its efforts to control land through legislation;
• the superimposition of the ancient rights of legitimate users of the same space and the resulting competition; and
• the privatisation of land, its entry on to the market and the emergence of new strategies of land control.

Beyond these common lines, however, there are also many factors of diversification, resulting in numerous specific local situations. Once again, it is a question of accounting for this diversity, while also endeavouring to elucidate some of the features that structure it.

SOCIAL SYSTEMS AND LAND SYSTEMS

The previous chapter showed to what extent the principles of organisation in Sahelo-Sudanese societies could vary. The way in which relations between humans and land were ordered in the past, and in which they are being reconstructed today, is intimately linked to this social order. Basic as it is, the typology of social systems established above can at this point help to clarify some of the differences observed with respect to land.

Land and the primacy of alliance

As already noted, lineage-based peasantries are above all organised according to principles of filiation and precedence (generation rank, primogeniture).[2] They perceive their capacity to survive as being strictly dependent on the preservation of fundamental equilibria that demand a consensus between the lineages of a community, and the maintenance of a lasting accord between humans and the supernatural entities that govern the world's forces. Their relationship to land is largely a reflection of these social representations. Guardianship over agricultural land is incumbent upon the lineage that first cleared and cultivated it. The responsibility of maintaining this alliance and sharing its benefits with those who have joined them is transmitted to the descendants of the ancestors who made the initial pact with the primacy divinities of the bush. To obtain the right to clear new fields on village lands, all newcomers must, therefore, seek authorisation from the member of the founding lineage who fills the function of 'land chief'. This authorisation will be granted to the newcomer after a process of social recognition, which is of varying severity depending on the society, but which always aims to ensure that the newcomer can be integrated into the community without disrupting its harmony. In any case, the newcomer is usually not completely unknown, but is coming to join a person or a family with whom he already has relations.

The role and power of the land chief can vary appreciably from one case to another. Among the Sénoufo of Mali, for example, he does not relinquish control over lands under his authority. Rights obtained from him are precarious and must be confirmed regularly. This is the meaning behind the offerings of millet and sorghum made to him during harvest by those to whom he has assigned a parcel of land. Sanogo summarised this as follows:

> The allocation of plots to different lineage chiefs is not definitive. Each time a field is moved on to new land a new request must be made. Furthermore, the cultivation of all land that has lain fallow for more than three years must be the object of a new request to the land chief of Ngary. In this way, a definitive appropriation of land is prevented.
>
> (Sanogo, 1989: 391)

Land organisation among the Sérer is very different. While the land chief is the supreme trustee of the land of his matrilineage, it is another individual, the 'guardian', chosen from another descent line who is in charge of its daily management and, more specifically, who divides it among its users. Usufruct granted in this way is confirmed through the payment of a fee, henceforth paid in cash, but it is transmitted by inheritance (sometimes through the paternal line) and over the course of generations acquires an ambiguous character, somewhere between rental and appropriation (Gastellu, 1981).

The Gurmantche of Burkina Faso exemplify a third type of situation, in which the institution of land chief has become a purely symbolic position, such that the assignment of usufruct has become virtually definitive, following a trend that is reinforced by permanent settlement and the stabilisation of cultivated lands (Rémy, 1967).

A final example of organisation is seen in the Bwa of Burkina Faso. Up to the middle of the nineteenth century they were organised according to principles very similar to those of segmentary societies. They then grouped themselves into defensive villages, the better to defend themselves against the threat posed by the expansionist aspirations of the state of Macina. This prompted them to reorganise, on a community basis, the distribution and exploitation of agricultural land, through a reallocation of the central ring of village land by neighbourhoods and by collective work on vast tracts of land situated in bush areas vulnerable to external attack (Savonnet, 1979).

Originally, apart from some local variations, a single vision informed the land organisation of lineage-based peasant societies. Divinities were the true landowners and humans beings were no more than the holders of usufruct rights. In this context, the role of the land chiefs was essentially to intercede with the supernatural world. Alliance, therefore, was the relation that constituted the touchstone of such a land system; it united the founding lineage with the agrarian deities, just as it united the land users with the land chief. The former alliance was periodically renewed by ritual sacrifices and the latter by the payment of a symbolic fee.

What were the operating principles of such a system? The founding lineage had no more right than anyone else to claim to own the land. It could, however, claim a privileged link with the divinities of the area. Without the consent that it alone was able to obtain from the gods, clearing the bush was a risky activity and efforts to make it productive could well prove fruitless. The mediation of the land chief was, therefore, essential to anyone who wanted to work the land free of danger and with any hope of success. What could have been the function of such an organisation at the time when it was established, a period when land was abundant (even taking into account long periods of fallow) and was not the object of competition? The thesis we wish to defend here is that what was at stake through such a network of alliances was not, despite appearances, the preservation of rights to land, but the exercise of control over human beings. In fact, the problem to resolve was not the control over land as a scarce resource (which it was not) but, on the contrary, the possibility of introducing a social mechanism to regulate the use of an abundant resource. In the absence of any risk of competition for land use, controlling access to land was the apparent role of the land chiefs. In fact, in granting or refusing permission to a newcomer to settle, and in assimilating a newcomer into a network of alliances from the beginning, they ensured the regulation of a process by which a local community was constituted. The principle at work was one of territory, not one of productive space. This assertion is in accordance with Meillassoux's statement regarding the status of land in a domestic community:

> However, with membership of a community being the condition for access to land, it is generally considered that this collectivity possesses 'communal property'. In reality, an awareness of 'appropriation', that is of an exclusive link with a piece of ground, does not derive from the exploration and occupation of land, nor from the work invested in it by past and present members of the group . . . Domestic society does not, in general, put up obstacles to the admission of foreign individuals or families once social relations are defined that will link them to the community.
>
> (Meillassoux 1982: 61–2)

With the increasing scarcity of land, the stakes have been reversed. Today, control over land has become an end in itself, a key condition in the functioning of systems of agricultural production and, consequently, of the material and social reproduction of peasant communities. The land organisation of lineage societies has, therefore, been profoundly transformed – even while an apparent stability has been maintained. This evolution has followed two principal directions. In many cases, the disappearance of the bush and the quasi-permanent exploitation of lands has progressively secured land use rights to the point that they have gradually become confused with permanent appropriation. The role of the land chief, when it has been retained, has become strictly ritual and now focuses on the performance of propitiatory rites for the benefit of all those who are working the land under his guardianship. Sometimes, however, as in the case of the Sérer communities studied by Gastellu, the chief's position has evolved to that of a holder of land rights, such that the symbolic offerings previously made to him (a handful of sand, pieces of wood) have been transformed into rental income, paid in the form of monetary fees (Gastellu, 1981: 185ff.).

Social stratification and land exclusion

The case of the warrior aristocracies is different. As we have already seen, they are characterised by a strict stratification into orders of social status. This manifests itself by fundamental splits between nobles, free persons and captives. It is common for this marked differentiation in status to be projected into the structure of landholding. This can be direct, as is the case of the Soninké of Guidimaka in Mauritania, among whom only the noble lineages hold rights to land, especially the most desirable land such as flooded basins, the levée along the river and the banks uncovered by receding waters.[3] Captives are, of course, permitted to cultivate such land (a right that can be withdrawn at any time) but only in small numbers and in return for the regular payment of a fee. The only areas that are freely open to them are the relatively unproductive dune areas at a distance from the village (Steinkamp-Ferrier, 1983). In this society, land has long been a social issue. This is largely related to the annual flood. By making the occupation of land in the flooded areas permanent, the annual flood has contributed to making land a valued factor of production, and the object of appropriation by dominant social classes.

This is not the case everywhere, even in the realm of the Soninké. Although in the Malian Diahunu, studied by Pollet and Winter (1972), as in Guidimaka, captives and artisans have never had permanent land rights, land was not (until very recently) the object of real appropriation by nobles and free persons. Again, power over people was a greater determinant than control over land. In fact, in this type of society, as with the preceding cases, the land question was not an issue until relatively recently. A strict social stratification made possible the subjugation of humans and the exploitation of their labour without it being necessary to establish private rights over land.

In looking at the case of Songhay–Zarma communities in Niger, Olivier de Sardan (1984) has analysed the recent emergence of the land question in an aristocratic society and the transition from a society in which differential status was of a political nature (domination exercised over people) to one of economic inequality based on control over land. He demonstrates how the notion of land appropriation is, in the final analysis, the product of colonial law and how that law has been appropriated and manipulated by the dominant classes of the local society. These classes replaced the captive/noble relationship, rejected by the foreign powers, with a relationship recognised by French law, that of tenant/landowner based on contract, and legitimised by the ownership of land (with tribute being replaced by a land fee). In order to accomplish this, families of the chiefly class and their allies employed all the strategies within their power to appropriate the lands over which they had previously only had political guardianship. The colonial administration colluded by producing a 'customary' land law, which although a complete fabrication, was more in accordance with its own ethical principles than the subjugation of captive to master. A Nigerien historian has forcefully summed up this evolution as follows:

> The chief, a warrior by vocation, did not work the land. With the loss of his weapons he wanted to have lands, to live off the land, following the example of his serfs and vassals. By the power vested in him by colonialism he appropriated those lands by force, an act distinctly divorced from custom. And thus a new cultural phenomenon was born among Africa's social strata: that of the battle for land, which in some instances continues today, bitter and sickening. Out of this struggle emerged the duality of chief and peasant, reproducing the former duality between master and serf.
>
> (Boubou Hama, cited by Olivier de Sardan, 1984: 219)

The evolution we observe here is misleading in appearance, and doubly so because, on the one hand, a major rupture is hidden behind the fallacious invocation of 'custom', and, on the other, this change has made the perpetuation of the former relationship of domination at least partly possible. While today it can be observed in aristocratic societies that social stratification is being projected on to the landholding structure, as illustrated above, this is sometimes the result of a complex and recent evolution that represents an attempt to perpetuate that stratification. Such an observation obliges us to be very cautious when considering 'traditional' land relationships in the Sahel because, as is the case here, they can be simultaneously the echo of an old social reality and the result of a radical change.

241

Whatever the case may be, and in spite of the sometimes quite recent emergence of the land question, the shadow of unequal social relations is usually a significant marker of the current conditions of land access in aristocratic Sahelian societies. There is a strong contrast between social classes claiming to control the use of land, if only in terms of recent appropriation strategies, and those who supposedly have access only within the framework of a relationship of personal subjugation or in the form of sharecropping. Such a situation generates demands, disputes and conflicts. These conflicts can lead to open clashes when the stakes are high, especially when there is external intervention of some kind – a hydro-agricultural development, for example, which gives the land a value it did not previously possess. In eastern Senegal, the creation of irrigated perimeters along the river enabled certain social classes to attempt to gain access to kinds of land from which they had previously been excluded (Weigel, 1982). Elsewhere, notably on the river bank in the Gorgol area of Mauritania, it is the dominant classes within the Moor and Toucouleur societies who are endeavouring to use their social position to seize the developed lands (Grayzel, 1988).

The superimposition of land authorities

In the former great Sahelian states, the land situation was generally complex and ambiguous; to a certain extent it preserved traces of previous forms of social organisation which were laid down in the process of state formation. Consequently, the lineage structure often continued to provide the basic framework that defined the ordering of relations to land. Among the Mossi or the Hausa, for example, opposition between the original occupants, the masters of the land, and the immigrants possessing political and military power constituted a founding paradigm from which the structure of the overall society was conceived (Nicolas, 1975; Izard, 1985). Within this general framework, many peasant communities organised themselves according to a model that had features in common with that of segmentary societies, where, as we have seen, those who originally cleared the land were the guardians of that land, and played the role of intermediaries with the deities of the forest and fields.

In these societies, however, the integration of communities into larger superstructures has subjected the organisation of the land relationship to distorting factors, rendering it significantly more diversified and complex than in the preceding cases. In fact, superimposed on the somewhat private relationship between peasant lineages and their land was the sovereign's public guardianship over the entire domain under his control, in accordance with an initial 'social contract' agreed between his ancestors and the original occupants.

Thus, in Hausa society, the sovereign was seen as the husband of the land,

242

the guarantor of its fertility (Nicolas, 1975). As such, he collected a fee when land was transferred through inheritance. He also had the right to oversee how land was used and could regulate its exploitation (see the example in Chapter 9, n. 4, p. 234, concerning the ban on the felling of *Acacia albidas* in the Sultanate of Zinder). Political organisation was projected on to the land by dividing it into a series of increasingly smaller units ranging from the land of the state in its entirety, to the village, via the intermediate levels of local chieftaincies. At each step, he who held power exercised his authority over a portion of land. Therefore, all individuals wishing to establish themselves in an unoccupied area to dig a well and lay the foundations of a new village had first to solicit the agreement of the political chief in charge of the desired area. Once permission was gained, the individual then had to perform propitiatory rites in order to obtain the benevolence of the divinities in the area.

In such a system, two legitimising powers constantly rubbed shoulders: one of a political nature, emanating from the sovereign, and the other, of a religious nature, originating in the link established with the occult masters of the land. While the chief of a Hausa village was usually a member of the founding lineage, he nevertheless derived his authority from the investiture of the chief of the territory upon which he depended. As such, he exercised certain functions of control and regulation which in other societies devolve upon the land chiefs. He could, for example, authorise a newcomer to clear a field in unoccupied lands within the village territory. He also had the right to decree, and to ensure respect for, common rules regarding the use of land and its resources (especially with respect to the circulation of livestock on cultivated lands). Originally, the relationship to the supernatural was not so much driven out, as dissociated from political power. It assumed an essentially ritualistic dimension in the form of periodic sacrifices, per-formed by specialised lineages (hunters, farmers), in order to open up the bush to human activity or to reassure the fertility of already cultivated areas. With the penetration of Islam, long-established in some areas, this dimen-sion gradually faded in favour of the sole power of administration and arbitration held by the village chief.[4]

In Mossi society, the institution of land chief existed, but its form was also much more complex than that observed in lineage-based societies. In a hierarchical manner, one to another, this institution presided over territ-ories that corresponded to units of command for the populations existing prior to the centralisation of power in the hands of Mossi immigrants. These territories became settlement and command units within a political struc-ture established by the Mossi, and the two boundaries did not always perfectly coincide (Marchal, 1983).

In these diversified social formations, where the political and religious dimensions of territoriality are in opposition, and where different levels of guardianship are found side by side, the complexity of the land structure is further accentuated by inegalitarian divides which, while not as rigid as those in aristocratic societies, affect social relations no less profoundly. Thus, belonging to the structure of the chieftaincy or being a notable usually went hand in hand with the possession of vast agricultural domains. After the colonial conquest, the servile labour force that was previously used to exploit those lands gave way to labour obligations furnished by subjects, by dependants and eventually by agricultural labourers.

In this type of society, as with those mentioned above, the land question arises only when land becomes scarce. Here, too, for a long time power over humans overrode the appropriation of land – all the more so in this case because a unified political structure, bolstered by a solid administrative apparatus, functioned as a powerful means to mobilise men and their labour power. As everywhere else, land has today become a major issue, but it seems to have evolved in a markedly different manner than in the other social systems. Starting from an initial situation characterised by a territorial structure divided into multiple levels, and by a duality between the power of a political apparatus and the religion-based legitimacy of the first lineages to occupy the land, this evolution created a complex and diversified land tenure situation. Traditional chieftaincies were weakened as a result of colonial conquest and the constitution of post-colonial states, while the structural and religious frameworks of the lineage-based aspects of these societies were progressively eroded (not least because of the advance of Islam). These changes created conditions for the breakdown of ancient rights, favouring the individual appropriation of land. In itself this phenomenon is obviously not specific to societies organised as states, because it is a trend wherever land has become scarce, but it has found fertile terrain in the complexity of their structure and in the potential contradictions created by a superimposition of different levels of control over land. This evolution is very clear in Nigerien Hausa country (Raynaut, 1988a), but can also be observed in a more subtle way in Mossi country (Marchal, 1983).

SYSTEMS OF PRODUCTION AND LAND SYSTEMS

Up to this point the land problem has been considered solely with regard to agriculture. However, as mentioned earlier, the Sahel is subjected to several forms of exploitation which are practised in dissociated, complementary or competing fashions, thus creating different land situations.

The herders' space

The major distinction here is again between agriculture and pastoralism. In Chapter 5 we saw how both activities shared the land of the Sahel. Three large zones were distinguished: the first where pastoral herding predominates, with virtually no sharing; the second where it is competing strongly with agriculture; and the third where livestock and agriculture have long coexisted, although with growing difficulty as land pressure increases. The sometimes violent conflicts that have arisen today in various areas of the Sahel cannot be understood without considering the specific traits of pastoral land organisation, both from the point of view of those internal contradictions that can be discerned, and from that of the potential confrontation with agricultural systems of exploitation.

In pastoral economies, the relationship with the land follows markedly different principles than in agricultural systems. Access to water represents the most important need:

> In the herder's mind, it is water that is the determining factor, that is foremost, in order for livestock to have access to a given area. Without water, it cannot survive. While fodder sometimes poses crucial problems, the herder feels that, in this domain, he can better cope with a difficult situation. . . .
>
> (Kintz, 1982: 44)

The social definition of the conditions of access to water is, therefore, one of the keys to pastoral land organisation. Wells that are shallow and easy to dig (as is the case in north Burkina) are not the objects of strong social control. However, when they require a large investment in labour and a certified technical specialisation (they can sometimes reach up to 100 m in depth (Bernus, 1989)) their use is subject to restrictions that are sometimes very strict. In pastoral societies, if there exists a clearly identified political authority, it is up to that authority to grant permission, but, if no such authority exists, the established procedure is that of negotiation and alliance between the different potential users. In the case of surface water, it is not appropriated as such, but can be subject to conditions of regulated access (especially in the valleys of the major rivers).

In arid conditions, the watering of livestock is the foremost condition for survival. Therefore the regulation of the use of watering places has long constituted the keystone of pasture management. Because a herd cannot stay long within the proximity of a forbidden well, the individual or group who controls its use also controls the surrounding pasture, without actually having to assert a specific right to the land. The appropriation of a defined area of land was of little interest in this context for, on the contrary, it was mobility which was the only adaptive response to drought and to the exhaustion of pasture land.[5] In these conditions, the most vital right to hold on to, the right upon which the survival of the group and its herd depended, was the possibility to move (Le Bris et al., 1991). This right was sometimes won by force but more often it was acquired through alliance. Here too, the management of social relations ultimately overrode the control over land as a material entity. This form of resource control, which was perfectly efficient for centuries, turned out to be doubly fragile in the face of the social and technological change that the Sahel has undergone in the last few decades.

Within the pastoral zone itself there has been a destabilising effect from the establishment, by state authorities or by organisations working on behalf of the state, of the drilling of public wells with unrestricted access. This freedom of access, combined with the emergence of new kinds of herders who have profited greatly from this hydraulic infrastructure (investors from the dominant political and economic classes, see p. 122) rapidly led to the anarchic use of pasture land. The result has been accelerated degradation. From crisis to crisis, triggered by droughts, countless traditional pastoralists lost their livestock, thus giving way to new public and private users, whose presence they could not contest because they had never exercised any rights over the land.

In addition to this relatively recent change, we must add the older threat posed by pressure from farmers on the southern fringe of pastoral land. Agriculturalists largely escape the regulation of land use by means of control over water because, not having large herds to water, they are much less dependent on it. They have, therefore, been able to adopt a 'nibbling' strategy which started with the establishment of temporary farming hamlets during the rainy season but, once land use rights were progressively established, led to the digging of a well and eventually permanent settlement. A 'zone of competition' was thus created where, today, two rival modes of resource exploitation are in conflict, as well as two land systems organised according to different principles.

Further south, where cohabitation has long existed, agriculture often dominates. It is, therefore, on the basis of agriculture's needs that ascendancy over land is ordered, with herders and agropastoralists fitting in between the spaces controlled by the peasant communities (Figure 5.1, p. 117). The cohabitation of two such different forms of natural resource exploitation is conducive to potential conflict. Access to water-holes, the use of fallow land as pasture and the consumption of crop residues left in the fields are all factors that cause numerous problems to be resolved, as already demonstrated (Chapter 5). Generally, the free circulation of animals, a necessary condition for extensive herding, must be assured without at the same time threatening crops. A lasting reconciliation of interests can only come about on the basis of negotiated agreements. These can be made at several levels, that is with individuals or small family groups, as is generally the case with manuring contracts, or with larger collectivities (village communities and pastoral groups) in the case of access to watering holes and fallow lands. State-based societies could formerly decree a regulation of general significance that applied to all partners involved and could establish arbitrating authorities that would intervene in the case of conflict. The great Hausa states, for example, included among their dignitaries a 'Fulani chief' and a 'Tuareg chief' responsible for managing relations with these two ethnic communities (Nicolas, 1975). The Dina, created by Shekhou Amadou in order to achieve a concerted use of the resources of the Central Niger Delta, addressed similar objectives, although in this case according to the priority interests of the Fulani herders (Gallais, 1975). In segmentary societies and small village aristocracies, the coexistence of farmers and herders gave rise to more unsettled relations, with episodes of confrontation, negotiation and mutual avoidance.

Circumstances today are very different. With the extreme scarcity of land in this most densely populated part of the Sahel the old rules of coexistence between shepherds and farmers are increasingly difficult to apply. Competition often degenerates into conflict, which is all the more difficult to arbitrate since those within traditional society who fulfilled this function, i.e., land chiefs and political or religious leaders, have been deprived of most of their authority. For a long time the modern state, through its local representatives, fulfilled this role (thus limiting the scale of confrontations), but the current weakening of the state has naturally been accompanied by a flare-up of often violent confrontations over land. The example of Niger is particularly illustrative in this respect. In fact, farmers and pastoralists have opposed one another for a long time, both in the zone of competition and in the south of the country, when herds come through by trans-humance or *en route* to Nigerian markets. These occasions have always caused local conflicts to erupt. Yet at the first sign of violence, they were quickly resolved by the administrative and judicial authorities. Arbitration was not always neutral and often depended on the political power of the parties present. Nevertheless, more serious situations were avoided. In recent years, however, numerous violent clashes have occurred.

Overlapping uses

While the evidence clearly demonstrates a contrast between pastoralism and agriculture, Sahelian systems of production are also divided along many other lines that are echoed in the modes of social control over resources. Thus, even within agricultural societies the relationship to water often constitutes an important criterion for differentiation, especially with regard to land rights. The example of riverside agricultural communities along the Senegal River (Toucouleur and Soninké) is very relevant in this respect. The strict appropriation of land liable to flooding (the flooded basins (*walo*) and the levées on the river banks (*falo*)) contrasts with the free access to interior dune lands (*dieri*) cultivable only with rainfall (Bradley *et al.*, 1977; Lericollais and Diallo, 1980; Steinkamp-Ferrier, 1983). In the Hausa country of Niger, the division takes other forms, bringing into opposition irrigated gardens, which have long been under private individual appropriation, and fields dependent on rainwater, which until recently were under a more collective type of control (Raynaut, 1989b; Yamba, 1993). In actual fact, all techniques involving the continuous use and development of the land tend to create a distinctive land status. This is particularly the case with manured lands which are cultivated in a quasi-permanent fashion and in which, regardless of the existing land system, the form of relation is often closer to that of appropriation than that of use pure and simple.

While the agricultural and pastoral production systems that currently predominate in the Sahel set the broad lines of control over space and the sharing of resources, other more specialised means of exploiting nature are employed for particular resources. These, too, are socially managed, according to their own code of appropriation. This is particularly the case with hunting, fishing, gathering and wood collection. In some situations, these rights are exercised over a specific ecosystem and, therefore, interfere very little with other systems of appropriation and exploitation. They are, therefore, organised entirely according to their own principles. This is the case, in particular, for artisanal fishing in the Central Niger Delta, where the means of social control serve a dual purpose, that of the long-term reproduction of the fisheries as a resource, and the organisation of its appropriation in a way that is compatible with the reproduction of Bozo society. Fay (1989) describes the organisation of this control and shows how it functions simultaneously through a division of resources into fishing territories allocated to distinct lineages, and through a decree of general prohibitions that regulate the means of capture by limiting both the techniques used and the size of the fish caught.

Several specialised means of appropriation specific to distinct resources usually overlap in the same area. A classic example, described by Schmitz (1986), has already been cited above (see Chapter 5, p. 115) and consists of the rotation of agriculture, herding and fishing in the seasonally flooded basins of the Senegal River. Gathering and hunting are also generally practised on land already under other forms of exploitation by herding and agriculture.

It is with wood collection that the coexistence of several forms of harvesting and numerous rules of appropriation pose the most acute problems. With the increase in demand for firewood and wood for construction, especially in areas close to large urban centres, the exploitation of the forest cover is becoming so intense that it is threatening the existence of the remaining forests. As Yamba (1993) clearly demonstrated in his recent work on Niger, the problem has been aggravated by the frequent discrepancy between traditional rules concerning land and those applied to self-regenerating vegetation. In fact, while land today comes under private control (the result of an evolution that we will return to later), this is not always the case for trees and bushes that grow naturally. These are generally considered to be common goods accessible to any individual who wishes to use them, even to a stranger to the local community upon whose territory they are growing. Village bush reserves are consequently being decimated by the axes of professional woodcutters. Residents often discover, without having any means of protest, that a tree on their property has been felled. This crisis is only resolved once the legal status of the tree is changed and its appropriation is linked to that of the land upon which it stands. When this occurs the effects on the regeneration of the vegetation cover can sometimes be striking (Yamba, 1993).

The superimposition and overlapping of rights of different natures and different levels on the same area is unquestionably the most common situation across the whole of the Sahel. This state of affairs can be the result of a political history in which there was a succession of guardianship rights, without one form ever fully replacing the others as was the case, for example, in Mossi society. It can also be the outcome of the coexistence of different communities, or of distinct social categories within the same community, where each exploits a different stratum of the ecosystem. This superimposition of uses and rights, be they recent or old, continued without any major obstruction as long as there was an abundance of resources which served to keep potential conflict at bay. While conflict was always nascent, it was limited and appropriately regulated by the existing mechanisms of arbitration. Therefore, it did not threaten the very principle of cohabitation. Today, the character of the problem has fundamentally changed, and with ecosystems now exploited to the limit of their reproductive capacity, maintaining the old conventions of use is no longer possible. New forms of social compromise, often supported by appropriate technical solutions, will have to be found so as to permit, once again, a management in common of land and its resources.

THE DYNAMIC OF LAND CHANGE

The title of this chapter, 'The birth of the land question', was chosen deliberately in order to show from the outset that land problems should be addressed from a dynamic perspective. Indeed, as well as the different local situations, which as we have just seen reflect the diversity of both social and technical systems, a number of general trends towards change are currently occurring across the Sahel. While their effects are modified according to the context within which they occur, they nevertheless contribute to a refashioning of the relationship to land, and more generally to nature, according to entirely new principles.

Major factors of transformation

A number of major factors of change can be identified. They are present everywhere, although to varying degrees of intensity. While they have already been cited a number of times in the above analyses, they are briefly recounted here.

1 *An increased scarcity of land and natural resources* as a result of an increase in population density. The demographic question was examined closely in the first three chapters of this book. Here we briefly revisit the issue. We have witnessed strong demographic growth throughout the Sahel over the last few decades, although this should not mask the existence of large disparities in the distribution and dynamics of settlement. Further-more, we have noted that, far from being primary, demographic con-straints were themselves the result of a combination of historical, econ-omic and natural circumstances (Chapter 2). It is necessary, therefore, to treat the demographic argument with care when it is used to explain the situation in the Sahel. As Thornton (1992) very rightly emphasised, the importance a society attributes to the possession of land does not depend simply on the amount of land available. We therefore reaffirm the fact that we must tread carefully if we wish to use the demographic argument to account for the situation in the Sahel. It is no less true, however, that wherever there is high concentration of population, factors related to land are profoundly disrupted because the land and its resources, which were once abundant, have become scarce and thus become the object of competition that previously did not exist.[6] Faced with new imposed constraints, the technical systems of production also had to evolve. The period of fallow has been shortened and maintaining soil fertility demands an investment of work or money. Relations with the land are, therefore, more permanent and more intense. The demo-graphic variable is certainly an important factor which influences the dynamic of land tenure systems. It must, therefore, be recalled that the duration and intensity of land pressure is far from uniform across the Sahel. While certain areas, from as far back as available records go and up to the present, were never exploited other than in a very haphazard way, other land has long endured a high level of agricultural exploitation – a particular example is the Mossi plateau (Marchal, 1983: 405). These historical situations must be taken into account when analysing present realities.

2 *The impact of the money economy* is another essential factor in the con-temporary reality of the Sahel. Today, a peasant or pastoral economy that can function without money no longer exists, even if certain sectors continue to follow principles other than those of a market economy. The consequences are many. Two are explored below.

(a) *The pursuit of money* is becoming a major determinant in the demand for plant and animal production, thus weighing heavily on the demand for land and the exploitation of resources. This point is well illustrated by the role the commercial circulation of wood plays in the ex-acerbation of competition between agropastoral production and the exploitation of woody plants, or by the rapid expansion of commercial cropping. The consequences are those described above in relation to the intensity of land use.

(b) *The possibility open to all to enter the market* and to negotiate individually the product of their labour profoundly transforms the objectives and the scale of values upon which systems of production are organised. Economic strategies that were previously collective and subject above all to the imperatives of social reproduction, are now tending towards diversity and autonomy. They introduce competition and rivalry to the appropriation of natural resources. The preceding chapter touched on the resistance by the Sérer social system to the introduction of market relations. Similarly, the Sénoufo of Mali have long been reluctant to use modern money (Diabaté, 1986: 300). These examples provide counter-proof of the threat of the market economy to the operational bases of such societies. It is clear that the effects of monetarisation are most striking where commercial speculation has been introduced on a massive scale (the groundnut basins of Senegal and Niger, Mali's cotton basin). However, this phenomenon occurs well beyond these specific geographic zones, for in sectors that fall outside these major 'development' schemes money is being intro- duced into local economies by way of migration (with significant consequences with respect to land).[7]

3 *Social and cultural changes* are also areas relevant to the evolution of land relations in Sahelian societies. These changes have been profound and multiple, and much research remains to be done on this subject. Three aspects will be examined here that have had a particularly striking impact on land organisation.

(a) *The disintegration of lineage-based structures* is an element common to all Sahelian social systems. Today, the link between the individual and the lineage is weakening, although to differing degrees depending on the particular situation; the process is generally slower in segmentary societies where kinship structures constituted the pillar of all social organisation. The subjugation of the young people to their elders is now tending to be relaxed and women's economic margin for manoeuvre is widening. Even when the individual maintains strong ties of solidarity with the family group to which he or she belongs by birth or marriage, he or she gains some autonomy, which is expressed not only in social and economic life (consumer practices, migration, matrimonial choices) but also in the relationship to land. The large lineage-based territories of an earlier time are breaking up because young people and women are increasingly taking an active role in agricultural life (Monimart, 1989). This change may be gradual and can be concealed behind an apparent preservation of family unity. This is the case with farms among the Sénoufo of southern Mali, where

251

there is a progressive shrinkage of collective family fields (under the guardianship of the elder of the lineage segment) in favour of the individual fields of the younger members. Eventually, the size of the collective domain is reduced to the point when the elder is no longer able to meet his social obligations (feed his dependants, pay their taxes, etc.), thus representing the final break-up of the farm (Diabaté, 1986). To varying degrees, this parcelling of land, which is a consequence of the disintegration of extended kinship structures, is a trend noted in the majority of studies on Sahelian agricultural societies (see especially Marchal, 1987).

(b) *The decline of agrarian religions* which sustained the representation of the relationship to land. As stated earlier, the alliance with the deities of the land was the very foundation of the land chiefs' power. Once faith in these divinities faded, to remain only as beliefs disengaged from the system of myths and rites that gave them their coherence, the operational principles of the land system crumbled because the symbolic barriers that ensured their respect had collapsed. Hence, as Marchal (1983: 358) has noted, among the Mossi, the land chiefs have lost their effectiveness in their religious functions. This change would have been more rapid in state and aristocratic societies, which have long been open to trans-Saharan influences and where Islam has long been present, even where it was only practised within certain social classes (merchants in particular). Niger's Hausa societies offer a particularly illustrative example. In the past they were characterised by a dualism between the universe of the chieftaincy and commerce, influenced by its old contacts with Islam, and that of the farmers and hunters whose entire world vision was organised around lineage cults (Nicolas, 1975). Very rapidly, that is within the last 30 years, Islam has made spectacular progress in these societies, to the point that it is becoming the dominant ideological and ethical model. By providing a much larger place for personal responsibility, it offers a conceptual framework more in accord with the current evolution of social relations than that offered by lineage religions. Thus, by relegating agrarian divinities to a secondary role, and by relieving land use of the necessity of an alliance with them, Islam fosters the emergence of both the concept of appropriation and the individualisation of land relations. Old beliefs survive only as relics cut off from their context and can no longer provide any meaning in today's world (Raynaut, 1984). In relation to Mossi society, Marchal also notes that Islam has favoured: 'The change from a right of usufruct of a social nature (itself founded on an essentially religious law) to a right of cultivation of an economic nature, along the lines of "the land belongs to him who works it"' (Marchal, 1983: 335).

In segmentary societies, resistance by agrarian cults is much stronger, although even they are tending to weaken, particularly under the effect of migration which is taking the youth off to the cities where they become familiar with other values and ways of thinking. There have, for example, been many conversions to Christianity and Islam among the Sérer (Gastellu, 1981).

(c) *The emergence of external political and economic structures* – first colonial power, followed by the national state. These new power structures, through their domination of communities under their authority, have significantly influenced these communities' relations with land and natural resources. The weakening of traditional political authorities or the reorganisation of decision-making structures have had crucial repercussions. In centralised or aristocratic societies the former chieftaincies, cut off from a large part of their power and often deprived of their legitimacy by the arbitrary actions of a new power which made and unmade title holders as it pleased, were no longer able to play fully their previous role of arbitrator and regulator (a role since usurped by representatives of the administration, albeit according to different criteria). In segmentary societies, on the other hand, the imposition on communities of village chieftaincies, where previously decisions were usually the outcome of a consensus between lineages, favoured the emergence of new notables and sowed the seeds of conflict over power and the appropriation of natural resources.

More generally, but still with respect to land change, the application of new legal principles to relations between humans and to their relations with land was a major change introduced by the colonial power. This heritage was generally maintained by the independent national governments that followed. Thus, as we have shown above, the abolition of bonds of servitude, coupled with the introduction of the notion of landed property, brought the Songhay–Zarma nobles to substitute power over land for the power they previously exercised over men and, thereby, passing from a master/slave relation to one that binds landowner and tenant (Olivier de Sardan, 1984). Of course, the government's recognition of these new legal principles did not immediately overturn the peasant land systems. The situation was still far from the kind of land reforms undertaken by the British colonial administration in Kenya. However, by authorising the administrative registration of private landownership, recognition of these principles allowed strategies of individual land appropriation to unfold. Although these appropriations are still small in number, they are nevertheless not without significance.

In a less direct way, externally imposed legislation occasionally interfered with traditional land rules resulting in new practices. Thus, according to Gastellu (1981: 194), in Sérer country, the maximum period of ten years, at the end of which the fee paid by the user to the land chief had to be renewed, was linked to the recognition of a colonial decree through which the continuous use of a piece of land for ten years opened up the right to permanent use. Sooner or later after independence, the new states enacted measures designed to affirm their pre-eminent guardianship over national land, for example, by nationalising almost the entire territory, as in the case of Senegal in 1964 and Burkina Faso in 1984, or by reserving the possibility of doing so in a selective manner in the event of development works, as in Mauritania and Niger.[8] Here too, this legislation generally had only a limited practical effect. However, by giving land access to social groups previously deprived of land rights (as with the hydro-agricultural developments on the Senegal (Weigel, 1982)) it had great local significance. It often also led to a suspicion of the government's intentions when it initiated land development projects. The government's involvement in reforestation or anti-erosion schemes are frequently interpreted by peasants as a prelude to a seizure of their lands (Yamba, 1993).

There is a final area in which the superimposition of the nation state on to old Sahelian societies has played a determining role: the appearance of social groups vested with new powers issuing from their links with the politico-administrative apparatus or their active participation in commercial speculation. These privileged positions were often the point of departure for strategies of accumulation of the means of production. These include land, but only to a limited extent due to the precariousness of agricultural activity in the Sahelo-Sudanese regions if control over water is not assured. Agricultural development schemes are, on the other hand, actively coveted by the privileged classes (Saul, 1988: 254). These strategies primarily involve livestock. The extent to which the emergence of investment stock-breeders has disrupted the use of land and pasture across the entire Sahelian pastoral zone has been considered above (Chapter 5).

The axes of change

The growing scarcity of land, the monetarisation of economies, the transformation of social relations and of the collective representation of nature – these are the three areas where the major causes of the profound change in Sahelian land systems take place, a change which, as was stated above, has today led to a real redistribution of the land 'cards'.

What are the trends along which the relation to land is changing, under the combined effects of these factors? We examine here three axes of change.

The advance of the concept of land appropriation

This leads gradually to ideologies of privatisation and individualisation. Within an historical context marked concurrently by competition for the control of scarce resources, by the weakening of social and political institutions that were the cement of communities and by the disintegration of belief systems that were the foundation of the community's relationship to nature, land becomes a good over which those who exploit it tend to affirm gradually a personal and exclusive right. As gradual as it is, this change leads to a major rupture. The regulation of resource use essentially based on the needs for social reproduction of the community (it was one's social position, a master's favour or an alliance that opened up the right to work the land) is replaced by an instrumental regulation defined, above all, according to the private agenda of the users/possessors, as a function of their individual needs as well as of requirements linked to the accomplishment of technical undertakings. In the final analysis, the privatisation of land rights leads to their individualisation. This change is furthered by the disintegration of production units and by their frequent reduction to a limited family size (see Chapter 11). Formerly managers of a collective patrimony founded on co-operation with other adults of an extended family cell, the heads of farms have now become the sole masters of the land which they cultivate with their children and, sometimes, with their wife or wives. Furthermore, the relative emancipation of those who traditionally occupied a position of dependency with respect to the chiefs constitutes an additional element of individualisation. There are even situations, such as in Hausa society, where women and youths can possess fields in their own name.

In the following chapter we shall return to the changes that have brought about this dismembering of family production structures. Here though we note that it goes hand-in-hand with the segmentation of land reserves. In this process, the modification of the rules governing intra-family transmission of land was an essential step. In the lineage-based system, the land of a domestic unit generally remained intact, placed under the authority of the individual to whom guardianship had been entrusted. Birthright and primogeniture (see note 2, p. 261) were the two principles by which rights of succession were determined. Now, the maintenance of the community is less and less the primary concern and, usually, when the head of the family dies the land is distributed among the beneficiaries, i.e., younger brothers when it is an older brother who dies, the son when it is the father (patrilineal system) and nephews in the case of an uncle (matrilineal system). The advance of Islam, accompanied by a decline in the worship of agrarian divinities, brings with it new notions of self and of individual rights, and has no doubt favoured these changes in the rules of transmission. However, it has certainly not been the direct cause, because this division of land is ultimately one manifestation of a widespread change in kinship relations, of which the main feature is a growing emancipation of the individual and of the power of his labour (see Chapter 11).

There are a number of consequences of this new organisation of land relations. First, it leads to an increase in potential contradictions between private interests and those of the community, with the emergence here and there of new regulations aimed at mitigating them. In the Maradi region in Niger, for example, a concerted defence of the last remaining bush reserves is being proclaimed by adjacent village authorities (Raynaut et al., 1988). In addition, under this new organisation, the possession of land gradually comes to include an exclusive right to all the resources within it (including natural vegetation in particular), in contradiction with the traditional overlapping rights of use on the same space. Such change is illustrated in the expulsion of pastoralists by agriculturalists (Mossi plateau) and by the appropriation of trees by field owners (Maradi). It is clear that these transformations do not occur as sudden changes. They are a slow and discontinuous development, punctuated with conflicts and resistance. They do not progress in the same rhythm everywhere, and while parcelisation and individual appropriation is very advanced in Mossi country (Marchal, 1983) or in Hausa country (Raynaut, 1988a), it is much slower in lineage-based systems such as those of the Gouin (Dacher, 1984). In intermediate situations, the strong pressure for change which often results from the introduction of commercial crops is counterbalanced by the capacity of the traditional structures and values to resist, as in Sénoufo–Minianka societies (Diabaté, 1986; Jonckers, 1987; Sanogo, 1989), and among the Sérer (Gastellu, 1981). Tightly stratified aristocracies present a third case, where the appropriation of land generally occurs within the framework of old relations of domination and reproduces the exclusion formerly experienced by the servile classes (Pollet and Winter, 1972; Steinkamp-Ferrier, 1983; Olivier de Sardan, 1984).

The entry of land into the market

In order for this phenomenon to occur there must be a convergence of three conditions on all or part of the land. These conditions are: (a) that land pressure has become such that demand can no longer be satisfied through traditional means of inheritance, loan or clearing; (b) that land rights have become individualised at the level of a limited production unit, or even at the level of the individual; and (c) that land has become a tool of market production through the cultivation of one of the major export crops, or through market gardening or the cultivation of fruit trees.

This convergence has reached a pinnacle in Nigerien Hausa country, because of a variety of reasons related to demography, to the long development of groundnut production and to the properties of the local social system. Precise data from fieldwork demonstrates that the purchase and sale of agricultural land is a common practice today and will in the future be a determining factor in peasant land dynamics (Nicolas and Mainet, 1964; Raynaut, 1988a; Yamba, 1993).[9] Meanwhile, on the basis of the available literature, it does not appear that there is a comparable situation elsewhere in the Sahel. The sale of land in the fullest sense of the term, that is the definitive alienation of all rights to a piece of land in return for a sum of money, remains a limited practice. Olivier de Sardan (1984) has noted that among the Songhay–Zarma of Niger it is an ancient practice in the rice-producing areas on the river banks, but it is an exceptional and very recent development in rain-fed agricultural sectors. It also exists sporadically in Burkina Faso, particularly in fields and orchards located in proximity to urban centres (Saul, 1988). Although these are limited observations they nevertheless indicate a change in progress. They show that the entry of land into the market does not occur as an abrupt upheaval, but rather as a progressive whittling away which can begin in particular land sectors. This process, therefore, gradually overcomes the resistance of residual social values which, even long after the disappearance of religious beliefs and the forms of social organisation engendered, continue to proclaim the inalienable character of the land. Moreover, this extension of market relations to land can take forms that are less abrupt than a straightforward sale. In Sérer society, for example, as described by Gastellu (1981), the monetarisation of the fees periodically paid to land chiefs has provoked a shift in collective ways of thinking. There is now some ambiguity, with those who make a payment perceiving it as a purchase, while the land chiefs still see it as a rental.

While as yet it is still discreet in its emergence, the entry of land into the market can nevertheless be considered to be in progress in most agricultural societies in the Sahel. Mathieu (1987) shows that where land has become an object of market production, if not a marketable object itself, the conditions exist for a 'managed transition' to capitalist agriculture. This change is slow and occurs in spurts, but in a determining way it contributes to the development of future relations between peasant communities and nature.

Land accumulation

The inequality of access to land is not in itself a new phenomenon in the Sahel. Today, however, it assumes a renewed form. In the past, it was an accurate reflection of social hierarchies and, in the last analysis, of a more fundamental reality, that of the control exercised over humans through relations of kinship and alliances, relations of subservience and political domination. Once land became coveted and appropriated due to its increased scarcity, it became subject to strategies of accumulation and now represents a source of investment. As was shown in Chapter 6, an increase in cultivated hectarage constitutes the most common response by farmers to the worsening of climatic conditions, as well as to the need to make technical equipment pay for itself as rapidly as possible. Consequently, control over land has become a major economic issue resulting in a 'race for land', which is all the more fierce and contested when land pressure is high. The movement towards land concentration in the Sahel is widespread, whether it be in the Senegalese Groundnut Basin (Gastellu, 1981; Lericollais, 1990), on the banks of the Senegal River (Weigel, 1982; Steinkamp-Ferrier, 1983), in the cotton zone of Mali (Bosc *et al.*, 1990; Raynaut, 1991a), on the Burkinabé Mossi plateau (Marchal, 1983; Saul, 1988) or in the dense agricultural fringe in Niger (Raynaut, 1988a).

This occurs at two levels, not only as an internal redistribution within peasant communities, but also as a seizure of land by external actors (merchants, bureaucrats, political officials) who generally reside in urban areas (Saul (1988) particularly stresses the latter phenomenon). This accumulation can take several routes. It can rely on the existence of pre-eminent social positions within the traditional social system. The most obvious example is that of strongly inegalitarian societies where former privileges of status are transformed into advantages over land. Somewhat analogous strategies can also be observed when dominant positions within a lineage-based structure are taken over to create advantages over land (Marchal, 1983: 357; Sanogo, 1989: 243ff.).

As long as the occupation of land is not total, however, the clearing of bush areas that have not been appropriated remains the most common means of accumulating land for the entire zone. This practice often takes advantage of the relaxation of coercive or symbolic controls formerly exercised over land by political chiefs or land chiefs. Occasionally, national legislation has added further confusion by nullifying traditional land rules, as in Senegal, or by decreeing that land belongs to those who develop it, as in Niger. The possibility of validating new rights over land before modern courts of law has resulted, in some areas, in veritable strategies of conquest, practised both by peasants and by wealthy urban dwellers, with the latter using their financial capacity in order to recruit wage labour, or their privileged access to power tools (Raynaut, 1988a).

In this accelerated race for agricultural land, which Marchal (1983) has termed 'land inflation', the principal asset everywhere has been the quantity and efficiency of available labour. Production units capable of mobilising many workers,[10] and especially those with access to animal-drawn agricultural equipment, have been most able to increase their landed property, as long as reserves of bush land have not disappeared. Once the land is fully occupied, and where money has permitted access to land, purchase (or a similar process such as a secured debt or mortgage) now forms part of land accumulation strategies. This phenomenon is best documented in Nigerien

Hausa country, for both rural and urban dwellers (Raynaut, 1988a; Yamba, 1993), although similar practices are reported locally in Burkina Faso (Saul, 1988).

The current trend towards land accumulation, fed through various channels, is certainly one of the major characteristics of the new profile of Sahelian peasant societies. Of course, the importance of internal adjustment mechanisms through the non-market circulation of land should not be underestimated (donations, loans) as these help to mitigate, here and there, disparities that are too extreme. As such, Marchal (1983) has noted that in Yatenga loans are involved in 35 per cent of the land. A similar phenomenon, although of lesser importance, has been observed in Niger (Raynaut, 1980b; Luxereau, 1987).

Of course, in contrast to other parts of the world, there are very few landless peasants in the Sahel. Nevertheless, it is no less true that unequal access to land has today become a major discriminating factor among farmers in this zone. It is clear that not everyone participates in the same way in the land inflation to which 'desertification' has been largely attributed. The various actors do not weigh equally heavily on the natural environment. This is due, of course, to differences in the surface area they exploit, but it is also because these disparities in landholding are often accompanied by marked differences in technical practices. On this point, the facts are very complex and require more meticulous study than that which already exists on the subject. Very generally, it is known that it is through agricultural extensification strategies that land accumulation occurs and, consequently, it is often the cause of environmental degradation (on this point, see pp. 152ff.). On the other hand, the technical options taken by those situated at the bottom of the land ladder can vary considerably. Some practise agriculture on their small plots with little care for using its resources sparingly, be it the fertility of the soil or its vegetation. Their principal objective is to reduce as much as possible the time required to farm their own land, in order either to be able to leave earlier for migration, or to be employed by their neighbours as agricultural labourers (Raynaut et al., 1988). Others, in contrast, put all their energy into farming and it is here that efforts of intensification based on fertilisation and better preservation of the vegetation first emerge (on this point see Yamba's (1993) observations for the region of Maradi, in Niger). The evidence, although incomplete, shows that it is not possible to look at the impact of peasant practices on the Sahelian environment without taking into account current socio-economic dynamics and, in particular, the uneven relationship becoming established today with regard to land, which as we have seen is itself becoming a scarce commodity.

CONCLUSION

The emergence of the land question, as we have shown in these pages, is the result of profound changes in Sahelian societies – considered both in their internal functioning and in their relations with the environment they exploit. While there are certain fundamental tendencies at work every-where, their dynamic has taken various forms, depending on the particular geographical setting, and has not assumed an identical rhythm in all cases. This heterogeneity in local situations is due as much to the original diversity of land systems as to the variability of the agents of change that have acted upon them. While the intensity of the exploitation of natural resources can be attributed to objective factors (demographic, economic, technical) the immaterial dimension of society/nature relations is essential to an under-standing of current environmental changes. The modes of control which Sahelian societies previously employed to ensure both their permanence as social systems and the preservation of the environment from which they derived their livelihood, have broken up, giving way to rivalry between competing users. These users can be of different origins. They may be members of the same community but belong to different social classes with different interests, or they may come from social groups which, up to that point, had shared the same space but exploited its various resources differently, or finally it may be individuals external to the rural milieu who, by exploiting their political or economic privileges, seek to enlarge their range of activities to include agriculture, herding or forest exploitation. It is often the state, through legislative action or through its coercive powers, that seeks to establish itself as the new regulator of natural resource use, but it has generally shown itself to be inefficient in this role, as demonstrated by Yamba (1993) with respect to Nigerien forestry policies. This is due to several factors: the inadequacy of the conceptual frameworks that guide its action (marked most often by the pursuit of strictly technical objectives), the weakness of control mechanisms available on the ground and often an absence of neutrality with respect to the parties involved, some of whom are particularly close to the state due to their ethnic membership, their economic functions or their political status. At the very most, the presence of strong power structures has for long made it possible to limit the scope of conflicts by repressing them as soon as they threatened the social order.

In a phase of radical mutation, where former systems of land regulation have collapsed without being replaced by new, clear and stable 'rules of the game' concerning the use of natural resources, individualised strategies have developed, guided above all by a concern either to preserve acquired entitlements to land or to gain new land. In the latter case, of course, the actors entering the competition are not all equally armed. As demonstrated above, an understanding of the changes occurring in landholding is essential in order to comprehend current practices in the Sahel with regard to the environment and its resources. These changes, linked to the contemporary history of the societies in this part of the continent, are none the less also inseparable from other major changes that the agropastoral systems of production have simultaneously undergone, especially with respect to the social control of labour and its organisation.

NOTES

1 This statement must be qualified because in some places the valorisation of land and restrictive forms of control resembling appropriation have long existed. In most cases, the relation to water has been involved, such as in recession agriculture and traditional practices of irrigated gardening.
2 The notion of generation rank reflects an individual's position according to the hierarchy of generations; primogeniture relates, in effect, to the chronology of births. In terms of rank, an uncle will always be superior to his nephew, even though he may be younger.
3 Pollet and Winter note the existence in Diahunu, Mali, of land chiefs as distinct from political chiefs, but these individuals nevertheless still belong to noble clans (Pollet and Winter, 1972: 315) and, formerly, only nobles could obtain from them the right to cultivate land (Pollet and Winter, 1972: 321).
4 We see later how the modern state, in its role as arbitrator and authority for land protection, has tried to substitute itself for the traditional political structures.
5 Thus, following the drought of 1973, certain groups of Fulani herders migrated to the borders of the equatorial forest, which enabled them to save their herds.
6 Or existed in a very localised way, both in time and space (especially where water was present to enhance its value).
7 On this subject, see in particular Mahir Saul's analyses of a Bobo village in Burkina Faso (Saul, 1988).
8 On this point see the summary of national land profiles presented by Riddell and Dickerman (1986).
9 These observations were confirmed by those made further to the south in Nigerian Hausa country (Hill, 1972).
10 One must not lose sight, however, of the fact that the privilege of large families is fragile, for while they can assemble enough workers to clear and cultivate large areas, the domain thus created has to be broken up among a much larger number claiming inheritance rights, as was shown in P. Hill's (1972) study of a Hausa village community in northern Nigeria.

11

THE TRANSFORMATION OF SOCIAL RELATIONS AND THE MANAGEMENT OF NATURAL RESOURCES

2 The Emancipation of the Workforce

Claude Raynaut and Philippe Lavigne Delville

As we have just seen, access to land has only recently become a truly limiting factor for agricultural and pastoral production, yet by contrast Sahelian societies have always been preoccupied with their workforce and its management – in other words its production, its social control and its efficient use aided by appropriate techniques. This is, therefore, a particularly relevant theme of analysis for those wishing to understand the workings and evolution of natural resource exploitation in its diversity.

Of course, this subject on its own can spark many long debates. In this chapter we limit ourselves to highlighting several particularly revealing features that help differentiate the organisation of modes of production in Sahelian societies and aid in identifying the changes which these societies are undergoing.

THE DIVERSE FORMS OF CONTROL OVER WORK

In the end, the social control of a workforce amounts to the control of potential workers by including them within relations of personal dependence. These relations lie at the core of the functioning of social systems, and in examining them we confront both what distinguishes them one from another in organisational principles and what inspires the transformations they experience. It is not surprising, therefore, that even a schematic typology of these major social systems constitutes once again a useful matrix for highlighting the diversity of local situations.

Domestic production relations

There is one important caveat to the application of this matrix, however. Regardless of the type of social system envisioned, kinship is the major frame of reference through which the relations of production are organised in a concrete manner. The domestic mode of production is dominant everywhere, therefore, even when other forms of organising production and controlling the workforce are superimposed. Consequently, although the following description of the structure and workings of domestic units of production is particularly relevant for lineage-based societies, it is also valid for all agricultural and agropastoral societies in the Sahelo-Sudanian zone.

In these societies, the concept of an 'ideal' or 'typical' domestic production unit consists of a large group corresponding to a lineage segment, assembling under the tutelage of a father or elder, and including several adult men and their spouses and offspring. The existence in the past of such extended communities is observable in various types of societies, such as the Lobi or Sénoufo–Minianka lineage systems (Savonnet, 1979; Jonckers, 1987; Sanogo, 1989), the aristocratic Songhay–Zarma or Soninké systems (Pollet and Winter, 1972; Olivier de Sardan, 1984), and also the large centralised Mossi and Hausa chieftainships (Raynaut, 1973b; Nicolas, 1975; Marchal, 1983; Izard, 1985). In these last two types of society, and for the predominant lineages, the community consisted not only of the extended family group, but also of the families of clients and dependants, the *griots*, the families of artisans supported by the chief and the domestic slaves.

We must take care not to view too monolithically these original communities which we know about more through the recreated image of an oral tradition rather than through direct observation. In fact, the various functions associated with the economic and social reproduction of the communities (residence, production, consumption) do not necessarily coincide and may be carried out by distinct social groups, as the present morphology of the domestic units shows. Ancey (1975) and Gastellu (1978) have analysed these economic units systematically. The largest is the familial residence unit, formed by individuals living in the same place and linked through kinship or marriage, along with any dependants (slaves or clients). It need have no apparent economic function, yet it often represents a privileged framework of solidarity and mutual aid, essential to the strategic battle against uncertainty.[1] The production unit groups individuals who actually co-operate in agricultural activities, often through participation in one or another of the operations conducted on lands worked in common. In reality, this definition is not easy to apply, since several production units of different levels can coexist within the same domestic group. The one which brings together the most manpower is based on co-operative work in collective fields, cultivated under the supervision of the head of the extended family. Others, smaller in size, are constituted around parcels of land allotted to family sub-units or even individuals, such as households, women helped by their children and young men working alone or in collaboration with younger siblings. More often than not, these different production units, which may include inactive members (children, old people, invalids), also constitute consumption units. Thus, products from a collective field are used to respond to an extended family's particular material and social needs, notably all or part of its food needs; accumulation of the necessary wealth for concluding matrimonial alliances or the accomplishment of certain ceremonies; and investments for the maintenance or replacement of production tools, real property and buildings. In principle, sub-groups and individuals who have access to their own plots benefit from the private use of their harvest.[2]

It is largely through this distinction between the different levels of production units that the allocation of the workforce is organised. The forms of organisation can vary considerably. In Hausa country, for example, the seven-day week is normally divided into four days reserved for the collective fields and three days when individuals work on private plots or the plots of the sub-group to which they are linked (Raynaut 1973b; Nicolas, 1975). The organisation is different with the Soninké from the Mauritanian Guidimaka and the Malian Diahunu (Pollet and Winter, 1972; Bradley *et al.*, 1977). Here, all the men gather in the morning on the collective fields; in the afternoon, and in successive segments of time, each individual owes work service on the private plots of his elders, and receives, in turn, labour for his own field from juniors. Thus an interlocking system of labour duties is created in which the higher one's place in the hierarchy the more labour one receives and the less one is called upon to provide.

In all cases, a lineage's social capital is measured according to the number of dependants, the size of its storehouses and the number of cattle. The community chief, who controls most of the group's production, is also the supreme controller of all the strategies of economic accumulation and of investment in the constitution of a social network. He also controls a major part of the group's physical reproduction (through food) and biological reproduction (through the matrimonial alliances of his younger brothers and dependants). He thus guarantees the existence and the cohesion of the community. By concentrating the surplus produced by the active members of the group in only a few hands, the domestic unit's capacity for accumulation is thus proportionate to its size.

Although the structure we have just described illustrates a model by which domestic production is organised in Sahelo-Sudanian agricultural societies, the specific forms this organisation takes at the local level vary considerably from one situation to another.

1 *The degree to which women participate in agricultural production* is one distinguishing characteristic. Situations can vary widely within this domain. With the Hausa of Niger (Nicolas and Mainet, 1964; Raynaut, 1973b), as with the Mossi of Burkina Faso (Marchal, 1983; Izard, 1985) and the Bambara of Mali (Lewis, n.d.), women participate actively in all phases of agricultural work (except clearing) both on family fields and on their own plots. But this situation is far from universal. The position occupied by

women in agriculture has for a long time remained quite limited through-
out many Sahelo-Sudanian societies of very different types, such as the
Songhay–Zarma of Niger (Olivier de Sardan, 1984), the Soninké of Mauri-
tania and Mali (Pollet and Winter, 1972; Bradley *et al.*, 1977) as well as the
Sénoufo–Minianka of Mali (Sanogo, 1989) and the Lela from Burkina Faso
(Barral, 1968).

As Meillassoux (1982) demonstrated, women play the key role in reproducing the workforce in domestic production. But they can intervene more
directly and with relative autonomy by incorporating their labour into the
production process itself. Clearly, the amount of available labour for
exploiting the environment and its resources increases according to whether
or not women are involved in the production activities of a given group.
The different role assigned to women's agricultural work in different areas
cannot therefore be considered an isolated cultural factor. On the contrary,
it must be linked to the working of the entire agricultural production system
and to the larger social and economic complex in which the system is
inscribed. Although we lack precise information for such an analysis, we
would like to formulate several hypotheses. In certain cases the minor
contribution women made to agriculture in aristocratic societies can be
attributed to the fact that sufficient labour power was previously provided
by slaves. Similarly, according to the model developed by Meillassoux
(1982), it seems logical that women's reproductive function was privileged
in lineage-based societies where men's essential energies were mobilised
around agriculture and their labour appropriated along kinship lines.
Finally, in communities that have long had diverse economic activities,
notably export crafts and commerce, putting women to work can be seen
as a compensation for labour shortages created by men's participation in
non-agricultural activities.[3]

2 *The respective status of collective and personal fields* is another domain
whereby concrete forms of the domestic organisation of agricultural
production can differ, especially with respect to their function within the
domestic economy. While these may appear to fall within the domain of
land tenure issues, in fact what is at stake is the pattern of labour allocation.

In certain cases, as with the example of the Hausa, the division between
responsibilities and charges is inscribed within the structure. It is incumbent
upon the collective domain, through the responsibility of the head of the
family, to take care of a certain number of expenses that are indispensable
not only to the extended family reproduction as a social unit, such as
payment of tribute to the political chief (later the payment of taxes to the
colonial or national administration), marriage payments, ceremonial expenses and so on, but also to the family's sustainability as a production unit.
By withholding a portion of the produce of the collectively worked fields at
the time of harvest, the amount of grain necessary to see the entire extended
family through the three-month growing season was ensured. In contrast,

during the dry season each household, often even each woman, lived almost exclusively from the produce of individual plots. Such a distribution of food responsibilities existed until very recently (Raynaut, 1978) and, like the women's work discussed above, is no doubt attributable to a long-standing economic diversity. Hunting, craft production and trade represented important spheres of activity, essentially practised by men in the dry season, often outside the domestic sphere and on a much more individual basis than agriculture.

In other societies, especially lineage-based ones such as the Sénoufo–Minianka of southern Mali, the needs of the extended family group were met exclusively from the produce of collective fields, which was controlled by the elder. Individual parcels were limited to elderly women who gained rights to them after being freed from all collective obligations (Sanogo, 1989).

Throughout the Sahelo-Sudanian zone, the principle of distinguishing between collective and individual fields has long existed in many agricultural production systems, yet the force that this distinction carries is not everywhere the same. It is even likely that until recently, individual plots were of only minor significance. The fact that they have become so much more important is the result of change which we will consider at a later stage.

Slave labour

While kinship relations represent a privileged social framework which exerts control over the labour force, for a very long time the same was true of slavery. Thornton (1992) clearly showed that this institution represented an essential element in the organisation of African systems of production. However, we must be careful not to make generalisations that might obscure the diversity of the social realities. Sahelian agricultural societies have not all treated slavery in the same way. Indeed, the particular place of slavery has long been a major distinguishing factor among them. In cases such as the Fulani in southern Mali (Amselle, 1990) or certain Soninké groups (Pollet and Winter, 1972), agricultural activities were almost entirely carried out by slave labour.[4] Elsewhere, especially in Hausa country but also with the Bambara (Pollet and Winter, 1972) and probably the Songhay–Zarma described by Olivier de Sardan (1984), slaves played an important, though less exclusive, role in production. They enabled certain members of the dominant class to exploit vast landholdings and represented a supplement to family labour for cultivators of more modest status. In some cases, slaves played only a minor role in agricultural production. This was the situation in many lineage-based societies such as the Sénoufo (Amselle, 1990) and Sérer (Gastellu, 1981).

In analysing the role of slavery in Sahelo-Sudanian societies, it is important to specify to what type of social relations this status corresponded and in what form slave labour was incorporated into the productive process. In this regard, the reality was certainly more complex than the simple term 'slave' suggests, evoking as it does an image of ancient societies and the colonial plantation economy. In nineteenth-century Sahelian societies, the status of slave may have been quite different from that which came to be associated with the human cargo that embarked for Caribbean and American destinations during the same period. In his study of the Songhay–Zarma, Olivier de Sardan (1984) proposes an analysis of captive status that may be generalised to most Sahelian societies who practised slavery (see the description given by Pollet and Winter (1972) for the Soninké and the memoirs of Baba of Karo retranscribed by M. Smith (1954) for the Hausa). He underscores the radical division between two types of captives. The family captives (*horso*), long integrated into the domestic structure, were subjugated to their master through unbreakable bonds,[5] yet were not entirely under their master's will, because *horso* captives could work for themselves and above all could not be sold as merchandise. The autobiographical work of Amadou Hampâte Bâ (1991), who speaks of the Fulani of Mali, illustrates the confidence that could be placed in the family slave under certain circumstances: for instance, being delegated the role of guardian of the master's children after his death. Traded slaves, in contrast, whether acquired through market or war, were nothing more than brute labour force and a readily negotiable commodity: '[The slave] is merely property while the *horso* is a human. The inferior slave is comparable to livestock, while the *horso* is more like kin' (Olivier de Sardan, 1984: 44).

While the generic category 'slave' often existed as such in Sahelian societies (designated by a specific term for the whole category) it nevertheless covers a profound ambiguity by grouping into one concept what appear to be two radically different realities. In the case of family captives, the relationship between master and slave implied a social relation based on reciprocity, where submission was exchanged for protection. In the case of slaves for exchange, we enter the realm of the instrumental, where the person owned was reduced to the level of object, to a tool of production, to merchandise destined for circulation.

Following Olivier de Sardan (1984: 59) we might deduce that two slave structures of different origins coexisted. The older of these corresponds to a 'soft' captivity, where prisoners of war gradually became integrated into existing networks of subjugation, particularly the lineage (but also the political structure, since 'slaves' with important functions could be found, especially in the Mossi and Hausa chieftainships). The other, that developed alongside the advance of Islam and the growth of interregional and intercontinental commerce, saw slaves above all else as choice merchandise. Ultimately, this historical explanation is not entirely satisfactory because the dualism between the domestic slave and the slave as chattel is found in a number of slave-owning systems, throughout the continents and across the centuries. Perhaps this is simply an expression of the difficulty that exists in sustaining the contradiction inherent in a social representation which denies the humanity of human beings. In order to sustain this contradiction, the commercial mobility of the slave would have to be maintained and slaves would have to circulate continuously and be exchanged. As soon as the slave remained long enough for a personal relationship with the master to develop, the fiction of the slave's non-humanity would become increasingly difficult to sustain and other ways of envisioning inferiority and submission would need to be substituted.

If such is the case, as Lovejoy (1986: 264) has emphasised, there is an inherent ambiguity in the very status of the slave. This provides the flexibility that allows for the differing treatment of slaves according to social system and historical circumstance. In systems which do not deny slaves their human status, the acquisition of slaves introduces women, and sometimes men,[6] into lineage structures to correct the demographic imbalances that can disturb their normal functioning. Slave status is, therefore, only transitory and can eventually dissolve into the fabric of kinship relations. At the other extreme, the slave may represent nothing more than an instrument of production and the accumulation of wealth. In this case, the slave is inscribed in a true relation of production under the slave system. These two 'solutions' are no doubt inherent in the status of slave and will be realised in different ways, depending on the circumstances and with, of course, an entire range of intermediary possibilities.

Historically, despite Thornton's (1992) reservations, it is probably the case that the astounding advance of the slave trade started in the seventeenth century (along with the development of an industry specialising in the taking of captives). The demise of the Caribbean and American trade are major events that had considerable effects on the function of slavery in Sahelo-Sudanian societies and led to its reorientation. The division of roles among them between predator and prey was consolidated during the period of heavy external demand (Bazin, 1975; Héritier, 1975), with intermediaries like the Soninké ensuring the transport of the slave caravans to the trading posts. It is an oversimplification to polarise these roles too sharply, at the very least because they could be reversed through the fortunes and misfortunes of war. In any case, there was none the less an observable tendency in certain groups, in particular the warrior aristocracies, to specialise in the capture of slaves, with neighbouring lineage-based societies being their favourite victims. Throughout this period, which was accompanied by an increase in conflict and insecurity, the 'production' of slaves appears to have been primarily geared towards supply for export channels. Certainly, some portion of slave labour was used at home, both for agriculture and craft production, yet it may be said that the generalisation of slave agriculture essentially dates back to the end of the trans-atlantic trade (Meillassoux, 1982: 90).

Given our current knowledge, it is difficult to proceed beyond this level of conjecture. All hypotheses suggest that when the colonial presence was first felt, with the exception of peasant lineage-based societies (autonomous or imbedded in other types of social formations), slave labour represented, whatever its form, an important source of production and accumulation of wealth in many Sahelo-Sudanian societies, at least for certain categories of the population.

Much work remains to be done in order to understand the impact this situation may have had on agricultural techniques and on the modes of natural resource exploitation. Following Raulin's (1967) reasoning, we can conclude that in the absence of land constraints, the abundance of slave labour encouraged recourse to extensive agricultural techniques. Extending this reasoning, it may be suggested that intensive practices, on the contrary, developed more easily in lineage-based societies that, without the use of slaves as an instrument of production, only had access to the human resources provided by natural demographic growth. Certainly, examples on the ground support these hypotheses,[7] yet systematic studies are necessary if we are to analyse the precise influence of ancient slave practices on the orientation of Sahelian technical production systems and to measure the impact of its suppression on their workings.

Labour service

Slavery was not the only method besides kinship for mobilising the labour force for agricultural work. In a more marginal yet significant way, other social relations could be activated as productive relations, giving rise to labour service. Included in this category are forms of horizontal co-operation that guaranteed an exchange of work for work, making tasks possible that demanded a great deal of labour (land clearing, for example), or through mutual aid, as a way of overcoming events that were disabling such as sickness. More often, however, the provision of labour took place within a clientist relationship which, in the widened code of gift-giving, exchanged work for another form of compensation. This was particularly the case with young men who worked the fields of their future in-laws, or for impoverished farmers who offered their labour to rich and powerful persons in the hope of obtaining protection. In many societies, young men were grouped into work associations that provided labour on demand in return for meals and payments to the age group's fund. With the Minianka, for instance, young men spent more time in work associations than in the fields of their own production unit (Jonckers, 1987). With the development of commercial agriculture in the Senegalese Groundnut Basin, the *navetanat* system represented another specific form of exchange that created work service. Coming from border regions, and later from further away (see Chapter 3), *navetanes* entered into contracts with local agriculturalists where part-time work (often half-time) for the landlord was exchanged for shelter and a plot of land that could be cultivated to meet their own needs.

In politically centralised societies, certain services were obligatory and thus took the form of tribute, where subjects were periodically obliged to cultivate the chief's fields. This practice was widespread in the Hausa system (Nicolas, 1975), and obligatory *corvée* led to a reciprocal obligation on the part of the beneficiary, who was thus obliged, in times of famine, to open his granary to those who had worked for him. In the Mossi example, the ambivalent character of such subjugation, that is its affinity with a servile status (emphasised in another context by Thornton (1992)), was made plain, because agricultural obligations to the chieftainship were concentrated on a specific social category, namely that of 'servant': 'Personnel formerly acquired through raids and moved from village to village, categorically assigned to field labour' (Marchal, 1983: 456–7). In this case, political domination took a form approaching that of master and slave. It is in this form, in fact, that slave 'emancipation' often took place in the early twentieth century. Slave relations were substituted with clientist relations along chief–subject (or elder–junior) lines following a process of ideological euphemism that disguised exploitative relations within a discourse of kinship or allegiance.[8]

As long as land was plentiful, the regulation of Sahelo-Sudanian agricultural production systems was primarily exercised along lines of appropriation and workforce management. In spite of diverse forms of social organisation throughout the zone, the familial community was the fundamental institutional framework through which this function was fulfilled. The domestic mode of production, a concept defined by Sahlins (1972) and elaborated by Meillassoux (1982), was therefore the base that ordered the exploitation of natural resources throughout the Sahel. This is not to suggest, however, that the specific conditions of organisation were everywhere the same.

Hence, the role accorded to women in agricultural production represents a particularly important axis of differentiation. The margin of autonomy left to individuals regarding work decisions, especially the balance between time spent on collective fields and on individual plots, was another characteristic that distinguished the different forms through which work was organised.

Yet while kinship structures, and more precisely the domestic cell, were the common structural base for all production systems, they were not by any means the only mode for appropriating and controlling the workforce. In many cases, slave labour supplemented and sometimes even replaced available family labour. Elsewhere, political subjugation was frequently accompanied by tribute obligations in the form of labour service, especially in highly centralised chieftainships. Finally, in a fairly general way, although at a variable degree of institutionalisation, broadened forms of co-operation and exchange of services brought about at least a partial redistribution of work among domestic production units, helping to ease inequalities arising out of demographic uncertainties, or, on the other hand, strengthening the accumulation capacities of the dominant families.

In conclusion, as we observe them today, Sahelian peasant societies exist in a continuation of forms of work organisation which, beyond their common structural kinship roots, are markedly different. This said, we do not necessarily refer to 'secular' traditions – to which each society may have remained loyal – but to their historical state at the beginning of the colonial period, the state on which their present form is modelled. The role slavery played in the economy and the degree of submission exerted upon subjects by the chieftaincy (features that are particularly relevant in the comparison of social and production systems) was not the heritage of an immutable past, but instead the product of a more recent history. By this we mean:

- the destabilisation introduced by the transatlantic slave trade;
- the rise and fall of political and military powers; and
- the division of roles between predators taking captives and potential prey which were found most frequently in segmentary societies situated on the periphery of military states, and whose fragmented structure was left especially vulnerable.

In exploring forms of work organisation (a key factor in the workings of production systems) we are once again confronted with a diversity that reflects those of the social dynamics. While we are still far from understanding all its causes, this diversity provides useful clues to our interpretation of the changes that are taking place today.

THE EMANCIPATION OF THE WORKFORCE

Beyond the various concrete forms of work appropriation and organisation found in precolonial peasant societies, and in spite of the sometimes profound transformations imposed by history, one characteristic feature remained unchanged: individual labour fulfilled social roles that were defined in relation to collective structures, structures that went well beyond the domain of production (kinship, servile relations, political submission, community co-operation). Inversely, what principally characterises the changes that have taken place throughout this century is the gradual emancipation of the workforce from these collective frameworks, at times in combination with a recomposition into new forms of solidarity and dependence.

The end of slavery

The suppression of slavery was the first and the most radical shock for those societies where slave labour occupied an important place. It is useful to recall briefly the circumstances within which this upheaval occurred.[9] The suppression of slavery was not undertaken by the colonisers as rapidly, or in as determined a manner, as one is led to believe by the proclamations of the great 'civilising' principles that inspired it (Suret-Canale, 1968–72). The French administration was aware of the social revolution that an abrupt emancipation of slaves could cause, and feared above all the reactions of local leaders whom it could not afford to confront directly before establishing its own authority. The colonial administration needed these local leaders to administer populations over which it had little control, at least until schools could produce an intermediary corps of '*évolués*' to take over that function.

Even though 'free villages' had been created at the end of the last century in Senegal and the Sudan (now Mali) to take in fugitive slaves, and although the right to return to one's country of origin was recognised for all those who still remembered it, administrative measures were taken to limit the massive departures that would disrupt agricultural production (Pollet and Winter, 1972: 254). Slave labour therefore declined gradually, especially because, once cut off from their origins and then completely integrated into their new community, domestic slaves often felt they had more to lose than to gain by abandoning the protection of their masters.

An administrative report of 1947, cited by Olivier de Sardan, discusses the continuation of slavery in Niger: 'There remain today only a few captives, referred to as "house" slaves, who are more in fact like servants, serving a family from generation to generation, rather than slaves who submit to a master's rule' (Olivier de Sardan, 1984: 196).

Until quite recently, for the Soninké of the Mauritanian Guidimaka and Malian Diahunu, the status of slave still, in some cases, implied the provision of labour service for a master (Pollet and Winter, 1972; Steinkamp-Ferrier, 1983).

As with all events that bring change, local societies sought their own means of adjustment. In the last chapter we saw how, with the more or less deliberate support of the colonial power, Songhay–Zarma notables attempted to transform the personal dependence of a slave on his master to a contractual arrangement between the owner of land and the share-cropper who worked it (Olivier de Sardan, 1984). Similarly, at Fouta Toro, access to floodplains was reserved for the aristocratic lineages, but they exchanged direct control over the labour force for indirect control, via landholding where former slaves were able to cultivate this land by means of a new social relationship – fifty–fifty sharecropping. Exploitation rights were thus maintained in the form of a land rent and no longer in the form of a direct levy as extra work (Minvielle, 1985).

It is certain, nevertheless, that even though the end of slavery was spread over the course of a generation, it represented a major upheaval, especially for the warrior aristocracies. At the beginning of the colonial presence, and in spite of the precautions mentioned above, the departure of a great number of slaves, who were traded, created an abrupt drain on the labour force (Pollet and Winter, 1972: 253). This therefore raises the question of the role played by the abolition of slavery in the great agricultural crisis that struck all of Sahelo-Sudanian Africa in 1914. As a result of abolition, lineages of free men were abruptly deprived of most of their workforce, and thus of their production capacity. In addition, many individual farms had been established by recently emancipated slaves, whose hold on the land was precarious, who had only small labour forces and, more especially, had no food reserves. Later on, gradually yet inexorably, what happened was the almost complete extinction of what had formerly been an essential mode of appropriating labour. The notions of master and slave survive today only as ideological categories expressing social identities, without relevance to the domain of economics.

Yet if we are to grasp fully the impact of the abolition of slavery, our inquiry must be directed to areas beyond agricultural production. The prohibition of an open slave trade and the institution of a colonial peace which dried up the supply of captives were in fact the death knell of the slave industry. Thus came the abrupt collapse of the very economies of whole societies, those in which the working of the soil held, in the end, only a secondary place. In both warrior societies, such as Segu, and in merchant societies like that of the Soninké, the elite classes were constrained to turn towards agriculture at the very moment when they were stripped of the right to appropriate the labour of those who performed degrading manual tasks. This change inevitably did not take place without resistance from those who saw the basis of their wealth and hegemony undermined. According to an administrative communication from the beginning of this century, cited by Pollet and Winter (1972), Soninké traders from the Bamako region expressed their recriminations on the subject in the most explicit terms:

> They were of free race and therefore unable to work the land or carry out the manual labours incumbent upon servants; and they hoped that the French would change their policies regarding slavery. If the French persisted, the country would be destroyed.
>
> (Pollet and Winter, 1972: 254)

In Sahelo-Sudanian Africa, many societies had thus to adapt to the new social and economic order imposed upon them. Yet it ran contrary to the primary vocation that saw the exploitation of nature as a secondary and devalued activity: 'From being fierce warriors, the Fula of Wasolu suddenly became reticent agriculturalists' (Amselle, 1990: 213).

As we have observed on p. 226, it is probably this loss of both their social and material bases of reproduction that caused the aristocratic and warrior societies, and to a lesser degree the market-based societies, to invest at a very early date (from the beginning of the century) in seasonal migrations; whereas the lineage-based peasant societies only resorted to this course of action much later (from the 1940s). While current development programmes consider them as simply peasant societies like so many others, the relationship with agriculture and herding maintained by these groups is, in fact, the product of a radical recomposition. Thus far, the relationship has not been made in the same spirit, nor with the same technical strategies, as in communities for whom these activities have long been the object of profound symbolic investment and a powerful mobilisation of the entire social body.

In an effort to generalise, we can take the distinction earlier developed by Pélissier (1966) in his comparison of the Senegalese Sérer and Wolof societies:

> The political organisation and social hierarchy remained alien – if not hostile – to the development of agrarian institutions, of which not the slightest trace can be found in Wolof society. In one of the apparent paradoxes of black Africa, it is, on the contrary, politically acephalous and socially egalitarian societies that developed these kinds of institutions.

> (Pélissier, 1966: 101)

Similarly, one may contrast the Fulani of Wasolu in Mali and the Bambara of Segu with the Sénoufo–Minianka or the Dogon, the Mossi with the Lobi or the Goin in Burkina Faso, and the Songhay–Zarma with the Gurmanche in Niger. In the end, this division suggests an ancient distribution of roles centred on war, trade and slavery that placed in confrontation throughout the Sahelian area a centre and a periphery which, to a certain extent, existed through each other. Yet, the reality is still more complex, as Marchal (1984) demonstrated with the Mossi, where the dividing line passes right through a single 'ethnic group', i.e., between the *têngbîise* (people of the land) and the *moose* (people of power) – two components of the same society yet with radically opposed value systems and technical strategies. Unsupported by such well-documented analysis, personal experience suggests that similar conclusions could be reached with respect to Hausa society through the contrast between farming concerns belonging to agricultural lineages and those with ties to commercial and urban interests.

What this type of comparison reveals is the existence within diverse Sahelian societies of radically different approaches to nature and its exploitation. How can we hope to understand the current evolution of the environment without taking into account these intangible relations that have guided concrete practices and given them meaning?

The break-up of the domestic unit

The end of slavery was one of the first events to mark the entry of Sahelian societies into a new era, with entirely new principles of social and economic organisation. For some, the adjustments imposed were radical. In particular, it ruined the major slave plantations to which the aristocratic societies had tried to turn when the slave trade was stopped.

With a massively reduced production capacity, the free lineages also lost their distribution capacity and similarly the means of maintaining the clientele networks on which they relied. Nor was it only the slaves who gained their autonomy – the craftsmen and the *griots*, who had previously looked to the noble families for their upkeep, now began to sell their work and to cultivate for themselves. The most enduring effect everywhere was the institution of the domestic unit as the major framework for the exercise of relations of production. Everywhere the control and management of the agricultural workforce became almost exclusively governed along kinship lines, although in some cases part of the slave labour force was reintegrated into this institutional context. It is for this reason that an analysis of the evolution of domestic production structures best illustrates the dynamic from which present-day Sahelo-Sudanian peasant societies are born.

In this regard, no matter what type of society is being considered, a single word summarises the general movement of the domestic production unit over the course of the last century, and that is 'break-up'. Whereas at the beginning of the century, agricultural activity was carried out in large social units, with lineage groups of several dozen individuals or even as many as a hundred, today one is faced with a large number of residential units, varying considerably in size, which in certain regions are now reduced to an individual household. This is a clear indicator of profound changes in social relationships and must have consequences for the way farming is organised. This phenomenon has already been mentioned in relation to landholding (see Chapter 10). It takes on its full meaning here to the extent that, as we have said, the conditions for allocating land have for a long time reflected social forms of control over humans above all. As long as land remained abundant, the allocation of fields to juniors and sometimes to women ultimately represented the right of relative economic autonomy (including the ability to dispose freely of part of their work time).

This widespread disintegration of domestic units combined different movements which, while being closely interlinked in reality, are in fact analytically quite distinct and which can have a different degree of importance from one region to another. At a structural level, there was a decline in the functional role of the all-encompassing family institutions (clans, lineages) of which these units formed part. From these, they often drew their social and religious reference, which in turn gave meaning to their own cohesion. Thus, whereas in many Sahelian societies the role of lineage

275

was one of land management (ensuring the occasional redistribution of land between units of production), this role has now often disappeared, thus pushing the land management function back down to the level of residential unit (or even of the production unit).[10] Similarly, the practice of work service or gifts of cereals for the lineage chief is now tending to disappear. At the level of individual practices, we shall see later that the autonomisation of dependants within domestic units and units of production is a phenomenon which is first a prelude, then an accompaniment, to the dynamics of segmentation which these units undergo.

The segmentation of domestic units is certainly a cyclical process linked to the very properties of this type of structure (Goody, 1958). In order to avoid reaching a size that would make it difficult to govern both socially and materially, a descent community must periodically divide itself. From generation to generation, as it grows, fragments have to detach themselves and become in turn new poles of demographic development. Unfortunate events (famine, epidemics), conflicts over succession and tensions between the paternal and maternal descent lines were for a long time the most common catalysts of these fissures. However, there were counter-constraints that limited this tendency to split. In particular, it was necessary to maintain a unit size sufficient to mobilise a critical mass of labour power for certain essential agricultural operations (such as weeding) and to cope with the many tasks that must be accomplished simultaneously. Moreover, an abundant supply of labour meant less vulnerability to accidents and disease. In addition, with the conditions of insecurity that have long prevailed in this region, the capacity to assemble a large number of combatants meant that a community could better defend itself. Besides these functional constraints, there were also structural features related to domestic organisation that checked fragmentation. Take, for example, the control exerted by the elders over access to women and to land. There was no point in leaving to establish another cell unless one was accompanied by a large enough group of individuals (for the reasons just mentioned). In other words, one had to have enough women and children able to work and, unless an alliance was made with another lineage, this was possible only for men who had already achieved a degree of seniority. For all these reasons, until the turn of the century, the situation that most frequently existed was that of large domestic production units, composed of a lineage segment, under the tutelage of an elder.

The tendency towards break-up could have begun in some areas even before the colonial period. This was notably the case in Mossi country, but even there the smallest production units observed at the beginning of this century matched the largest recorded in 1970 (Marchal, 1983: 464ff.). Be that as it may, the process of segmentation in family structures has accelerated during the past hundred years. It has acquired an entirely different dimension in that the size of production units often tends to

276

approach that of the nuclear family. While this phenomenon may not occur to the same extent everywhere, it has certainly begun to take hold everywhere.[11]

This evolution has occurred gradually and in several stages. In many cases, it began with a modification in power relations within production units, creating a growing individualisation of the work process. For those juniors and women who already had access to personal plots of land (with the Hausa and Mossi, for example), this evolution modified power relations between the head of the family and his dependants, thus facilitating an eventual fission. These changes did not fundamentally challenge the organisational principles of domestic production. However, this was not the case in lineage-based societies, where, to a large extent, the social order depended on a cohesive family cell, united around a collective field under the chief's tutelage. The remarks of one Sénoufo family head from southern Mali express well the force and depth of this obligation: 'In the lineage of which I am head, there is a pact. This pact is what links us to our ancestors. The link is forged in the permanence of the common field' (Diabaté, 1986: 429).

Yet this idealised model was progressively eroded by the new aspirations of women, and even more so by young men, who sought emancipation. Jonckers (1987) describes well how the Minianka, neighbours and relatives of the Sénoufo, granted individual fields to juniors as a means of guarding against their departure. However, in the end, this partial autonomy actually reinforced their desire for independence, and thus began the disintegration of the domestic unit. Little by little, the size of the common field shrank while the time individuals spent working it fell. As Diabaté (1986) has demonstrated, when its size relative to individual parcels becomes too unbalanced, the common field is reduced to a merely symbolic plot, and each person takes his independence; the unit breaks up.

The causes of this development are numerous. Overall it can be said that it reveals a profound change in the representation of the individual in society. We raised this issue in our discussion on land problems, for it is this change which drives the movement towards individualised land appropriation. Whereas in bygone days, from birth via marriage to death, it was impossible to conceive of a person's social destiny outside the lineage,[12] options have now gradually multiplied (thanks to school, migration, salaried work and commercial agriculture), making the achievement of individual success both conceivable and desirable.

Meillassoux (1982) demonstrated how the circulation of subsistence goods between generations used to constitute the fundamental principle of membership in a domestic community:

> To participate in the production cycle, that is, to belong to the community ... as well as to contribute to his own perpetuation,

each producer must: 1) return to the community that part . . . consumed during his non-productive years so that it can be reinvested into the development of a future producer; 2) advance that part . . . that he will consume once incapacitated; 3) produce that part . . . necessary for his current needs.

(Meillassoux, 1982: 91)

These requirements, which used to cement the community, are nowadays felt much less urgently or can at least be satisfied without resorting to co-operative work, although family bonds have by no means been broken. On the contrary, these bonds can remain extremely powerful, although now ordered by an entirely new principle. It is no longer necessary for each person to work towards the reproduction of a collective entity whose existence must take precedence over his own. The perspective is now reversed and individuals use family membership and its possibilities for mutual aid in resolving individual problems.

Even if the more profound significance of these observed developments actually resides in the transformation of collective representations, it is often material circumstances that spark the break-up of production units. Often mentioned is the burden of a head tax.[13] Instituted by the colonial administration and continued by independent states, it was one of the principal constraints that drove the heads of extended families to separate from their younger brothers and/or sons, splitting up the familial domain and leaving each individual the responsibility of paying his own taxes. This is effectively what happened in countries where taxation was especially oppressive. Thus in Niger, heads of families devastated by taxes gave their married sons autonomy, disassociating themselves from all economic responsibility for them. Then, so as not to lose the benefits of their labour, they employed them as salaried workers on the fields they had kept for themselves (Raynaut, 1973a). Clearly this is an extreme case, yet what better illustration could be found for the changes in productive relations within the family?

In any case, the impact of the tax on family structure went way beyond its financial constraints. Its initial function, and a deliberate strategy on the part of the colonial administration, was to introduce money into the local economy. To meet their tax obligations that had to be paid in cash, Sahelian populations had to enter the marketplace to get the cash by negotiating the sale of either their products or their labour. Thus the tax was not only the touchstone of policies for introducing commercial agriculture,[14] but also of those for attracting migratory labour power towards employment centres (see Chapter 3). The tax was, therefore, a vector for a genuine upheaval in economic relations within the family. This explains its powerful destructuring effects on the family. Izard captures perfectly this dynamic in the case of Mossi society:

For the payment of the tax, a purpose of an individual character over-determines the atomised nature of the pursuit of money, received outside the community in return for work foreign to traditional ways, turned over to the administration, thus to outsiders. . . .

(Izard, 1985: 213)

As a result of the imposition of the tax, the use of money gradually penetrated every domain of the family's material and social life. Whether it be for the satisfaction of daily needs (such as food or clothing), for the accomplishment of social rites marking an individual's life cycle (marriage in particular) or even for the pursuit of a productive activity (the acquisition of seeds or tools), it became impossible to do without money. This development had two consequences.

On the one hand, it increased the heavy burdens on family heads, all the more so as decreases in the prices of agricultural products frequently reduced their income (Chapter 4). (This represents a generalising of the situation concerning taxes.) On the other hand, for individuals in dependent positions within the family, particularly the young, it created the potential to earn money to satisfy, at least in part, their individual needs and aspirations. One arena that was particularly significant was that of marriage expenses (not only the dowry but also the many other associated expenses which today can amount to a considerable sum). More and more frequently, young men provide most of this money themselves, which puts them in a stronger position when choosing a bride. In seeking greater autonomy, many younger sons have tried to maximise their monetary income (Ancey, 1983) by devoting their personal lands, whenever possible, to commercial agriculture. In many cases, women too now enjoy real economic autonomy, particularly thanks to commercial activities (Raynaut, 1978). This development in familial economic relations carried within it the seeds of the destruction of the large domestic unit, which became less and less able to control the centrifugal forces created. Increasingly, therefore, it became less viable socially and economically.

One last factor, which is often not taken sufficiently into consideration, favoured the dynamic of fragmentation. This was the transformation of inheritance rules (see Chapter 10). Usually, in a lineage-based system, succession to the head of the domestic community and to the overseeing of collective possessions went according to the double rule of generation rank and primogeniture. Where three generations were grouped together this encouraged communities to remain intact. During the present century, and partly as a result of the emergence of Islam which favours this form, inheritance by lineal descent has increased, which leads to the systematic segmentation of landholdings among male beneficiaries (see Chapter 10). At Fouta Toro (Senegal valley), Minvielle (1985) showed that the division of the land paralleled the family break-up, and that the split in the inheritance rules (which had triggered the process) coincided with the first migrations. Islam may have laid the legal foundation for land division, but it was emigration alone which made it possible to obtain the income needed for the break (Minvielle, 1985: 97).

This disintegration of production structures has deep-seated repercussions on the very conditions of production. As small production units are unable to mobilise a large workforce at crucial times in the agricultural calendar, they are forced to stagger their sowing, they then fall behind with the weeding, resulting in a decline in productivity, which in turn means that they must adopt more extensive cultural practices. The growing season is short in the Sahelian or Sudano-Sahelian climate (see Chapter 1) and therefore any delay in sowing involves serious agricultural risks. Finally, being unable to produce any surplus, these small units find it more difficult to equip themselves with farming materials. Furthermore, the loss of just one worker is felt much more keenly. In fact, small family farms suffer from an increased economic fragility, especially as they are subject to the demographic cycle characteristic of the peasant economy described by Tchayanov (1990) and to variations in the number of workers.

Marchal (1989) provides a particularly striking illustration of the repercussions that the size and structure of a farm can have on its agricultural productivity. He compares two farms in Yatenga: one conforms to the 'old' model (*Nayiri*) and the other is reduced to a single household (*Warma*). Both have a similar surface area per family but, because it lacks the workforce, the workers on the second farm must each cultivate on average an area twice the size of that cultivated by their equivalents on the first farm. *Nayiri* finished sowing in 21 days, whereas *Warma* had to wait for a second good rainfall in order to finish his (bringing the total time to 39 days). As he was completing the sowing, *Nayiri* was able, at the same time, to begin weeding and to practise earthing up. Each of *Warma*'s workers devoted four times as much time to sowing as *Nayiri*'s workers, and then encroached on the time that he should have spent weeding. At the end of these two phases in the farming calendar, each of *Warma*'s workers will have worked on average 46 days, still without bringing the weeds under control, whereas each of *Nayiri*'s workers will have worked only 27 days. Not only have *Nayiri*'s workers drawn most benefit from the climatically favourable days, but the work of the group was more efficient.

An example such as this shows clearly how the fragmentation of production units can wipe out productivity and strengthen the principles of extensification. By a paradox which has very serious consequences, the search for autonomy can result in a decrease in the results of production and an increase in vulnerability, along with the threats to material and social reproduction that this implies. In fact, the current trend towards fragmenting landholdings here comes up against functional needs which had for a long time counterbalanced the action of the centrifugal forces inherent in any economic and social structure founded on the basis of a domestic unit. For this reason, we must be careful not to overgeneralise.

Although the fragmentation of landholdings has been fairly widespread, there have been other trends favouring the safeguarding of solid family-based structures. The combination of these opposing forces, each exerting a different degree of pressure and each at a precise moment and on distinct social systems, could only result in the great diversity of situations that we see today which are themselves in a constant state of flux. A comparative history of this process of differentiation remains to be written. One thing, however, is certain: this phenomenon of breaking up land into parcels is not simply the result of the mechanical interaction of the various factors that we have identified and described. It is the product of a social dynamic, a complex interplay of actors, each trying to jostle for position. Within a single region or a single village, different sizes and forms of domestic unit coexist today with different combinations of residential unit and production unit and a different balance between common fields and individual plots.

Paradoxically, it is precisely where we would expect the forces of disintegration to be strongest, in zones of massive emigration, that large domestic units have at times been best preserved. In fact, the ability to depend on a cohesive and large family cell can be an important fall-back for migrants. It was, therefore, common for Soninké men of the Mauritanian Guidimaka to act in common with brothers and cousins within the domestic unit in planning their departures for France. By belonging to an extended family, migrants were able to take turns at migration, placing their families under the authority of trustworthy men. They could therefore absent themselves for several years without abandoning the women and children who remained in the village (Bradley et al., 1977). Moreover, given that he controls most of the savings of his younger brothers who are emigrants, the family elder may find that his function and even his position in the family is strengthened, even though he no longer has responsibility for the material needs for family reproduction. Mathieu (1987) shows that such a situation has to rely on a tacit agreement between elders and younger emigrants as to the non-productive use of the income from migration. According to Quiminal (1991), migrants continue to see their future as being in their village, as eventual heads of families, and they do not therefore want to see any diminution in the status of the family elder.[15] The same concern over the preservation of social and economic functions within extended kinship structures was observed with Bambara migrants from Mali (Lewis, n.d.).

The continuity of older forms of organisation in such cases, far from being some blind attachment to 'tradition', is on the contrary a manifestation of precise adaptive strategies to current realities. While the break-up of family structures appears to be general throughout the Sahel, we should not overlook the capacity of certain groups or individuals to respond independently, at times contrarily, to the constraints placed upon them. What is emerging therefore in the Nigerien Hausa country are new ways of dealing with land shortage. In families with few landholdings it is now common for brothers to maintain joint ownership after their father's death. In this way they keep the farm large enough to make animal traction equipment profitable (Yamba, 1993). For the Soninké from the Diéma district and Gadiaga (Mali), in the high emigration zone, the crisis in food production and the shortage of manpower through migration has forced the family elder to concentrate all his workforce in one large common field. The plots allocated to the younger family members are now reduced in size or have even disappeared and the women work more on the common family field (Coulibaly (ed.), 1985). Without doubt, the variability of family strategies constitutes a powerful engine of diversity.

Nevertheless, even when co-operative work structures are maintained, the emancipation of individual labour is still observed. In fact, the present forms of family work organisation are no longer dominated by the dual requirements of, first, ensuring the physical continuity of the community as such, and, second, reproducing the configuration of roles and status which defines it in social terms. Co-operation is now in effect the result of negotiations (albeit tacit) between all those involved. In a reversal of perspectives, although the agreements between individuals to which such negotiations lead preserve the collective nature of work and living, this is no longer an end in itself, but a guarantee of efficiency and security for the benefit and in the interests of each individual.

The rise of market relations

Where it existed, the disappearance of a slave class along with the breakdown of production units now makes recourse to outside work necessary. In part, the response to this need is found in the work exchanges that take place outside the framework of the domestic unit but still depends on the initiation of social relation networks, most often in clientist form. Hence, older forms of personal allegiance, especially those that linked slave to master or subject to chief, continue to furnish support for labour service (Pollet and Winter, 1972; Nicolas, 1975; Steinkamp-Ferrier, 1983; Olivier de Sardan, 1984). Often, an important role is played by agricultural associations of young people, as well as by the *navetanat* (Amselle *et al.*, 1982; Jonckers, 1987). Currently, new types of clientist relations can also be observed and can sometimes create highly structured, collective forms of work organisation, as in the case of the Mouride brotherhood of Senegal which has demonstrated its great capacity for mobilising agricultural labour (Cruise O'Brien, 1971; Rocheteau, 1979).

The existence of such forms of co-operation suggests that we should be wary when considering the notion of individualism, which is so often referred to when describing the present changes in social relations in African societies. We cannot simply transpose into this context the essence of this notion in the ideology of European societies.[16] In Africa today, as in former times, the principles of behaviour still continue to resist the complete dissociation of economic practices from the social structures and the systems of representation that give them meaning. The continuing erosion of former statutory rights is not about to give rise to the kind of individualism which would simply be a turning in on oneself. On the contrary, it reactivates and enhances social relations that already existed, such as clientelism, patronage and co-operation between peers. In the context of the town, always described as the ideal melting pot for individualism, the permanent nature of community solidarities has been shown together with their role in the success of household subsistence strategies (Raynaut, 1988b). Without a doubt in the Sahel, and probably throughout Africa, economic relations continue to be deeply rooted in the fabric of social relations. This should not mask the fact, however, that a very real and major evolution is taking place, an evolution that increasingly ascribes an exchange value to work and makes it, more or less openly, a commodity.

We are thus confronted with a complex dynamic, quite similar to that observed when we considered the emergence of a market for land. Even if it may have existed since the precolonial period in certain highly monetised Sahelian societies such as that of the Hausa, wage-earning only became general when the colonial administration had systematic recourse to it, whether on construction sites (after the abolition of compulsory labour) or in migration areas. However, only slowly, and sometimes indirectly, did it find a place in agriculture.

According to an administrative communication cited by Pollet and Winter, it is from the beginning of the century that the Soninké of the Malian Beledugu (around Bamako) 'had ... the idea of hiring for the season the labour of their Bambara neighbours' (Pollet and Winter, 1972: 253), to compensate for the departure of thousands of slaves. The method of payment adopted remains unclear, although it was perhaps similar to the *navetanat* contract. What is clear is that a view of work as a simple factor of production, and no longer the realisation of social relations, was already inspiring this strategy. Much later we see this at work with the Soninké from the Mauritanian Guidimaka and the Senegalese Gadiaga, in their efforts to attract wage-earners from northern regions or neighbouring Mali in order to compensate for the drain on their own labour pool as their young men left for France (Bradley *et al.*, 1977; Weigel, 1982; Steinkamp-Ferrier, 1983). It is significant that even today people outside the village, or even the ethnic group, are often called upon when wage-labour is needed. Thus, Sénoufo cotton agriculturalists from Mali recruit large numbers of seasonal Bambara and Dogon workers (Sanogo, 1989). Such behaviour underscores these communities' reluctance to allow market relations to penetrate the fabric of their interpersonal relations. When the resistance weakens, it is often in an indirect fashion. As Jonckers (1987) noted with the Minianka, it is the young people's work associations that tend to transform themselves into paid work crews for the service of rich agriculturalists. Nevertheless, it does not always stop here. Gradually, the practice of agricultural wage-earning occurs within communities. This phenomenon has appeared in social systems like that of the Sénoufo whose otherwise strong attachment to their ancient values have been mentioned several times (Sanogo, 1989). In the Hausa peasant's world, it has become a daily reality that a segment of the population can no longer survive without selling its labour (Raynaut et al., 1988).

Diverse social categories can provide this labour. Many junior members are finding that selling their labour is more profitable than cultivating their small personal fields. In the end this is a variation of the strategies of emancipation mentioned above (p. 277). The phenomenon is entirely different when it involves the heads of farms, who are forced to it when their production, and the income from it, are no longer sufficient to meet family needs. They then enter a cycle of impoverishment that becomes difficult to escape, whereby they are forced to hire themselves out as daily workers to the detriment of their own fields (Raynaut, 1984). In many regions of the Sahel, the practice is not limited to a small proportion of the agricultural population. Thus, in Mali over 40 per cent of the production units in the groundnut zone are involved (Amselle et al., 1982), 28 per cent in the Beledugu (Teme, 1985) and in the Nigerien Hausa country the proportion is 30 per cent (Raynaut et al., 1988). In other regions the phenomenon is also observable, albeit on a smaller scale, and some are barely touched by commercial agricultural development. This is the case, for instance, in eastern Senegal (Lavigne Delville, 1988). Production deficits linked to the recurrent climatic crises of the last two decades have accentuated this phenomenon by bankrupting the most vulnerable farms.

Who benefits from salaried labour? Generally, all those who are in a position to resist successfully the increasing constraints on agropastoral production. First, of course, are those producers who by various means have acquired enough land to shield themselves from the effects of land shortages and thereby to maintain a satisfactory level of production (but with a resultant shortage of labour). There are also those producers (and sometimes they are the same ones) who, with the benefits of non-agricultural revenues, are able to withstand production shortfalls and sometimes to invest in modern production equipment. In this way a gap slowly widens between a more or less limited minority of units that are able to accumulate land, labour and tools,[17] and a growing proportion of producers who maintain equilibrium only with difficulty or, in the most severe cases, founder and are forced to negotiate their labour – if not their land, as in the case of the Nigerien Hausa (Raynaut, 1983).

In a more marginal but nevertheless extremely significant way, from the point of view of the development of production structures, women also are now employers of paid labour. In a way, women are compensating for their poor access to modern means of production, but, more importantly, they are also realising and reinforcing their economic autonomy within the domestic cell, thus having less need for other family members' co-operation and gradually creating their own autonomous production unit.

CONCLUSION

Throughout this chapter we have seen once again that it is impossible to analyse properly the difficulties experienced in society/nature relations in the Sahel (problems that take the form of a severe environmental crisis) without taking into account the upheavals in their social systems. The overexploitation of soils or the degradation of the vegetation are not simply the result of demographic excess or of destructive technical practices. These are proximate causes that make sense only when placed in relation to the evolution of social frameworks (modes of thought, values system, human relations) that order the management of natural resources and human energies.

How can we understand the relative inability of some Sahelian societies today to establish sustainable relations of use within their environment, without understanding the road they have taken in the past, and without taking account of the recent changes they have undergone and those that they are undergoing today, changes through which they have remained faithful to a vocation that embodies their identity while at the same time adapting, sometimes in radical fashion, to incentives? It cannot be ignored, for example, that warrior aristocracies have adopted the role of peasant societies against their will, and if we are to understand the present we cannot close our eyes to the upheavals of the recent past. This is exactly what Pélissier (1966) is saying when he notes the original 'hostility' of Wolof social organisation to agrarian institutions and shows how this collective attitude persisted, *mutatis mutandis*, well after agriculture, by force of circumstances, had become their dominant activity. It is also what Marchal (1983: 810) is saying when he attributes the 'evil-doing to the landscape' of the inhabitants of Yatenga to the *moose* society's specific intentions for nature – intentions formerly marked by political objectives (to control the maximum territory) and which continue today in the form of unrestrained consumption of space. Amselle (1990) is also saying this when he analyses the present characteristics of Fulani/Bambara/Malinke production systems as the result of a slippage from the 'warrior–slaveholder' model towards the 'Islamic–merchant model', both having in common a devaluation of working the earth.

285

We are faced with a complex alchemy in which the principle of social reproduction and the demands of change come face to face. Often, it is in such confrontations and in the contradictions that give rise to them that we must seek the origin of the incompatibilities between, on the one hand, production strategies (and their technical practices) and the demands of ecological reproduction on the other. When a society directs the same predatory strategies that formerly inspired its military or commercial activities towards working on the land, the beginnings of land inflation denounced by Marchal (1983) is plain to see.

Such an analysis is too broad, however, and therefore remains incomplete. But even at this stage, it certainly helps us to understand the diversity of approaches to nature hidden behind the unifying label of 'peasant societies'. Yet it does not identify the internal dynamics at work within communities and their effects on concrete conditions for managing natural resources. The major phenomenon at this level resides in the break-up of the social production structures and in the progressive individualisation of the work-force – a phenomenon that parallels the land control phenomenon. It is the social and economic foundation of production systems that is therefore being transformed. These transformations occur in two ways.

In all types of society, along with the disintegration of old collective frameworks comes a sharp rise in competition, which leads to increasing disorder in the conditions for appropriating nature. Individuals must enter into a struggle dominated by short-term objectives. All those who are able to do so accumulate labour, production tools and, of course, land without which neither of the other two would be of any use. Given these conditions, concern with sustainability and resource renewal becomes secondary. The struggle becomes all the more bitter as control over agricultural or pastoral production becomes more necessary for subsistence and the satisfaction of social aspirations, especially when a major portion of revenues is generated from the commercialisation of production. Similarly, the struggle is more intense when land grows scarce, or when a climatic crisis reduces the production capacities of the ecosystem. More than ever before, individual accumulation takes place at the expense of the rest of the community. Thus new forms of inequality are created, favouring the multiplication of divergent technical and economic strategies. A study of the Nigerien Hausa clearly shows the socially destructive effects which new socio-economic relations have engendered, particularly the upsurge in sorcery as a means of expressing interpersonal rivalry and conflicts (Schmoll, 1993).

Liberating the workforce from the collective framework that once surrounded it considerably increases its possible applications. The creation of a labour market integrates peasant communities into wider networks of circulation permitting part of the familial workforce to turn outwards towards neighbouring production units or more towards distant poles of attraction located in other Sahelian agricultural regions, coastal countries, or sometimes even Europe. Through a gradual amplification, migration has become both a means for, and a result of, the emancipation of the workforce. Under the combined effects of factors described throughout this chapter, certain social groups and geographic regions specialised in the exporting of labour, while others received wage-earners or *navetanes*. How can we analyse the environmental strategies of groups like the Mossi of Burkina Faso, the Bambara and Malinke of central Mali, the Soninké of south-eastern Mauritania and western Mali or the Zarma of western Niger without taking into account the fact that migration has for these rural populations become the principal source of revenue? As Marchal (1983) aptly notes, these rural systems now function within a 'double space', and it is impossible to understand the 'desertification' of Yatenga without acknowledging that the territory under consideration represents only one side of a double entity that also includes the regions that receive its emigrants. When agropastoral production becomes a mere supplement to profit-making activities conducted at a distance, it is hardly surprising that village-level efforts to improve techniques and to protect the environment are so modest:

> Thus, if the inhabitants' idea of well-being is a forest clearing with cotton fields (south-western Volta) or coffee and cocoa plantations (Ivory Coast), we will be a long time deploring the degraded state of 'old Yatenga' and waiting, even if the agricultural administration is stepped up, for destructive cultural practices to become more pro-ductive by restoring to the environment what it takes out.
>
> (Marchal, 1983: 812)

To a large extent, this double movement feeds the dynamic by which Sahelo-Sudanian agrarian arrangements are gradually diversifying, with a growing economic disparity between members of the same community accompanied by a divergence in ecological strategies and a division of roles between those regions that have long focused on exporting their labour power and those where agriculture represents a real dominant activity. Such diversification could not have occurred without a mobilisation of the workforce, itself made possible by profound changes in social production structures.

Nevertheless, this does not amount to the proposal of a simple framework for the analysis of environmental situations. There is no evidence for stating that 'desertification' is more intense where migration has outstripped agropastoral production. What is true for Yatenga is by no means true for Mauritanian Guidimaka or for western Mali. Besides, just because humans take from their surrounding environment the essentials for their subsist-ence and their social existence does not mean that they have given particularly special attention to the renewal of exploitable resources. The groundnut basin, for example, provides evidence to the contrary.

The reality is much more complex. Many other factors interact to give

287

rise to a particular situation. They have been described throughout this study and it remains to tie them together. In addition, local realities are not static because communities can develop new strategies in response to the challenges confronting them, which are at times counter to the major trends that have been described. Thus we have seen the beginning of a return to extended domestic structures in the Hausa country of Niger according to the observations made by Yamba (1993).

In this chapter, as in the preceding one, our specific purpose has been to demonstrate that it is impossible to explore the way in which the environment is utilised without taking into account the dynamic of cultural representations and of social relations which organises such use. Such a conclusion is far from original, yet it is nevertheless useful to restate the fact that the Sahelian crisis is as much a manifestation of social upheavals, themselves a part of the history of this part of Africa, as a disturbance in the material relations between societies and their biophysical environment.

NOTES

1 We must not underestimate the functional importance of the network of social relations founded on neighbourhood, which do not openly mobilise institutionalised co-operation, but rather rely on the free expression of individual affinities. Thus, in Hausa country, not only were we able to observe the importance of food sharing and exchange in the structuring of social relations within a village community (Raynaut, 1978), but also its role in food security (Raynaut, 1988b).
2 In some cases cereal production, the key to a group's physical reproduction, is monopolised by the family elder. Control and distribution are therefore reserved for the principal production unit.
3 An objection could be raised in the case of Hausa society, the archetypal example of just such a system, where frequent recourse to slave labour should have rendered the mobilisation of women's labour unnecessary. Yet such an argument ignores the point that slavery was not a structural principle of the society but only one of many instrumental forms of control over the workforce, quite limited within the lineage-based part of the society and differing in this regard from the organisation created in the territories conquered by the Fulani in the nineteenth century (present-day northern Nigeria). There, slavery acquired great importance and the sequestering of women for religious reasons removed them from agricultural activities.
4 Administrative reports from the turn of the century put the slave population at 60 per cent in certain Soninké groups in Mali (Pollet and Winter, 1972: 238). It is estimated that one-quarter of West Africa's population fell in this category (Meillassoux, 1982: 94).
5 Theoretically, emancipation was possible by payment of compensation to the master – a possibility beyond the reach of the slaves unless the master encouraged it by allowing them a chance to accumulate the necessary wealth. With respect to modes of repayment, see the details given by Lovejoy (1986) concerning the caliphate of Sokoto.
6 For example, in the genealogy of a Hausa chieftain in Niger we found a marriage of a daughter to a slave, in order to provide descendants in the reigning line

otherwise threatened with extinction with no male heir (Raynaut, personal observation). In the Samo, Burkina Faso, Héritier (1975) noted that marriage to slave girls was one of the methods used by free lineages to increase their numbers.

7 See, in particular, Amselle's (1990) comparison of the Fulani and the Sénoufo of southern Mali.

8 Thornton (1992) and Lovejoy (1986) rightly emphasise the similarity between certain forms of slavery in Africa and clientele relationships.

9 As Lovejoy (1986) has clearly demonstrated with the example of the caliphate of Sokoto, slavery has always met resistance from those forced to endure it, as slaves would run away and there would even be attempted uprisings. Colonial law introduced radical change to the slave's subjugation by his master, by condemning the very institution of slavery.

10 Semantic shifts in meaning highlight these changes. For the Soninké, the terms *jamu* and *kâ*, which previously meant clan and lineage, are now used to denote, respectively, lineage and residential unit.

11 Thus between 1969 and 1990, in a Hausa village in Niger (Tahoua department), Faulkingham observed a splitting of residence units and households reduced their average size by half – from 15.5 people to 8.9 in one instance and from 9.35 to 5.3 in another (personal communication).

12 As always, such a clear-cut statement is too reductive. Here, as elsewhere, history reveals that individual destinies have always been attainable by adopting flexibility in relation to ethnic identities, by negotiating political allegiances and by opting for deviant forms of behaviour or transgressions. Such examples were only exceptions, however, which never compromised the principles which underlaid the functioning of the institutions and the definition of the social roles within them.

13 This imposition consisted of collecting a lump-sum payment for each person who was theoretically active, i.e., all healthy men and women old enough to work in the fields (from the end of adolescence to old age).

14 This is evident by the fact that, for a long time and in many Sahelian agricultural societies, plots devoted to commercial agriculture (groundnut or cotton) were called 'tax fields'.

15 This consensus is not necessarily reached once and for all, however. With the *Haalpulaar* of the Senegal valley, there are indications that splits are beginning to emerge. These take the form of concentrating migrants' remittances within their own households and could eventually lead to a major split between the large families which are still in existence (Lavigne Delville, 1994).

16 In our societies this essence is indeed ideological, rooted in the paradigm of the market, which itself forms the basis of the prevailing representations of the rationality of economic relations. In practice, however, there is every indication that, on the contrary, and even within these societies the behaviour of individuals derives ultimately from reference points of a collective nature, i.e., family, social class, local identity, etc.

17 Between 15 and 20 per cent of farms in Mali, according to various studies (Amselle *et al.*, 1982; Benhamou *et al.*, 1983); 25 per cent in the Maradi area of Niger (Raynaut *et al.*, 1988).

12

CONCLUSION: RELATIONS BETWEEN SOCIETY AND NATURE – THEIR DYNAMICS, DIVERSITY AND COMPLEXITY

Claude Raynaut

At the end of the journey, we return to our initial question, which has been the guiding thread of the entire study: what are the dynamics that are masked by the hackneyed term 'desertification' as it is applied to the entire Sahelian zone, often without real thought for its meaning? We have attempted to address this question both by emphasising the diversity and the specificity of situations encountered in this part of the African continent (that portion, at least, that has been explored here) and by clarifying the role played in their genesis, not only by the actions of human beings as a link in the food chain but also by the social logic that orders the exploitation and transformation of the natural environment.

Thus we have posed the problem in terms of relations between nature and society. In so doing we have adopted, one after another, several analytical directions representing different angles of approach to the question. Drawing on the wealth of scientific literature concerning the five countries included in our study, and often forced to confront gaps in the research, we have taken pains to identify and to set out current knowledge and to derive from it some of the elements of description and analysis of the Sahelian reality. Thus we have successively considered the role played by demographic, economic, technical, social and cultural factors in determining the conditions for land occupation and natural resource use.

Now the time has come to gather these partial views into a more synthetic understanding, and to attempt to draw from this understanding some conclusions that may be pertinent to strategies for development intervention – since now, more than ever, the Sahel is the object of great concern.

OBSERVING DIVERSITY: CONFIRMATION

The impossibility of reducing the diversity found in the Sahelo-Sudanian zone to one single analytical model constituted our initial hypothesis. This hypothesis is, at the end of our study, abundantly confirmed. The Sahel cannot be viewed as one homogeneous entity.

Situations coexist that are very different, sometimes in sharp contrast. This observation is not new. Although it is sometimes eclipsed by media images that are too often simplistic and reductionist, it is obvious to all those with some knowledge of the region. Nevertheless, we have gone beyond merely reminding the reader of a self-evident truth, and we have tried to systematise the image of this diversity. To accomplish this, we have reduced it to several major environmental situations, which are identified within the geographical limits of the Sahel, delimiting zones to illustrate their geographical distribution (Figure 8.2, p. 189). We must emphasise again that this categorisation is, and remains, schematic. The areas identified are only indicative of the broad pattern of diversity, their borders represent only a coarse delimitation of their extent. Moreover, these situations correspond only to a dominant dynamic. They do not take into account the nuances and more specific local realities that can be encountered within them. This division into zones should, therefore, be used as a frame of reference to help us formulate questions on the ground, knowing that the reality is incomparably richer and more varied than is suggested by the image drawn on the map. However, as we have already stated, such a level of generalisation permits us to go beyond a simple observation of the singular individuality of each local situation, and with the distance thus gained we are able to identify several major axes of differentiation whose combined effects generate diversity.

The agro-ecological facts

Throughout this study, we have amply shown how the constraints and potentials of the Sahelian environment vary greatly from one location to another. However, these variations can only partly explain the population distribution in the region and the way natural resources are exploited.

Thus we cannot satisfactorily account for the alternation, from Senegal to Chad, of areas of demographic concentration and vast, unpopulated regions by focusing on bio-physical characteristics, such as climate, soil type or water resources, or on the distribution of major parasitic disease vectors (Chapter 3). Similarly, with respect to the geographic distribution of production systems, at most, only two observations are possible. These are that irrigated agriculture is concentrated along the major rivers, and rainfed crops are absent where aridity makes only nomadic pastoralism possible. Outside these extreme cases, the respective places of agriculture and herding in the Sahelian region, and the geographic distribution of technical practices, are only partially illuminated by a comparison with the pattern of climatic and soil constraints. Agricultural and pastoral communities often live together within a single territory, exploiting in different ways (at times complementary, at times competitive) the same ecosystems (Chapter 5). The spatial distribution of agricultural tools, the hoe in one location, the iler close by and animal traction somewhere else, has more to do with a local community's technical options and the localisation of development programmes than to the influence of the physical environment. The same is largely true regarding the diversity of agricultural systems (Chapter 7).

This is not to suggest that anything is possible anywhere. Obviously, natural characteristics permit or rule out some practices, and favour or prohibit some developments. Numerous examples could be cited. The location of the major groundnut production zones, for instance, coincides with the presence of deep sandy soils while, for the most part, cotton production could only really take hold under favourable climatic conditions. The distribution of karite and *Acacia albida* groves is governed by their water and soil requirements (pp. 176–7). Agriculture has advanced into increasingly arid sectors only at the price of an increased vulnerability of production systems (p. 50), a fact which shows clearly that some natural boundaries cannot be crossed in any sustainable way, at least not without radical modifications to production techniques.

In the end, the major environmental parameters in the Sahelo-Sudanian zone (the degree of aridity, topography, soil type) create a general framework for determining the range of technical solutions available for obtaining from nature the resources necessary for human survival. The same parameters determine the vulnerability of a natural resource base to human action. It is in these terms, and these terms alone, that the 'rigour' of the Sahelian environment can be affirmed. This 'rigour' must not be seen as a 'fatality', however, but as a strict framework (adaptable to many local variations) which fixes the rules of the game for a sustainable relationship between societies and the resources they exploit.

The role of demography

As we have seen, the role of demography is a thorny subject that must be approached without the unfortunate simplifications that are so often made when dealing with this topic. One observation is undeniable and this is that, for agricultural production techniques, population density is only one variable that influences the level of impact on ecosystems. The greater the degree of human concentration within a limited space, the more the soil and vegetation are likely to become degraded. For this reason a major parameter in the division into zones is the pattern of population densities. However, the use of the demographic variable is subject to certain limits.

First, demographic phenomena can in no way be considered as primary causes, for they are themselves the product of a prior social and economic dynamic. The devastating local effects of extreme population density are, therefore, largely the result of a variable distribution of people in space, by no means dictated by some 'natural' law which is an expression of the intrinsic demographic properties of the local societies. It is, in fact, the outcome of a history that shaped territorial settlement (Chapter 3). It is in the context of this history, and not in so-called collective mentalities and behaviours influencing fertility, that an explanation for the present situation is to be found.

Second, with regard to the effect of the demographic variable on resource use, we should not lose sight of the obvious, that it never acts independently of co-factors that can substantially alter its impact. Two such co-factors appear fundamental.

Land consumption

Given identical population pressure, some technical practices are much more destabilising than others. This is notably the case with animal traction agriculture. Sahelian farmers often use this method to increase the area of land under cultivation, thus accelerating the clearing of land with the resulting consequences for vegetation and soils (p. 172). In this regard, pastoral systems represent an extreme case, because it is the number of animals and not humans that determines the level of impact on resources.

In reality, the effect of human density on the evolution of ecosystems is doubly complex. First, as we have just noted, it is mediated by technical practices that can alter its nature and intensity, but, above all, it does not take a single direction. As Mortimor (1989) clearly showed, the larger the population (and, therefore, the more labour that is available), the better the people can govern the applicability of techniques that are less consumptive of land and resources and more labour intensive. This is the case, for example, in agriculture with organic fertilisers, with the early weeding and thinning of plants, with irrigation and with erosion management. Where labour is insufficient or where, because of migration, it is invested mostly outside the rural area, the only applicable technical practices are those that are not labour intensive, however damaging they may be to the environment.

Therefore, even in the harsh natural conditions of the Sahel the concept of overpopulation is relative. An increase in the number of people does not only lead to a greater demand for primary production, it also represents the increase of a resource (labour) that can be used to manage and preserve nature. Of course, certain economic, material and social conditions must come together for this to occur, but it is clear that a simple equation that assumes a mechanical and necessary relationship between population growth and 'desertification' is inadequate.

CLAUDE RAYNAUT

Level of demand

The degree of impact that a Sahelian agricultural or pastoral community currently exerts on the lands it exploits depends less exclusively than ever on the fulfilment of fundamental biological and social needs. Rather, this impact is now closely linked to the amount of cash necessary for a community to sustain itself in a largely monetarised economy open to the external world. As a result, fiscal pressure, the evolution of consumption practices and the terms of exchange between local products and manufactured goods are variables that weigh heavily on the level of natural resource exploitation and are independent of the size of the population.

In conclusion, while population density continues to be an important parameter for defining a local environmental situation, its effect is not direct and mechanical, it interacts with the effects of other, very different kinds of variables. This is why, for example, given almost identical levels of settlement, the Yatenga Burkinabe and the groundnut basins of Senegal and Niger have distinct environmental situations (Chapter 8). Similarly, while human densities in southern Mali's cotton zone are lower than in regions to the north, the use of space is at least equivalent because of the mechanisation of agriculture and the dynamism of commercial agriculture.

Modes of resource use

It is clear from our previous discussion that concrete environmental situations throughout the Sahel depend, to a large degree, on the specific conditions under which different human communities exploit their natural environment. In this respect, several major criteria for distinguishing these conditions are used in the zoning we have proposed.

Herding practices

As we have seen, a sharp distinction between 'pastoral' and 'agricultural' zones is meaningless. The northern edge of the Sahelian region is indeed devoted almost entirely to livestock. It is also true, though, that livestock are also present almost everywhere else.

In fact, pastoralism can take many diverse forms, from the different ancient models of pastoralism to a rapidly developing form of agro-pastoralism and to the ever increasing herds accumulated by owner-investors. The analysis of, and distinction between, these local situations must not be based on static geographical divisions, but should take account of the way these different forms of stock-raising are superimposed and competitive with each other in a single space, and the way they interact with other productive activities, especially agriculture (see above pp. 110–18).

294

Agricultural practices

In discussing this topic, several factors need to be considered. With respect to the spatial distribution of agricultural systems, there are several major divisions that create diversity in Sahelo-Sudanian agriculture otherwise dominated by cereals, millet, sorghum and maize.

First, there is the opposition between irrigated and rain-fed agriculture. This opposition highlights the vast areas traditionally devoted to rice or floodplain sorghum. Partially superimposed on these areas are locations where major hydro-agricultural projects are located, as well as several areas where water management is achieved by means of hillside dams. Hydro-agriculture has long been the focus of grandiose development projects, representing a major strategic option for combating the climatic risks found in Sahelian countries. Problems with these projects have been diverse, their impact on the environment is imperfectly managed and they have caused extreme upheaval in the traditional modes of natural resource exploitation (pp. 126ff.).

Another differentiating factor is the existence of large agricultural basins devoted to the commercial crops of groundnut and cotton. There has long existed a concentration of strong forces for technical and social change in these basins, particularly the intensification of monetary exchange, diffusion of modern agricultural practices and the development of new collective structures (co-operatives). Even when the position of commercial agriculture has become modest, as in central Niger, the current state of relations between humans and their environment cannot be understood without reference to the history of these regions as centres of production for commercial crops (see pp. 189, 196–7, 203).

Finally, another important differentiating factor in determining resource use is the diversity of peasant technical strategies. Traditionally, of course, all of these strategies tended strongly towards the valorisation of the labour force, often at the expense of that of the land. Therefore, extensive practices, with their high level of land consumption, dominated (pp. 152–4). Nevertheless, these practices most often coexisted with more active techniques for increasing labour efficiency through the use of appropriate tools (pp. 167–70), management of the environment (pp. 174–81) and the restoration of the soil's potential fertility (pp. 138–48). Hence, an important criteria for characterising and distinguishing local situations is the relationship between intensivity and extensivity in peasant practices, as it may be expressed by such indicators as the surface area cultivated per person (Figure 6.2).

Integration into an enlarged technical and economic space

On numerous occasions throughout the preceding chapters, we have emphasised the fact that Sahelian societies have now been firmly penetrated by a monetary economy. The process does not cover all areas to the same degree. In the majority of cases, land and labour are not yet totally integrated into the market, but there are examples where both of these have become nothing more than a commodity (for example, in the Nigerien Hausa country). In daily life, however, money is now indispensable everywhere, even though some sectors of social life continue to follow other principles than the market economy, such as exchanging gifts.

It is access to money, in particular, which has provided the opportunity for individual strategies of emancipation from the family and political tutelage systems on which social order was formerly based. Thus the pursuit of money has become a key element in current technical and economic behavioural strategies.

Also, a large proportion of agricultural and herding production is destined for the market (the majority in the case of commercial cropping areas) whereas almost all the technical factors of production (seed, fertiliser, pesticides, tools, etc.) have to be bought outside the rural areas. Given this situation, the agropastoralist production systems have become largely dependent on a whole series of all-encompassing economic conditions (outlets, prices, supplies) over which they have virtually no control. The major factors here are local or regional development projects, national economic policies and fluctuations in world markets.

Now, more than ever, it is no longer possible to consider rural societies as closed systems, guided solely by their own principles. From now on, the most common acts of material and social life (eating, dressing, even marrying), most technical decisions, often even decisions relating to the best use of the workforce (on the farm, in agropastoral production; away from the farm, through migration) are subject directly to market needs and constraints. As we have seen, Sahelian producers are very sensitive to variations in their technical and economic environment (Chapters 6 and 7). Thus the differences in technical practices and the different ways of using resources that we note from one locality to another are very often due much less to the diversity of the societies themselves than to differences in the overall conditions to which they are subjected.

In the end, data from the natural environment, demographic factors, modes of resource use, the influence of the market and development policies are all groups of variables that work jointly to differentiate environmental situations found throughout the western Sahel. When considered alone, none of them can provide the key to comprehension of this diversity. This means that we are confronted with an order of reality that cannot be grasped through an analytical model built on the principle of linear causality. The diversity of these local situations allows us a glimpse of the complexity of the multiple interactions that make up this dynamic. Figure 12.1 is a schematic representation of one such network of interactions, reduced to several core factors and strong associations.

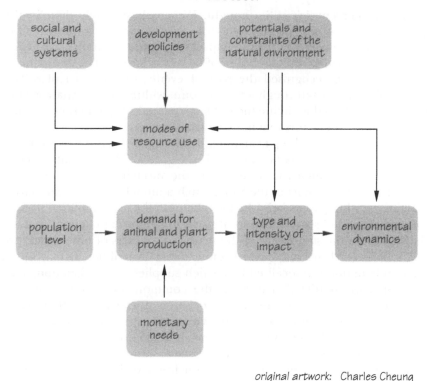

original artwork: Charles Cheung
computer graphics: Phil Bradley

Figure 12.1 Key elements in the analysis of modes of resource use in the western Sahel, and their articulation

What can be gained from this image of reality? To begin with, greater prudence when it comes to attempts to explain the current environmental crisis that the Sahel is experiencing. The harsh environment, population growth, monetarisation of the economy, insufficient development programmes – none of these, considered in isolation or accorded a primary explaining role, allows us to account for the concrete realities on the ground in all their variety.

In addition, there is the role of chance and event as sources of diversity. The complex systems model is at the heart of current efforts to understand the relationship between society and nature (Jollivet and Pavé, 1993). Extending an approach initially developed in the field of physical and biological sciences (Atlan, 1979; Prygogine and Stengers, 1991), the social sciences are now exploring the role of chaos in the life of societies (Balandier, 1989; Dupuy, 1990). We will not dwell here on a theoretical vision that, behind its current fashionableness, no doubt represents a major advance in current scientific thought. We emphasise only that, in the evolution of complex systems, the role of chance is significant. One modification, however tiny, or the impact of a single event on a particular element can disrupt the equilibrium of an entire system. The discipline of history has long recognised the role of events, of the particular. The epistemological breakthrough was to account within a single analysis both for the regularities that define the stable properties of system function and for the contingent factors that induce variable states.

The overall view resulting from a small-scale study shows the existence of certain major parameters (summarised in Figure 12.1), the multiple combined effects of which can account for the variability of the dominant environmental situations of the Sahel. Such a model, however, can hardly explain everything, especially when we abandon the oversimplified scale of our analysis and observe actual local situations. Thus, in contexts that are identical from the standpoint of the major parameters of our model, two neighbouring villages may present quite different realities. It may be an apparently minor and localised fact which supplies the explanation, such as an ancient conflict that disrupts the common use of resources, the personality of a village chief or of an association official, a migration stream that drains off the youth of a particular village. Many more examples of this type could be cited. Observations such as these do not at all invalidate the effort to understand the larger picture since, although events and local features play their role, they do so within the framework of a general context that circumscribes or orients their consequences. The effectiveness of individual initiative or the force of constraints inherent in a local situation notwithstanding, the reality found in one village in the old groundnut basin of Senegal (sector 1 in Figure 8.2, p. 189) cannot help but differ markedly from that observable in the Malian or Burkina regions of high population density (sector 7), or in the Interior Delta or the Bend of the Niger (sector 9).

The global and the local correspond to two distinct levels of observation of the same reality, levels that may reveal faces that are often quite different but ultimately indistinguishable one from the other. Each concrete situation encompasses a unique combination, contingent only up to a point, of elements that are always identical in nature yet variable in their place in the hierarchy of causal relations that bind them together.

Nevertheless, beyond these remarks lies a series of questions more general in scope:

- How are we to make use of these observations of interaction and variability?
- How can we create an interpretation that allows us to overcome the somewhat futile observation that 'everything is interrelated'?
- How are we to introduce the order necessary to understand and, if possible, to promote action?

ORDER IN COMPLEXITY

These questions cannot be answered merely by constructing a typology of concrete situations and a description of the interrelationships that characterise them. There is a general problematic here, that can only be addressed adequately by anchoring empirical facts in a more abstracted construct, through which questions of more general significance and application can be tackled. Local observations must also be resituated within the broader historical context that makes it possible to grasp their dynamic.

To explore the crisis occurring throughout the Sahel is, in fact, to apply to a particular reality a much broader problematic – one that deals with the nature and sustainability of relations between human societies and their environment. Superimposed on to the network of interrelationships, as proposed in Figure 12.2, is another, less descriptive but more powerful tool for understanding the situation. It posits a single all-encompassing system that may be designated the ecosphere, links two subsystems (nature and society) which interact and share some common features, yet are organised according to their own structural and dynamic principles (Jollivet and Pavé, 1993; Zanoni and Raynaut, 1994).

The use of this analytical model can help us to clarify the diversity that

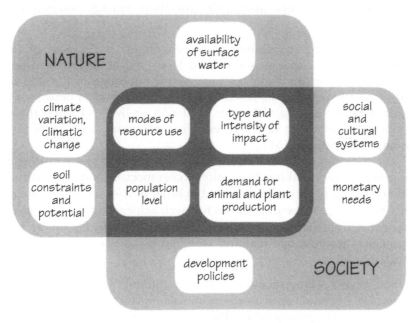

original artwork: Charles Cheung
computer graphics: Phil Bradley

Figure 12.2 Key elements in the relations between nature and society in the western Sahel

we have found in Sahelian environments and help us to understand the dynamics that underlie its variability.

Nature

The natural system encompasses all the biological and physico-chemical components that interact within the major spheres of biological organisation: atmosphere, pedosphere, hydrosphere and geosphere. This system includes a fringe that, although it can be analysed from the point of view of natural processes, is highly artificial due to its reconstruction by man. This is true of all lands occupied and exploited by humans. Humans are, therefore, included in the natural system as agents of change.

As long as man has been present, he has played an integral role in the dynamics of the natural system. This suggests that in analysing the Sahelian environmental crisis and its various local forms, one must begin with the interactions that operate between the systems of production and their specific natural environment. Without entering into a detailed analysis, one may say that a major characteristic of these systems has been their capacity, throughout the centuries, to maintain a sustainable environmental balance with only limited intervention from outside elements.

From this general operating principle there emerge two important consequences:

1 A direct relationship between plant productivity and a poor and uncertain rainfall, characteristic of arid lands, makes security a primary objective of agricultural and pastoral productive strategies. The choice of plant varieties, of farming practices and the movement of herds, are not geared to maximising yields per land unit or by head when there is sufficient rainfall but rather to ensuring minimum production in the event of drought (p. 151).
2 The critical importance of maintaining the soil fertility potential which is generally poor and fragile requires that, especially in these arid zones, any depletion of the soil be restored in order to re-establish equilibrium. For the most part, restoration was formerly achieved through extracting elements from the surrounding ecosystem, through fallows or transhumance, from animal manure, or by crop rotation or intercropping (pp. 138–49).

Under such conditions, and with the exception of particular cases such as irrigation, recession agriculture and horticulture, one of the constants of Sahelo-Sudanian agrosystems was the use, with respect to the environment, of extensive, light and mobile practices that favoured and took advantage of natural regeneration. Of course, such an adaptation to the conditions of reproduction of the natural surroundings did not prohibit the existence of a wide variety of possible techniques, evident in the diversity of agricultural and livestock systems, implements and modes of fertility renewal described throughout this study (Chapters 5, 6 and 7). It is none the less true that technical strategies had always to deal with a harsh climate and the physical and chemical fragility of the soils. Many examples can be cited, but let us simply recall the role of pastoral mobility in response to rainfall variability (pp. 111–12) and gradations in cultivated surface per individual according to the level of aridity (p. 150).

In exploring the recent evolution of these agropastoral systems, it is necessary to stress the obvious. The margin of flexibility available for increasing the volume of production was narrow in proportion to the severity of environmental constraints to which they were subject (see Chapter 1). This is not to be taken as an attempt to reintroduce some notion of natural determinism, for it is well known that such limits are not insurmountable. To exceed them, however, implies radical changes in the structure and function of production systems – what G. Bertrand (1980) calls the crossing of agro-technical thresholds. In this case, that means agropastoral systems have had to draw increasingly on elements from outside their local environment in order to restore a balance that internal mechanisms could no longer maintain. Fertiliser, pesticides, soil treatments, tools, agronomic know-how and food aid are all external elements intended to shift the threshold of equilibrium between a production system and its environment. This approach to the problem enables us to understand the evolution of relations between society and nature that is today most dramatically manifested in the Sahel in the form of 'desertification'.

There is, in this regard, a sharp division between two major situations. The first is when local populations act as a labour reserve for external poles of attraction and the second is when labour is locally invested in the development of agricultural and pastoral production, mostly for export. The environmental crisis is often most extreme in commercial agricultural areas because of certain major constraints:

- A rapidly and constantly growing demand for plant and animal production due to the combined effects of demographic growth and the development of a market economy, each potentially reinforcing the other.
- The low rate of internal capital accumulation that would be necessary to implement technical changes required by this increased demand; a low rate that is caused by the direct and indirect monetary demands placed on rural populations (p. 95) and the inadequacy of 'development projects' intended to enable them to invest in technological modernisations (see pp. 209–11).
- An increase in aridity throughout this century, with a marked acceleration in the last 25 years, which has significantly reduced the productivity of ecosystems, making them more vulnerable to human action (pp. 12–18).

Confronted by shortages resulting from an explosive demand, economically weakened, and facing a severely aggravated climatic risk, local production systems have had no choice but to react by continuing their dependence on time-honoured strategies. These have been pushed well beyond the limits dictated by a harsh and fragile natural environment, that is into a contradiction with the specific properties of the natural system. The result is the phenomena of environmental degradation and reduced productivity that are the most obvious features of the current crisis.

The example of Mali's southern zone (see above pp. 196–8) provides us with an alternative outcome to this breakdown of a balanced relationship between Sahelo-Sudanian societies and their environment. Over the last 20 years, technical systems in this region have changed radically, resulting in a huge increase in cotton production, accompanied by a significant rise in the production of food crops. Such an evolution was made possible only by an exceptional combination of favourable factors: the natural environment's potential made new equipment profitable and enabled labour productivity to be optimised; there was a supportive technical framework; and stable prices and effective market organisation made the economic environment secure. While disturbing phenomena of environmental degradation have indeed started to occur, especially in the heart of the cotton basin, they appear much later and thus far are still less severe than those observable elsewhere in the Sahel. This case highlights the fact that, depending on their natural environment and the technical and economic means at their disposal, not all the rural communities of the Sahelian countries have benefited equally from the same 'margin of flexibility' that would allow them to make critical adjustments to their relationship with nature.

It is incontestable that the Sahelian crisis is the manifestation of a disturbance in the biological and physico-chemical processes that interact within the natural system. It must, therefore, be analysed as such. In this regard, the exceptionally severe and long-lasting deterioration in rainfall conditions over the last 20 years has certainly contributed to the situation. Were it not for human impact, though, these climatic events would not have been as disastrous. However, this situation cannot be understood purely and simply as an 'increase in the rate of extraction of natural capital necessary for survival' (FNUAP, 1988). By definition, all resource exploitation constitutes an extraction. What makes a technical system sustainable is its capacity to restore to the ecosystem, directly or indirectly, what has been taken out. In order to frame the question of the disruptive nature of human intervention properly, therefore, one must explore not only the amount taken out but also the modes of restoration. The sustainability of the society/nature relationship is threatened when what is removed exceeds what is put back.[1] To approach the question in this way requires that we introduce into any analysis of the working of the nature system the dynamics of organised human action, that is, the functioning or disfunctioning of production systems. This cannot be achieved by examining only their technical aspects; we also need to take account of their social principles of organisation, resituating them within another organisational whole – society.

Society

Society is composed of all the elements and processes related to the organisation, reproduction and evolution of social relations and culture. Here, the coherences we seek to identify are rooted in the processes of the production and circulation of meaning (representations, values, norms), processes that are the product of a particular history and that remain largely independent of biological and physico-chemical determinations. Clearly, no social organisation can exist without a material base, composed primarily of, first, the people that make up the society; second, goods and objects (the production, circulation and consumption of which allow its reproduction); and, third, the necessary technology for creating those goods and objects. However, this material basis is an integral part of society only inasmuch as it contributes to the production of social relations and culture, or results from them.

In other words, this means that in analysing the interaction with eco-systems man cannot be considered purely and simply as one living organism among others, but must be analysed as the element of a social system, outside of which human behaviour has no meaning. To recognise this suggests that the immaterial, the '*idéel*' to use Godelier's (1984) formulation, is as much a part of the real as the material and, therefore, constitutes an equally legitimate inquiry for scientific analysis. Thus in this book, in an attempt to account for the social dimension of the variability of environmental situations found in the Sahel we have particularly emphasised two sets of factors.

The diversity of Sahelian social and cultural systems

Not all societies in the region are governed by the same principles of organisation. To restate, this diversity is reflected in the relationship a society has with nature and, therefore, the way it exploits the resources on which it depends for its subsistence. Beyond the obvious differences in ways of life, the most evident being the opposition between agricultural and pastoral societies, what are at work are criteria of differentiation having to do with the functional properties of social and cultural systems. In analysing this diversity, we have demonstrated the irrelevance of ethnic divisions, even though they offer convenient labels for description and constitute references at play on the contemporary political chessboard. In fact, ethnic divisions are largely arbitrary, for they are not the simple registering of stable cultural identities inscribed within the permanence of ancient tradition, but artificially freeze a continuous historical process of construction and deconstruction (pp. 215–19).

In order to understand the diversity of Sahelian social organisation, we have found it more fruitful to begin with the fundamental principles that govern and determine their specific functional properties. Thus, by crossing three axes – the degree of centralisation of power, the level of social stratification and the degree of tolerance of economic inequality – we have identified three major models of organisation to which all the concrete social formations found in the Sahel adhere more or less faithfully (pp. 230–2). A very different internal logic is at work when we compare lineage-based peasant societies with warrior aristocracies or large trading states, a logic that is based on social relations but is also reflected in the ways that nature is managed. Landholding relations (pp. 237–44), modes of controlling the workforce (pp. 262–71), the position of women and juniors in agricultural production (pp. 265–6), and even implements (note 7, p. 182 can vary according to the principles that govern the whole social and economic organisation. Sahelo-Sudanian societies are also distinguishable, one from the other, by the presence or absence of a centralised power that can impose collective rules for resource use (p. 221). Ultimately, the very role played by the exploitation of nature in the reproduction of social systems may be called into question. Thus it would require a deliberate ignorance of the social values that previously governed them (values that they continue to respect in other forms) to consider as 'agricultural' certain societies in which, as recently as the early twentieth century, war or commerce were the primary activities and in which farming was perceived as a degrading task to be performed only by slaves (p. 285).

Without taking into account the value systems, modes of thought and the collective norms that govern social practices, in other words the meaning they express, we cannot understand their deeper logic. Nor can we make sense of the opposing technical strategies, some wasteful of natural resources and others more economical, used by sometimes neighbouring groups, as is illustrated by the classic comparisons of the Sérer and the Wolof in Senegal or the *Moose* and the *Têngbîise* in Burkina (p. 274).

The force of change

Although, by their diversity, social systems contribute to the differentiation of Sahelian environmental situations, we must avoid creating a fixed image of them. These social systems are often composites and have never ceased evolving and transforming. Evidence of this comes from the permeability of ethnic borders, the rise and fall of great empires and the inclusion, either temporarily or permanently, of lineage systems into political entities governed by different principles. All of these changes are part of the long history of the Sahel, and have been amplified and accelerated since the beginning of the century by the creation of new nation states open to world political and economic influences.

Some of the major upheavals that Sahelo-Sudanian societies have had to confront this century include the end of wars, the abolition of slavery, the collapse of ancient commercial networks, the marginalisation of former political powers, the monetarisation of economies and the development of migration. Today, to a lesser extent than ever they can be seen as rooted in some ancestral 'tradition'. To understand properly the problems that these societies now face, as well as the character of their relationship with nature, we must view them as systems in the midst of profound change. These societies are engaged in a radical process of reconstruction. They do what they must to respond to new constraints and incentives, while attempting to preserve, to the extent that this is possible, their core identity. The adjustments that had to be made by Sérer society in order to protect its animating principle of equality at the same time as it embraced commercial agriculture is one example among many (p. 230). This process of change is made all the more complex by its exposure of the tensions and potential fractures that have long been present in these social systems, for example, between paternal and maternal lineages, or between dominant and subservient, and can often fuel conflicts of interest or power. Moreover, this dynamic is also associated with the weakening of collective structures, which in turn permits the emergence of individual strategies that profoundly alter older social frameworks, such as the individual quest for wealth and the beginnings of emancipation of women and the young.

With respect to the evolution of the social conditions for the exploitation of natural resources, two major areas are affected by these developments.

(1) Land relations The major phenomenon here is the growth of the idea of appropriation, which brings with it notions of privatisation and individualisation. These transformations take distinct forms, according to their respective social systems. Hence, very different landholding strategies are found in former slave societies compared to lineage societies (Chapter 10).

Nevertheless, all are following the same basic trend, which can briefly be described as a shift in the priorities for organising production. Formerly focused on control of the workforce, these priorities are increasingly dominated by the objective of control over land and the means of working it.

The land tenure question thus begins to take shape and is manifested at two different levels. Within a single social community, the seizure of available lands gives rise to confrontations between rival individual strategies. Between different communities, accustomed to exploiting jointly the same space, land possession tends to confer on the dominant group that takes control of it an exclusive right through which they can then oppose the use-claims of their former partners. Competition is particularly heated when agricultural and pastoral communities oppose one another, since their different conceptions of space are difficult to reconcile, giving rise to major and sometimes bloody conflicts.

It should be noted that land in the Sahel continues to enter the market, even if the manifestations of evolution are localised and discreet. Where this has taken place (notably in the Hausa country in Niger) land transactions become an essential element to landholding strategies, and it is largely through this that the 'new deal' of land operates. In the future, and in a determining fashion, this evolution will certainly contribute to a reformulation of the relations peasant communities maintain with nature.

(2) Labour relations This is another critical domain in the recent evolution of agropastoral production systems. The major trend here has been the emancipation of individual labour and at least partially its liberation from the institutional frameworks that formerly guaranteed its control. In this regard, three critical phenomena have marked production systems since the beginning of this century.

The first is the abolition of slavery – a major upheaval for many Sahelo-Sudanian societies in which servile labour had been of paramount importance, and where the dominant classes were forced against their will to become farmers or pastoralists, although without completely abandoning their ancient values and making every effort to preserve their privileges (pp. 272–4).

The second is the disintegration of domestic units which, in spite of history's many vicissitudes, had furnished the most solid base for exercising production relations. It has allowed the young to break free, at least partially, from the control of their elders. It has also started women on the path to greater economic autonomy (pp. 275–82).

Finally, within the context of economic monetarisation, the emancipated portion of the workforce can now offer itself anonymously on the market, detaching itself even more from the fabric of social relations. In some cases, wage-labour is used within a local community. Most often though, it is exercised externally and is accompanied by short- or long-distance migrations (pp. 282–5).

With respect to older forms of social control, this movement, with its three phases, towards the individualisation of the labour force and its emancipation from ancient forms of control has reached its ultimate stage, where the possibility now exists for substituting, at least partially, mechanised work for manual work (thanks to draft power from animals and, as in the case of southern Mali, even to motorised agriculture). Even though this change is still in its initial stage, it carries with it the germ of a profound upheaval. In the conduct of production strategies, the management of social relations is gradually being replaced by the management of machines.

Internal changes and the state dimension

The radical modifications in land and labour relations just summarised (and described in more detail in preceding chapters), are the most obvious manifestations of the new face of agricultural and pastoral communities in the Sahel. Of course, throughout their long history they have never ceased to change, but today they are facing a very different kind of change, on a very different scale. Until the recent past, no matter what adaptation they were forced to undergo, one principle was never threatened and remained constant, the principle that underpinned all the organisational configurations that the communities were likely to adopt. While lineage-based societies may have been absorbed into more centralised empires and these empires, in turn, split into small local aristocracies, and while some communities moved alternately between phases of agriculture and pastoralism, and war or trade played a changing economic role, never under the pressure of all these vicissitudes did the symbolic relationship of domination between individuals ever lose its place as the touchstone of the social system. The younger brother remained subservient to the elder, the woman to the man, the subject to his prince, the slave to his master. This submission comprised, above all, a shared adherence to a common system of representation, to the same social order. The changes we observe today are of a quite different order. Polanyi described a similar evolution in ancient societies as the 'great transformation', a development which affects the priority given to interpersonal relations over relations with objects (Polanyi, 1957). We shall look at two examples where this deep-rooted transformation can be seen.

(1) The internal recomposition of existing social systems Clearly, as described above, each society has interpreted the constraints it faces in the light of its own intrinsic properties. Nevertheless, through this diversity of reactions, the same powerful trends are at work. They foster the emergence of new stratifications where the social and the economic are combined and where inequalities are no longer reinforced by personal domination but by appropriation of the means of production, namely land, labour and equipment.[2] Although this transformation is not complete in all cases, the means of production are gradually becoming marketable commodities to which access depends not on one's social position but on a monetary transaction that may be divorced from any personal relationship (Chapters 10 and 11). New hierarchies are thus

307

springing up which tend to be free of any obligation to reciprocate (exchanging protection for submission), the principle on which to a large extent the old social order was based. As a result, the gap is slowly widening between a minority of producers with some prospects for the future and those who cannot overcome the obstacles confronting them. From this perspective, the successive crises in the Sahel over the past two decades have created a 'selection' process, whereby the most vulnerable go under and the strongest take advantage of their loss. We cannot comprehend the current disruption of the social organisation for resource use without taking into account the new forms of competition fostered by such a differentiation. A vision of these profoundly changing local communities as homogeneous and 'consensual' is artificial and far from the present reality. As noted earlier (Chapters 10 and 11), it is often competition such as this, inspired by short-term, individual strategies, that can be blamed for the disorganisation of the modes of natural resource exploitation.

(2) *The state dimension* Although extremely important, these internal transformations could not, alone, account for the dynamics that currently govern the reconstruction of relations between Sahelian societies and their natural environment. This is so because the competition already mentioned extends much further than the limits of a local society, whatever form it may take. More and more often, it brings into opposition different social groups exploiting the same space. Such rivalry has always existed but it is now developing in an entirely new context – that of the nation state.

Serious competition takes place at this level not just between long-standing partners (such as pastoralists and cultivators), but also among new arrivals from urban areas (politicians, merchants) who seek to enlarge their activity base through agriculture, pastoralism or by the exploitation of wood (pp. 122, 249, 258). These new users of natural resources have played a direct part in the damage done to the environment. Supported by new land legislation, they have also contributed to the disintegration of the old ways of controlling space and their presence has triggered a race to occupy the land.

In a broader context, among the factors which play their part today in reshaping the pastoral and agricultural communities and in the transformation of their relations with nature, one of the most important is the political and economic environment created by the actions of the state.

Through prices, taxes and many other direct and indirect forms of levy, the vast amounts of wealth that have been withdrawn from the rural areas over several decades have contributed extensively to the destabilisation of social relations and provoked a frantic rush into production. This predatory relationship was set up by the colonial powers and, following independence, was then maintained and strengthened by the new dominant classes, who thus found a way to enhance their own expansion (p. 95). Moreover, as we have been able to observe on many occasions, state policies, both colonial and post-colonial, have directly stimulated or curbed the material transformations in which local societies have been involved since the beginning of the century. We have also seen that it was impossible to understand such essential facts as demographic phenomena or the spatial distribution of production systems without taking into account the major politico-economic patterns which produced, in some places, areas devoted to cash crops and, in others, areas meant to serve as migrant labour reserves (Chapters 3 and 5).

Launching development programmes (some of which are extremely ambitious, like those along the Senegal River and in the Niger Inner Delta), forming a safe and stable economic environment and creating outlets for products are all beyond the reach of agricultural and pastoral societies. Yet such programmes have played a determining role with respect to the technical strategies applied by producers and, hence, also in the emergence of the environmental 'crisis' in the Sahel. We are therefore unable to assess the current state of relations between local societies and the land they exploit (some would say overexploit) without first taking into consideration the fact that these communities are now components in a social system of national dimensions, dominated by the state and its urban agents (merchants, bureaucrats). This larger system is itself subject to incentives and constraints from beyond national borders. International aid organisations share with national powers the heavy responsibility for the misconceived development policies carried out during the last 30 years and the inordinate debt that these policies have generated (Chapter 4).

SUSTAINABILITY AND CHANGE

The diversity of Sahelian situations therefore helps us to understand the extreme complexity of the interactions at work in the dynamics of the relations between society and nature. This is a fundamental question, for, more urgently than anywhere in the world, this part of Africa faces a challenge which it cannot ignore. This is the challenge of creating a form of development that allows societies, whose very survival is threatened, to preserve a future in a fragile environment without exhausting its resources. In a local context, this is the question of sustainable development. Formulated in the light of the preceding analyses, the question therefore becomes: what dynamics operate at the interface between nature and society? While we do not claim to have addressed this question in our study, several points may help advance a discussion of the topic. As a preliminary conclusion, we focus on two theoretical points that seem particularly relevant.

Reproduction and change

An inquiry into the issue of sustainable development suggests that we examine the conditions for the linked reproduction of both systems: nature and society. Here, the paradox resides in the fact that if a system is to endure, it must transform itself, change being the very condition for reproduction. This is true in ecosystems that are constantly adapting. As G. Balandier (1989) has shown, this is also true for social systems, even those where 'tradition' appears to govern strictly social life and to fix history. The example of Songhai society is a perfect illustration in this regard, where a major change in the status of land (the creation of a 'pseudo-customary' right) perpetuated the domination of the aristocratic class over the rest of society, the principle of hierarchy having been integral to this social system (p. 241).

The transformations of natural and social systems are closely interrelated. As we have described, the soil and vegetation degradation in the Sahel cannot be understood without taking account of the evolution of social strategies for the use of these resources. Conversely, such phenomena as the exacerbation of tensions between agriculturalist and pastoralist, or the emergence of new economic stratifications within peasant communities, cannot be analysed without reference to land scarcity or a climate that is becoming increasingly arid. Understanding these changes and the reciprocal principles that govern them is thus a necessary precondition for any conception of the future of Sahelian society. Here, as elsewhere, sustainability does not mean that history stops, with reproduction endlessly seeking an impossible equilibrium. It involves, on the contrary, a capacity for constant change, a constantly renewed conflict between the reproduction demands of nature and societies.

Material and immaterial dimensions of reality

In order to understand the dynamic of social practices, whether in the environment or in other areas, we must accord humanity its full dimension – that of the production and manipulation of meaning. Contrary to what is reflected by the behavioural models imbued with a simplistic rationalism that too often inspires development actions, in the final analysis the driving force behind technical and economic choices is the defence of ideas, images of identity and secular or religious values. Once basic biological reproductive needs have been met, a material advantage is perceived as such only in terms of the social value attributed to it, or of its utility in pursuing social objectives. We have demonstrated how the perception of wealth, the value placed on labour and the position held by competition and inequality among individuals can vary significantly from one Sahelian society to the next, with evident repercussions for the forms taken by man's relationship with nature (Chapters 10 and 11). Hence, the different values placed on working the land and the contrasting technical practices observable in Sérer and Wolof societies, the Sénoufo and the Bambara or the *Têngbîise* and *Moose* are the expression of very distinct systems of representation and values.

The major question that arises here is that of the way in which change operates in the domain of the immaterial. As a general rule, a great stability is ascribed to forms of thought in African societies. However, the appearance of stability is often illusory, reinforced by the long-standing image given by an ethnology more concerned with reconstituting a permanent 'tradition' than with describing its evolution.

There is abundant evidence of transformations in behavioural norms, in values and even in religious beliefs. The example of the break-up of older lineage cults or the massive push by Islam today furnish convincing examples. Nevertheless, evolutions can also occur within ancient belief systems as evidenced, for example, by the advance of beliefs and practices linked to sorcery (Schmoll, 1993).

Ideas are therefore subject to change. Yet, at times, they can also show a great capacity for resistance. How often are taboos or prejudices that 'obstruct' the spread of 'progress' put forward to explain the difficulties of development programmes! Even in this work it has been shown how certain Sahelian societies still remain attached to identity references and principles of organisation that, in spite of the upheavals they have had to endure, continue to form the stable core of an ancient heritage of norms and values.

Faced with these contradictory observations, one question stands out: why do certain ideas rapidly disappear while others persist? The classic Marxian analysis responds to this question by considering culture as a superstructure determined, in the end, by the material conditions that reproduce the social system (see, for example, the analyses by Meillassoux (1982: 57–96) concerning domestic reproduction and kinship subsistence structures). The less rigid interpretation adopted by Godelier (1984) suggests that an institution (familial, religious, political) occupies a dominant place in society to the extent that it structures production relations. If this is so, we can deduce that values, social norms and religious beliefs are stable to the extent that they are closely aligned with the organisation of production, and change only when production is transformed. This interpretation is confirmed by several observations. As was mentioned earlier, the decline of ancient agrarian cults in favour of Islam can, to a large extent, be explained by their inadequacy in the face of new conditions of production imposed by land scarcity and the growth of competition among individuals. Yet this analysis is not entirely satisfactory, since if production relations and technical practices are to change, ideas and knowledge must also have begun to evolve. Thus we have

CLAUDE RAYNAUT

seen that Sénoufo and Sérer societies did not adopt commercial crops and agricultural changes without a long period of maturation that allowed them to make these changes socially acceptable.

In fact, relations between a social system's material and immaterial aspects should perhaps be conceived as the product of a reciprocal process of adjustment, with each taking its turn in the dominant role. The key notion here is undoubtedly that of compatibility. When change is provoked, when a new technique is proposed or a new idea emerges, it will take hold more rapidly when it is not in direct contradiction with the order of material things or that of social relations. Sahelian peasants and pastoralists were quick to adopt technical objects from the industrial world, such as radios, bicycles and watches, although they were quite resistant to the plough and to agricultural improvements developed by agronomic research. Such reticence does not reflect an a priori rejection of modernism, but rather a certain prudence when faced with changes liable to disrupt the coherence of social and technical systems or to contradict high priority options, especially that of extensivity. In the realm of the immaterial, although the acquisition of new knowledge by the young through schooling was rapidly accepted and even sought out for its apparent utility regarding social and economic ambitions, opposition to the emancipation of youth, women and former slaves has been long-standing and fierce. This emancipation is perceived as a threat to the power of the dominant social categories of men, elders and nobles.

In general, however, even when these new elements were rejected, they did not completely disappear from the social landscape. Technical objects were pushed back to the periphery, so that, for many decades, animal traction agriculture, which has been promoted throughout many parts of the Sahel since the first quarter of this century, remained the prerogative of only a small minority of 'pioneer' cultivators. Many nonconformist ideas were taken up by a fringe consisting of groups or individuals, such as migrants, 'free' women, political militants or operatives, who more or less openly contested the dominant social order.

Things can remain static for a long time. Then the moment comes when the balance between power relations changes, for example, women become increasingly integral to production operations or the earnings of young people become indispensable to a family's survival. Then incompatibilities are reduced and change accelerates. Practices that had thus far appeared destined to remain marginal suddenly take off and increasing numbers of young people seek out wage-labour and more women start acquiring fields. In the domain of production, often it is clearly the constraints engendered by a monetarised economy and a scarcity of land that make desirable certain changes that were previously considered incompatible with the established order.

It is, therefore, a complex dynamic that drives the joint transformation of the material and the immaterial functional bases of a social system. New norms, values and techniques only become accepted after a process of adjustment through which they are made compatible. This process can be rapid when the effect of change remains localised or, again, when it benefits a social category in a position of power (as was frequently the case in the evolution of land structures in aristocratic societies). Or it can be very slow when it threatens the coherence of well-established social or technical organisations or when it is likely to favour a subservient social category.

Thus, a change in the relations between a social system and the environment it exploits acts on various levels: organisational and material levels, as well as that of ideas. Often, it also has a political dimension, using the word 'political' in its broad meaning of a clash of powers. Thus it has to do with a complex dynamic that cannot be captured by mechanistic models. In turn, this implies that change may contain within it a large element of chance, to events, to local initiative – isolated elements that allow a given situation to move on, while tension and dysfunction continue to accumulate elsewhere.

If we are to advance our understanding of this reality, we must proceed beyond the level of generalisation to which we have so far limited ourselves. Among the obstacles to a synthesis of the relations between society and nature in the Sahel, the most important has certainly been the absence of systematic multi-disciplinary studies that could explore this subject in depth, based on analyses of specific situations. Mostly we have had to settle for isolated, heterogeneous and disparate information. There is a need for studies of this sort if we are to arm ourselves with the analytical tools necessary to bridge the gap between the general principles of interaction between nature and society, on the one hand, and, on the other, the conditions that give rise to specific local situations, which can never be completely accounted for by the application of these general principles. Our reflections throughout this study and our use of the zones we have established as an instrument for describing diversity provide the basis on which several such studies could be founded. Each could be conducted in a different locality and illustrate a distinct situation. Such studies would make possible a comparative work which would test the explanatory capacity of the variables we have identified, the most important of which are summarised in Figures 12.1 and 12.2. It is not our intention to elaborate such a research programme in this study, although we hope our effort will encourage others to take on the task.

ACTION: SEVERAL AXES FOR REFLECTION

To close this scientific analysis, we cannot rest content with a conclusion calling for more research. The need for deeper knowledge is crucial if we are to act less blindly. Nevertheless, the little that we already know allows us to make some recommendations. Our formulation of these recommendations is here deliberately brief, as we believe that the evidence of the preceding pages provides adequate argumentation.

CLAUDE RAYNAUT

1 The first practical lesson to be drawn is the necessity to abandon the dominant conception of development as an intervention intended to introduce change into societies that are stuck in traditions that are responsible for their economic and material stagnation.

As we have seen, the situation is quite different; it is not a matter of tradition but of rapid transformation. Not one of the situations we have described is exempt from the tensions, the social or material contradictions that are rooted in history and that must be overcome if a lasting relationship between human beings and their environment is to be established. The Sahelian 'crisis' is the manifestation of a change in the relations between society and nature and all interventions must be conceived as participating in this dynamic. There are two practical consequences in taking this approach:

- First, is to start from the efforts made by local societies themselves to find answers to their own problems, in conformity with their conceptions of themselves, with the social forces that shape them and with the options that direct their technical practices. Thus, what is called for is a complete reversal of perspective. The much called for 'participation' must be not only the participation of 'populations' in the projects proposed to them, but especially the participation of 'developers' in the ongoing dynamics within which their actions will necessarily be inscribed.
- The second consequence, arising out of the first, suggests that these 'developers', regardless of their origin (national or foreign) or level of intervention, are not the neutral and disembodied agents of an immanent economic or technical rationality. They are necessarily actors who, willingly or not, are party to conflicts and demands that accompany the present economic and social changes. This fact, long hidden by the monolithic bureaucracy of the political systems in place, is now plain to be seen, given the weakening of the state and the rise of popular movements.

2 Natural resource use is inscribed within a shared space, which is subject to diverse strategies of control and appropriation. Nothing can be grasped about the current crisis or its various manifestations without consideration of the rivalries that confront (and the solidarities that unite) the many competitors and partners on the environmental scene, not only nomadic pastoralists, agropastoralists and urban livestock investors, but also woodcutters and, locally, fishermen and hunters. We have described in much detail the movements of conquest and withdrawal found throughout the Sahel, the phenomena of exclusion and those of territorial overlap. No sustainable project for the use and development of land can be attempted without a straightforward approach to this problem, not in order to propose 'miracle' technical remedies for reducing tensions but to initiate

314

negotiation among the parties and to ensure that development actions that benefit one category of users are not undertaken at the expense of their potential consequence for others. For example, developing hydro-agriculture in the large valleys or promoting animal traction agriculture that favours the use of land spared by manual cultivation will only succeed in exacerbating latent conflicts if no consideration is given to the problems this creates for pastoral communities who exploit the same spaces. The precolonial state of Macina took steps to reconcile the different modes of resource exploitation practised by the many populations who occupied the Central Delta of the Niger. Of course it accomplished this as it saw fit, that is according to the hegemony of Fulani pastoralists. The challenge to which they attempted to respond remains, nevertheless, even more real today. In the Sahel, the relations between nature and societies must be conceived in the plural which means that, in order to be 'sustainable', development must not be blind to the diversity of users present within the same space. On the contrary, it must help them to search for solutions to the rivalries that cause conflict. When reconciliation is not possible, arbitration is necessary, and this is where the problem can take a political dimension. Here, once again, the solutions are far from strictly technical. Social relations and conflict are often the problems that need to be managed if environmental problems are to be resolved.

3 Within countries and even more at the level of the Sahelian region as a whole, contrasts among situations are extreme. Ample evidence of this is found throughout our study. One of the challenges today is to manage this diversity. We are not confronted with the simple juxtaposition of irreducible local realities. This demographic, economic, social and en-vironmental mosaic also reflects complementarities, inequalities and relations of dominance and dependence between zones – a dimension primarily manifested in the movement of people and goods. While in some cases these disparities are the result of natural causes, more often they are the product of history – a history more or less recent, yet the effects of which are generally not irreversible. Of course, the consequences of the high population densities found in some of the major settlement zones is certainly cause for concern. However, this should not obscure the exist-ence of vast areas in which human presence is sparse and all the more so because of the emigration that drains them. We must direct our inquiry to the reasons for this heterogeneous population distribution pattern and to the dynamics that must be set in motion in order to change it. Certainly, success will not come about through the planned displacement of popu-lations – we saw how ineffective this strategy was in Burkina Faso and eastern Senegal. Our reasoning must proceed in terms of land manage-ment, the creation of basic infrastructure and of finding ways to facilitate the movement of people and merchandise. The scale at which these efforts

are conceived must also be larger than those of the nation state. Moreover, a framework must be created that allows community and individual strategies to be carried out autonomously, as is currently the case in the southernmost part of Mali, where intense population movements are taking place. Beyond the question of the spatial distribution of people, a territorial view of development is necessary in order to conceive a reasoned use of natural resources, taking local specificities into account. This means rejecting actions based on a standard 'technological package', while at the same time gaining enough distance to integrate the greater scale on which the strategies of Sahelian populations, largely founded on mobility, are deployed. Having said this, we want to stress the limitations of a *gestion de terroirs* approach, which by focusing on the micro-level can often lose sight of the essential larger picture.

4 Conflict management, arbitration between divergent interests, territorial development – these functions require the action of a state that is exercising its prerogatives to the full. Certain analysts have questioned the reality or 'legitimacy' of the African state, although in a number of countries in the region the recent disintegration of the mechanisms of authority and the accompanying disorder eloquently demonstrate the existence of this structure. In fact, to a large extent, current problems can be attributed to the economic and political processes that have contributed to the construction of the state apparatus. Sahelian states have certainly been predatory, often repressive and even violent. Yet almost everywhere, year in, year out, they fulfilled some of the basic functions without which a country cannot exist; functions such as the development of infrastructures in the collective interest (we need mention only the extension of road networks between 1960 and 1980), health and education, market controls and land use regulations and the distribution of agricultural equipment, to name but a few. Of course, these accomplishments were often poorly conceived and their profits frequently diverted to personal interests. Nevertheless they corresponded to a level of intervention that only the state can achieve and which it is impossible to do without.

In our analysis, we have strongly emphasised the diversity of local situations and the force of the dynamics that make peasant and pastoral communities, for better or sometimes for worse, the agents of their own transformation. We have stressed the necessity of taking into account the complexity and specificity of these dynamics. One reality cannot be forgotten, though, and that is that popular initiative, regardless of its potential, is always governed by specific aspirations that, should they diverge, no 'invisible hand' can reconcile. Essential though it may be to take into account the problems experienced by the populations involved, it is rare to find a solution that does not entail interventions at a higher level in the form of equipment, supplies, regulation, etc. The decisions and execution of such interventions are out of the hands of the local community.

This means that one must not, as is still done all too frequently, draw a sharp distinction between 'bottom-up' development, which is based on community dynamics and supposedly the only truly effective form of development, and 'state or top-down development' which is a synonym for waste and inefficiency. The time of the great projects is undeniably over. We must loosen the grasp of the burgeoning bureaucracies, whose only real objective inevitably becomes their own reproduction. Yet the need remains for a level at which more global and long-term problems can be addressed, a level at which, when divergent interests come into conflict, arbitration and choices could be effected. In this regard, a major characteristic of the current social evolution resides in the appearance or the strengthening of peasant forms of organisation, such as the cotton and cereal producers union in southern Mali or the peasant federation in the Bakel region of Senegal. In the context of openness created by the recent 'democracy' movement, these organisations appear to be the basis of new power relations between the centralised power and a rural world previously disorganised in its relations with the central state. This new socio-political reality is likely to create a new foundation on which to build the restructuring of relations between the societies of the Sahel and their environment. This is an area of study not dealt with in this work and it could legitimately be suggested that this is a serious omission. However, we believe that the general descriptive and reflective framework we have provided will be useful for others who might take up this analysis.

BEYOND THE SAHEL EXPERIENCE

The facts analysed throughout this work raise questions that reach far beyond the Sahel in their scope. They represent the fuel for a much wider-ranging debate, one that touches on the ways in which rural African societies have been able to respond to the major challenges facing them during this century. The main challenges have been the following:

- coming to terms with new cultural models;
- submitting to new political and administrative forms of organisation;
- integration into a market economy that is very much opened up to the exterior; and
- an ever-growing increase in subsistence needs, imposed by a very fast rate of demographic growth.

Given these profound changes in the very basis of the functioning of society, one is inclined to ask to what extent have they been able to maintain their collective entities and respond to the new aspirations of their members and, in order to achieve this, how have they been able to reshape their relations with nature from which they continue to draw most of their means of existence. One also wonders what perspectives for the future are now opening up. These questions are relevant for almost all rural societies in Africa, but particularly so in relation to the semi-arid zones, as a result of their very rigorous natural constraints and the high degree of risk they impose on human activity. It is for this reason that the case of the Sahel is a good starting-point for reflection.

The Sahel, even when reduced to the dimensions we have given it for the purposes of this study, is vast and varied. All our reflection has drawn on this diversity and what it showed us of the dynamics involved in the transformation of relations between society and nature in the context of what is normally called 'development', but which in fact corresponds more particularly to an historical process. This process has drawn the agricultural and pastoralist communities into the movement to construct new political entities (colonies, then independent states), while at the same time ensuring, in different ways and in distinct positions, their integration into a market economy. Studies in other geographical contexts have posed the same questions as ours, in particular, a recent study on a semi-arid region of Kenya, the Machakos District (Tiffen *et al.*, 1994). By bringing together and comparing the conclusions presented here and those in the study just mentioned, we can usefully stimulate reflection and reach a more general basis of questioning.

The scale of the work carried out was not the same in the two studies. In our work we covered a portion of the African continent whereas the Kenya study covered an area of a few tens of thousands of square kilometres. The methods used were also dissimilar. The great variety of localised situations, cultural and social realities that are contained in the western Sahel provide a wide basis for a comparative approach, whereas the research concentrated on Machakos District enabled an in-depth historical analysis to be made of the evolution of the Akamba society and their practices. The contexts also display considerable differences: first the natural context (relief and climate), next the colonial past (in Kenya, a policy of 'reserves' and direct cultivation of the land by the European colonisers), and last, population density, which in Kenya can reach 400 inhabitants per square kilometre and bears no comparison with numbers in the Sahel. Despite these differences, a certain number of common problems emerge, offering the possibility of some enlightening comparisons. Moreover, the two studies deal in depth with distinct aspects of the relations between society and nature and complement one another in analysing factors that contribute to their dynamic.

One of the major contributions of the Kenya research is an extremely detailed analysis of the complex network of interactions linking demographic growth and the conditions for exploiting natural resources. In our own study, we compared local situations which enabled us also to show that population densities and the intensity of the exploitation of the environment are by no means associated in a direct and linear fashion. Their relationship is influenced by other factors such as production techniques, economic constraints and opportunities and the importance of migration. The Machakos study, because of the wealth of factual data on which it is based, enables a much deeper analysis to be made of this nexus of interrelations. In particular, it allows us to dismantle the concrete mechanisms through which population growth exerts a positive role on the evolution of technical practices to achieve better productivity from work and from the land, and a better conservation of natural resources. A most stimulating contribution, from this point of view, is the way it shows how emigration (often presented as an adaptation strategy to the saturation of an area and as the consequence of a situation of environmental deterioration) can contribute to improving the exploitation of natural resources in the home base, by favouring an initial accumulation of capital and skills, indispensable to improvements in production systems.

In the Kenya study, we also find an eloquent illustration of a dynamic which has only been felt to date in very localised areas in the Sahel, i.e., the role of demographic growth as both a stimulus to increases in productivity (see the outline suggested by Boserup (1970)) and as a supplementary resource which makes feasible long-term resource management practices requiring a heavy investment in labour. One simple observation proves the point. In the 1930s, when population density was less than 100 inhabitants per square kilometre, observers noted that peasant production systems were in crisis and that the environment was under severe threat. Sixty years later, with population density in some areas reaching levels four times as great, there are more trees; successful soil conservation actions are more common and are carried out almost spontaneously; and both agricultural and non-agricultural incomes have increased considerably. Note that an earlier study, also in Kenya but in ecologically different regions, pointed out a positive link between population density and the volume of woody biomass (Bradley, 1991). In northern Nigeria (Kano region), Mortimore (1989) also showed that very high population densities were compatible with sustainable forms of natural resource management. In the Sahel, the examples of the Sérer (Pélissier, 1966) or of the Dogon (Gallais, 1965) have long supported the same conclusion. In the Maradi region of Niger, various forms of technical innovation can be seen which indicate a search for a new environmental equilibrium in densely peopled areas (Yamba, 1993). It was many years ago that Raulin (1967) first suggested that the transition from one agricultural implement to another, from the iler to the hoe could have been due to a population increase.

A major contribution of the Machakos study is that it documents a technological mutation in a detailed fashion and identifies the conditions that made it possible. In particular, it brings to the fore the role that the wider economic environment now plays in the context of monetary exchange, namely collective infrastructure (roads, schools, clinics), the level and stability of agricultural prices, guaranteed supplies and the possibility of non-agricultural income. We found we were making the same observations when analysing the only example in the Sahel where a massive intensification of rain-fed production systems is under way on a scale similar to that in Kenya, i.e., the cotton-growing zone in southern Mali (see above pp. 196–7). However, this area differs markedly from those mentioned above. The climate is not semi-arid and it therefore enables better productivity to be achieved from modern technical equipment (ploughs, fertilisers). Here, the technical transformation is not the result of great demographic pressure. On the contrary, it is partly the existence of a large area of available land that has contributed to the best use being made of animal traction techniques. Technological change, even though possible only with the support of local communities, received its impetus from a voluntarist intervention on the part of a development body which created an environment favourable to its emergence, through prices, marketing networks, infrastructure, passing on know-how and adequate supply of the factors of production.

The example from Mali confirms some aspects of the 'virtuous circle' that links population increase to an intensification of technical practices, as proposed by the authors of the Machakos study, but it also suggests that this model should be considered according to context. Certainly, Tiffen *et al.* (1994) approach the matter with a degree of flexibility, but their demonstration (although totally convincing in the clearly defined context in which they are working) must not be inappropriately generalised in order to sustain a new 'dogma', inspired by evolutionist theories, which would then replace the 'vicious circle' model within which, until now, the link between population growth and environmental degradation has been confined. Our observations on the wide diversity of situations in the Sahel can perhaps help to maintain the flexibility needed when applying the 'virtuous circle' model, and so we stress a few of the more important points.

If, as seems to be the case in the Mali cotton-growing zone, major changes other than population pressure can affect the relationship between society and nature, then conversely, the existence of high population densities does not necessarily lead to intensified technical practices. The example of the Burkinabe Yatenga is particularly relevant here. Indeed, Marchal has earlier pointed out that this example contradicted Boserup's argument (Marchal, 1983). There is clearly major migration out of the area, but this is not accompanied by any significant investment in the agriculture of the home base. In contrast to the Machakos case, Marchal observes an acceleration in extensive, land-consuming practices. Perhaps there is simply a time lag in the process, which will develop differently at a later date. In Niger's Maradi region we observed a similar move towards extensive practices in conditions of rapid demographic growth (Raynaut, 1984), but it is now possible to record the beginnings of an intensification. This variability in the phenomenon, whereby the move towards intensifying technical practices begins, raises a certain number of questions.

1 It is essential constantly to bear in mind the context into which the local situation being analysed is integrated. This context is more than simply a question of predominant prices, outlets and infrastructure. It also extends to cover the place occupied by the area under study in the context of a geographical distribution of work which can encompass vast stretches of the continent and form part of a long-term historical context. Thus we have seen (Chapter 3) that, since the colonial period, the exploitation of western Africa has produced a spatial fragmentation (l'Office du Niger developments, cocoa and coffee plantations in the Ivory Coast, groundnut regions of Senegal and Niger) which has defined regional specialisations where, in one area, we see the development of crops for export while, in another, we find the maintenance of a labour reserve which can be drawn upon when extra hands are needed elsewhere. For a long time, investment distribution and the geographical distribution of infrastructure were to a large extent dictated by this type of geo-political role-sharing. Even though much has now changed (especially with road infrastructure – Chapter 3), the past has left a legacy which still touches the present. Thus Burkinabe emigrants, who are accustomed to dividing their labour between their village of origin (where they still maintain their social roots and their cultural identity) and their Ivory Coast cocoa plantations (where they put most of their capital), invest as little money and labour as possible in the land of their forebears. In order to become more widespread, the complex model of interactions which links population growth with the transformation of conditions for exploiting resources in a given region must broaden its horizons. It must even be prepared, if necessary, to 'delocalise' in order to take into account the way the given area is integrated into a much wider political and economic domain.

2 The differences in the options available to social communities in relation to the exploitation of nature, even when they are in similar natural and economic conditions, show that the universalist paradigm of the market only takes partial account of the true conditions in which practices evolve. Indeed, this paradigm tends to reduce collective behaviours to a combination of individual choices, where each actor is guided by his will to ensure his own survival and that of his family, and by the wish to maximise his profits by taking advantage of all possibilities and overcoming any obstacles in the way. Whatever the degree of disintegration of political and family structures, no Sahelian rural community can currently be considered as a group of individuals simply organising the fluctuating equilibrium of their

321

interrelations, whether complementary or competitive. Each carries with it a certain number of values and representations which are common to all its members, which constitute their social identity and which, despite the particular objectives that individuals may have, guide their choice in technical and economic matters.

Different examples have shown that, according to the type of social system of which they are part (lineage-based peasantry, warrior aristocracy, major centralised market state) the farmers and herders of the Sahel view their relations with nature and whatever they can draw from it differently. According to different societies and places, attitudes to working the land, to trade or to the accumulation of wealth can vary. These differences may help to explain why a dynamic observed in one group is not confirmed in the neighbouring group and why, for instance, the Sérer and the Wolof, the Sénoufo and the Bambara, the Lobi and the Mossi have not developed the same forms of agrarian civilisation. Each social and cultural system has its specific qualities which tend to be reproduced through individual behaviour. This dimension of reality should be taken into account if we are to hope to understand the divergent dynamics observed in local situations.

3 In the course of this work, we have often discussed the relevance and the limitations of the concept of 'social reproduction'. We have insisted on the fact that collective values and representations are not imposed by an immutable tradition. On the contrary, they are constantly subject to change. However, except in the extreme case of total social collapse, these changes give rise, in turn, to new forms of organisation and new collective practices that cannot be fully understood from the sole standpoint of individual behaviour. Sahelian societies, and probably many societies in Africa as a whole, are undergoing an internal upheaval. As we have shown, the social conditions for controlling production factors (land, labour, tools) are a key arena in which this reshaping is taking place. It is for this reason that, when we note phenomena such as a global increase in productivity and improvements in the state of the environment, we should also consider the transformation in the social and economic relations that have accompanied these changes. In the Sahel we have seen that the emergence of salaried workers and land transactions is gaining momentum. In some cases, these market relations already play a fundamental role in the emergence of new social configurations. To a certain extent these new social configurations are based on ancient values and social divisions, but they also incorporate new hierarchies, new forms of co-operation and new conflicts of interest. This renewal is all the more marked in that one of the features of the recent changes in the Sahel is the increase and the diversification in natural resource users, such as peasants, pastoralists, urban investors and woodmen, who often find

themselves all in the same space and entering into competition. It has often been the case that the succession of environmental crises suffered over the last decades have been overcome as a result of a redistribution of control of the land, of the livestock and of the workforce. This movement has operated to the advantage of some categories of producers and to the disadvantage of others. Up to a certain point, the exodus towards the towns and the increased urban growth thus produced were the consequences of this evolution. Such social recompositions cannot be ignored in constructing a model of interactions, or its descriptive value might be considerably diminished and its relevance in the formulation of a scenario for the future might be reduced.

These are just some of the questions that we ask when comparing the results of studies carried out in different areas, and they prove the importance of a comparative approach. Only by observing the diversity of realities on the ground can we create conceptual tools which will enable us to understand and assess the dynamic of society/nature relations. In each case, these relations must be analysed individually, not only taking into account their purely practical and concrete dimensions, but also the extent to which they have been able to draw on the ability of human societies to picture themselves and to represent the world that surrounds them.

We conclude with one final remark. It concerns the underlying question of this entire work: what contribution can scientific analysis and, in particular, the social sciences make to the creation of a strategy of sustainable development, which is nothing less than the Promethean project of intervening in society/nature relations in all their diversity and complexity? The danger would be to fall prey to what we have called elsewhere the 'scientific illusion' (Raynaut, 1989a), the belief that more science, if only it is placed in the service of a holistic approach to the facts of development and the environment, would be able to supply the scientific basis of a consensus or a tool for manipulating reality. Only in exceptional cases can more knowledge, or broader knowledge that takes account of social dynamics, lead to the identification of a single solution to a problem. What it can provide, however, is a clarification which will help to identify the natural, economic and social dimensions of a situation and the means for all concerned partners to grasp them. On this basis, negotiation becomes possible – negotiation based on validated arguments that can best harmonise the interests and aspirations of all parties. In this sense one can say, with implications far beyond the Sahel, that democratic procedures, enlightened by shared knowledge, can give new flexibility to the resolution of the problems of environment and development, thus contributing, however imperfectly and in ways that must constantly be reinvented, to the search for a lasting balance in the relations between the systems of nature and society.

NOTES

1 In industrial economies, there is also the problem of removing what has been introduced into nature and disturbs its ability to function. This is a question of waste disposal, which is not at all relevant to the case under study.
2 Even if, from the point of view of the beneficiaries, the new inequalities simply prolong those that existed before, which is not always the case, their basis is to be found in entirely new principles; hence the former aristocrats have replaced domination over their slaves with landownership.

BIBLIOGRAPHY

ACCT (1987–) *Médecine traditionnelle et pharmacopées*, Bulletin de Liaison, Paris: ACCT.

Adams, A. (1977) *Le long voyage des gens du fleuve*, Paris: Maspero.

Afrique-Agriculture (1990) 'Hydro-agriculture au sud du Sahara', *Afrique agriculture*, no. 177, Etude spéciale no. 8.

Amin, S. (ed.) (1974) *Modern Migrations in West Africa*, London: Oxford University Press.

Amselle, J.-L. (1990) *Logiques métisses. Anthropologie de l'identité en Afrique et ailleurs*, Paris: Payot.

Amselle, J.-L., Baris, P. and Papazian, V. (1982) *Evaluation de la filière arachide en pays mossi, Haute-Volta*, Paris: Ministère des Relations Extérieures.

Amselle, J.-L. and Grégoire, E. (1987) 'Complicités et conflits entre bourgeoisies d'état et d'affaires au Mali et au Niger', in E. Terray (ed.) *L'Etat contemporain au Mali*, Paris: L'Harmattan: 23–47.

Ancey, G. (1975) 'Niveaux de décision et fonction objectif en milieu rural africain', note no. 3, duplicated paper, Paris: AMIRA/INSEE.

—— (1977) 'Recensement et description des principaux systèmes ruraux sahéliens', *Cahiers ORSTOM, Série Sciences Humaines*, XIV, 1: 3–18.

—— (1983) *Monnaie et structure d'exploitations en pays mossi, Haute-Volta*, Paris: ORSTOM.

Atlan, H. (1979) *Entre le cristal et la fumée*, Paris: Le Seuil.

Bâ, Amadou Hampâté (1991) *Amkoulell, l'enfant peul*, Arles: Actes Sud.

Ba, Cheikh (1980) 'Le Nord du Sénégal', in L. G. Colvin (ed.) 'Les migrants et l'économie monétaire en Sénégambie', duplicated paper, University of Maryland: 120–46.

Baier, S. (1980) *An Economic History of Central Niger*, Oxford: Clarendon Press.

Balandier, G. (1989) *Le désordre, éloge du mouvement*, Paris: Fayard.

Barral, H. (1968) *Tiogo (Haute-Volta)*, 'Atlas des structures agraires au sud du Sahara' no. 2, Paris: ORSTOM-Maison des Sciences de l'Homme.

Barth, H. (1965) *Travels and Discoveries in North and Central Africa*, new edn, 3 vols, London: Frank Cass & Co.

Bayart, J.-F. (1991) 'La problématique de la démocratie en Afrique noire', *Politique Africaine*, no. 43.

Bazin, J. (1975) 'Guerre et servitude à Ségou', in Cl. Meillassoux (ed.) *L'Esclavage en Afrique précoloniale*, Paris: Maspero: 135–81.

Beauvilain, A. (1977) 'Les Peul du Dallol Bosso et la sécheresse de 1969–73 au Niger', in J. Gallais (ed.) *Stratégies pastorales et agricoles des sahéliens durant la sécheresse*

1969–1974, 'Travaux et documents de géographie tropicale', Bordeaux: CEGET, CNRS: 169–98.

Becker, J. and Martin, V. (1982) 'Kayor et Baol: Royaumes sénégalais et traite des esclaves au 17e et 18e siècles', in J. E. Hinikori (ed.) (1982), *Forced Migration: The impact of the export slave trade on African societies*, London: Hutchinson: 100–25 (first published 1975).

Bellot, J.-M. (1982a) *Commerce, commerçants de bétail et intégration régionale. L'Exemple de l'Ouest du Niger*, Bordeaux: Centre d'Etude d'Afrique Noire, Institut d'Etudes Politiques.

—— (1982b) *La question du bois dans les pays sahéliens. Bilan, perspectives et propositions*, Paris: Club du Sahel.

Benhamou, J., Raymond, H. and Zaslavsky, J. (1983) *Evaluation des filières coton et maïs au Mali*, Paris: Ministère des Relations Extérieures (Coopération et Développement).

Benoit, M. (1975) 'Espaces agraires mossi en pays Bwa (Haute-Volta)', duplicated paper, Ouagadougou: ORSTOM.

—— (1977) *Introduction à la géographie des aires pastorales soudaniennes en Haute-Volta*, Paris: ORSTOM.

—— (1982) *Nature peul du Yatenga. Remarques sur le pastoralisme en pays mossi*, 'Collection travaux et documents de l'ORSTOM', no. 43, Paris: ORSTOM.

Benoit-Cattin, M. (1979) 'Projet technique et réalité socio-économique: les exploitations des colons sur les terres neuves du Sénégal oriental', in Collectif, proceedings of the Colloquium, *Maîtrise de l'espace agraire et développement en Afrique tropicale*, 'Mémoire' no. 89, Paris: ORSTOM: 30–9.

Bergeret, A. and Ribot, J.-C. (1990) *L'arbre nourricier en pays sahélien*, Paris: Editions de la Maison des Sciences de l'Homme.

Bernardet, Ph. (1984) 'Pour une étude des modes de transmission. La technologie du manche court en Afrique noire', *Cahiers ORSTOM, Séries Sciences Humaines*, XX, 3–4: 375–98.

Bernus, E. (1970) 'Espace géographique et champs sociaux chez les Touareg Illabakan (République du Niger)', *Etudes rurales*, 37–38–39: 46–64.

—— (1981) *Touaregs nigériens. Unité culturelle et diversité régionale d'un peuple pasteur*, Paris: ORSTOM.

—— (1989) 'L'eau du désert. Usages, techniques et maîtrise de l'espace aux confins du Sahara', *Etudes Rurales*, 115–16: 93–104.

Bernus, E. and Hamidou Sidikou (eds) (1980) *Atlas du Niger*, Paris: Editions Jeune Afrique.

Berthé, A. L., Blokland, A., Bouare, S. (eds) (1991) *Profil d'environnement Mali-Sud. Etat des ressources naturelles et potentialités de développement*, Bamako: Institut d'Economie Rurale/Amsterdam: Institut Royal des Tropiques.

Berton, S. (1988) *La maîtrise des crues dans les bas-fonds: petits et micro-barrages en Afrique de l'Ouest*, Paris: GRET/AFVP/ACCT.

Bertrand, G. (1980) 'Pour une histoire écologique de la France rurale', in G. Duby and A. Wallone (eds) *Histoire de la France rurale*, Paris: Seuil: 37–113.

Bigot, Y., Binswanger, H. P. and Pingali, P. (1987) *La mécanisation agricole et l'évolution des systèmes agraires en Afrique Sud-Saharienne*, Washington: World Bank.

Blanchard de la Brosse, V. (1989) 'Riz des femmes, riz des hommes au Guidimaka (Mauritanie)', *Etudes rurales*, 115–16: 37–58.

Bonte, P. (1975) 'Esclavage et relations de dépendance chez les Touareg Kel Gress', in Cl. Meillassoux (ed.) *L'Esclavage en Afrique précoloniale*, Paris: Maspero: 49–76.

Bosc, P.M., Calkins, P. and Yung, J.-M. (1990) *Développement et recherches agricoles dans les pays sahéliens et soudaniens d'Afrique*, 'Synthesis', Montpellier: CIRAD.

Boserup, E. (1970) *Evolution agraire et pression démographique*, Paris: Flammarion.
Bosma, R., Bergoly, D. and Defou, T. (1993) *Pour un système de production durable, augmenter le bétail: rôle des ruminants dans le maintien du taux de matière organique des sols*, Bamako: IER/DRSPR.
Boulet, R. (1964) *Etude pédologique du Niger Central*, Paris: ORSTOM/Niamey, Ministère de l'Economie Rurale.
Boulier, F. and Jouve, Ph. (1988) 'Etude comparée de l'évolution des systèmes de production sahéliens et de leur adaptation à la sécheresse', duplicated paper, Montpellier: CIRAD/DSA.
Bourgeot, A. (1975) 'Rapports esclavagistes et conditions d'affranchissement des imuhag', in Cl. Meillassoux (ed.) *L'Esclavage en Afrique précoloniale*, Paris: Maspero: 77–97.
—— (1977) 'Rapport de mission d'étude sur les agro-pasteurs twareg et buzu de la région de Maradi (Niger)', Bordeaux: GRID–Université de Bordeaux 2.
—— (1979) 'Structure de classe, pouvoir politique et organisation de l'espace en pays touareg', in *Production pastorale et société*, Cambridge: Cambridge University Press/Paris: Editions de la Maison des Sciences de l'Homme: 141–53.
—— (1990) 'Identité touarègue: de l'aristocratie à la révolution', *Etudes rurales*, 120: 129–62.
Boutrais, J. (1992) 'L'élevage en Afrique tropicale: une activité dégradante?', in G. Pontié and M. Gaud (eds) *L'Environnement africain*, special issue of *Afrique contemporaine*, 161: 109–25.
Bradley, P. N. (1991) *Woodfuel, Women and Woodlots. The Foundations of a Woodfuel Development Strategy for East Africa*, vol. 1, London: Macmillan.
Bradley, P., Raynaut, Cl. and Torrealba, J. (1977) *The Guidimaka Region of Mauritania: A Critical Analysis Leading to a Development Project (Le Guidimaka Mauritanien. Diagnostic et propositions d'action)* London: War on Want.
Bruntland, G. H. (1967) *Our Common Future*, Oxford: Oxford University Press (published in French (1988), *Notre avenir à tous*, Quebec: Edition du Fleuve).
Cahen, M. (1991) 'Vent des îles, la victoire de l'opposition aux îles du Cap-Vert et à São Tome et Principe', *Politique Africaine*, 43.
Caillé, R. (1980) *Voyage à Tombouctou*, new edn, Paris: Maspero/La Découverte.
Capron, J. (1973) *Communauté villageoise Bwa, Mali–Haute-Volta*, Paris: Musée de l'Homme.
Cebron, D. (ed.) (1990) 'Etude des coûts de production du paddy à l'Office du Niger. Campagne 1988–89', duplicated paper, Bamako: IER.
Chassey, F. de (1972) *L'Etrier, la houe et le livre. Les sociétés traditionnelles au Sahara et au Sahel occidental*, Paris: Anthropos.
Chastanet, M. (1984) 'Cultures et outils agricoles en pays soninké (Gajaaga et Gidimaxa)', *Cahiers ORSTOM, Série Sciences Humaines*, XX, 3–4: 453–9.
—— (1991) 'La cueillette de plantes alimentaires en pays soninké, Sénégal, depuis la fin du XIX° siècle. Histoire et devenir d'un savoir-faire', in G. Dupré (ed.) *Savoirs paysans et développement*, Paris: Karthala–ORSTOM: 253–87.
Chleq, J.-L. and Dupriez, H. (1984) *Eau et terres en fuite, Collection terres et vie*, Paris: L'Harmattan/Dakar: ENDA.
Clanet, J.-C. (1989) 'Systèmes pastoraux et sécheresse', in B. Bret (ed.) *Les hommes face aux sécheresses*, special issue of *Travaux et mémoires de l'IHEAL*, no. 42, Paris: 309–14.
Claude, J., Grouzis, M. and Milleville, P. (eds) (1991) *Un espace Sahélien. La mare d'Oursi, Burkina Faso*, Paris: ORSTOM.
Cohen, M. N. (1977) *The Food Crisis in Prehistory. Overpopulation and the Origins of Agriculture*, New Haven and London: Yale University Press.

Collectif (1968) *Atlas international de l'Ouest africain,* Organisation de l'Unité Africain–
ORSTOM.

—— (1975) 'Enquête sur les mouvements de population à partir du pays Mossi
(Haute-Volta)' vol. 1: 'Les migrations internes mossi', vol. 2: 'Les migrations de
travail mossi', duplicated paper, Ouagadougou: ORSTOM.

—— (1980) 'L'Arbre en Afrique tropicale: la fonction et le signe', *Cahiers ORSTOM,
Série Sciences Humaines,* XVIII, 3–4.

—— (1984) 'Les instruments aratoires en Afrique tropicale. La fonction et le signe',
Cahiers ORSTOM, Série Sciences Humaines, XX, 3–4.

—— (1989) 'Sahel 89, Colloque Etat-Sahel', *Cahiers de géographie de Rouen,* 32.

—— (1991) *Savanes d'Afrique, terres fertiles? Actes des Rencontres Internationales, Focal
Coop.,* Paris: Ministère de la Coopération/CIRAD.

Collins, J. (1974) *Government and Groundnut Marketing in Rural Hausa (Niger): The
1930s to the 1970s in Magaria,* Ann Arbor: University Microfilms International.

Colvin, L. G. (ed.) (1981) *The Uprooted of the Western Sahel: Migrants' Quest for Cash in
the Senegambia,* New York: Praeger (first published 1980 as 'Les migrants
et l'économie monétaire en Sénégambie', duplicated paper, University of
Maryland).

Comité Information Sahel (1974) *Qui se nourrit de la famine au Sahel,* Paris: Maspero.

Copans, J. (ed.) (1975) *Sécheresses et famines du Sahel,* 2 vols, Paris: Maspero.

Copans, J., Couty, P. and Delpech, B. (1972) *Maintenance sociale et changement
économique au Sénégal, travaux et documents,* Paris: ORSTOM.

Coquery-Vidrovitch, C. (ed.) (1988) *Atlas historique de l'Afrique,* Paris: éd. du Jaguar.

Coste, J. and Egg, J. (eds) (1993) *Echanges céréaliers et politiques agricoles dans le sous-
espace ouest (Gambie, Guinée Bissau, Mali, Mauritanie, Sénégal). Quelle dynamique
régionale?* Ouadagoudou: CILSS – Club du Sahel.

Coulibaly, A. (ed.) (1985) *Etude sur le stockage des grains au Mali,* Paris: GRET.

Coulon, Ch. (1992) 'La démocratie sénégalaise: bilan d'une expérience', *Politique
Africaine,* 45.

Courel, M.-F. (1983) *Analyse des changements biogéographiques dans le Sahel à partir des
mesures des satellites,* IBM: Centre Scientifique de Paris.

Couty, Ph. (1989) 'Risque agricole, périls économiques', in M. Eldin and P. Milleville
(eds) *Le risque en agriculture,* Paris: ORSTOM, 561–7.

—— (1991) 'L'agriculture africaine en réserve. Réflexions sur l'innovation et
l'intensification en Afrique tropicale', *Cahiers d'études Africaines,* 31, 1–2: 65–81.

Couty, Ph. and Hallaire, A. (1980) *De la carte aux systèmes. Les études agraires de
l'ORSTOM au Sud du Sahara (1960–1980),* brochure no. 29, Paris: AMIRA/INSEE.

Crousse, B., Le Bris, E. and Le Roy, E. (eds) (1986) *Espaces disputés en Afrique noire,
pratiques foncières locales,* Paris: Karthala.

Crousse, B., Mathieu, P. and Seck, S. M. (eds) (1991) *La vallée du fleuve Sénégal.
Evaluation et perspectives d'une décennie d'aménagements,* Paris: Karthala.

Cruise O'Brien, D. (1971) *The Mourides of Senegal,* Oxford: Clarendon Press.

Dacher, M. (1984) 'Génies, ancêtres, voisins: quelques aspects de la relation à la
terre chez les Goin du Burkina Faso', *Cahiers d'études Africaines,* 94, 2: 157–92.

—— (1987) 'Société lignagère et état: les Goin du Burkina Faso', *Genève-Afrique,*
XXXV, 1: 44–57.

Déat and Bockel (1987) 'Identification des situations agricoles en Afrique sahélienne
en vue d'améliorer leur adaptation à la sécheresse', duplicated paper, Montpellier: CIRAD.

Démarets, O. (1991) *Motorisation dans la vallée du fleuve Sénégal. Stratégie et dynamique
d'équipement des différents prestataires de service,* Montpellier: CNEARC/GRDR,
Mémoire pour le Diplôme EITARC.

Déramon, J., de Conneville, G. and Pouillon, F. (1984) *Evaluation de l'élevage bovin dans la zone sahélienne au Sénégal*, Paris: MRE/CODEV.

Descola, Ph. (1986) *La nature domestique. Symbolisme et praxis dans l'écologie des Achuar*, Paris: Editions de la Maison des Sciences de l'Homme.

Deshayes, Ph. (1989) *Le système agraire d'une petite région du Boundou*, Paris: AFVP/AISB/GRDR.

Dewispelaere, G. (1980) 'Les photographies aériennes, témoins de la dégradation du couvert ligneux dans un géosystème sahélien sénégalais: influence de la proximité d'un forage', *Cahiers ORSTOM, Série Sciences Humaines*, XVII, 3–4: 155–266.

Dewispelaere, G., Noël, J. and Pain, M. (1983) 'Application des données de simulation SPOT à l'étude de la végétation sahélienne du Ferlo sénégalais', in Van Praet (ed.) *Méthodes d'inventaire et de surveillance continue des écosystèmes pastoraux sahéliens*, Colloque de Dakar, November: 333–5.

Dewispelaere, G. and Toutain, B. (1981) 'Etude diachronique de quelques géosystèmes sahéliens en Haute-Volta septentrionale', *Photointerprétation*, 1, 3: 1–7.

Diabaté, D. (1986) 'Analyse des mutations socio-économiques au sein des sociétés rurales sénoufo du Sud Mali', unpublished PhD thesis, Paris: EHESS.

Diarra, S. (1979) 'Les stratégies spatiales des éleveurs-cultivateurs peuls du Niger central agricole', in Collectif *Maîtrise de l'espace agraire et développement en Afrique tropicale*, 'Mémoire' no. 89, Paris: ORSTOM: 87–91.

Djouara, H., Lavigne Delville, Ph. and Brons, J. (1994) *Profits de l'intensification: une analyse économique du processus d'intensification dans les villages de recherche de DRSPR*, Paris: Club du Sahel/OCDE/DRSPR.

Dobremez, J.-F., Jollivet, M., Hubert, B. and Raynaut, Cl. (1990) 'Pour une pratique de l'interdisciplinarité Sciences de la nature-Sciences de l'homme', duplicated paper, Paris: CNRS.

Dory, D. (1988) 'La production socio-culturelle des paysages tropicaux. Réflexions à partir d'un cas africain, in IVe Colloque franco-japonais de géographie', *Travaux et documents de géographie tropicale*, Bordeaux: CEGET.

Doumenge, J.-P., Mott, K. E. and Cheung, Ch. (eds) (1987) *Atlas de la répartition mondiale des schistosomiases*, Bordeaux: CRET, University of Bordeaux 3, 47 maps and accompanying commentaries.

Downs, R. E. and Reyna, S. P. (eds) (1988) *Land and Society in Contemporary Africa*, Hanover, NH: University Press of New England.

Dumont, L. (1966) *Homo hierarchicus. Essai sur le système des castes*, Paris: Gallimard.

Dupire, M. (1962) *Peuls nomades. Etude descriptive des Wodaabe du Sahel nigérien*, Paris: Institut d'Ethnologie.

—— (1975) 'Exploitation du sol, Communautés résidentielles et organisation lignagère des pasteurs Wodaabe (Niger)', in Th. Monod (ed.) *Les sociétés pastorales en Afrique inter-tropicale*, London: IAI/Oxford University Press.

Dupriez, H. (1980) *Paysans d'Afrique Noire*, Brussels: Terre et vie.

Dupriez, H. and de Leener, P. (1987) *Jardins et vergers d'Afrique*, Brussels: Terre et vie.

Dupuy, J.-P. (1990) *Ordres et désordres. Enquête sur un nouveau paradigme*, Paris: Seuil.

Durufle, G. (1988) *L'Ajustement structurel en Afrique (Sénégal, Côte d'Ivoire, Madagascar)*, Paris: Karthala.

Egg, J. and Igué, J. (1993) *L'Intégration par les marchés dans le sous-espace est: L'Impact du Nigéria sur ses voisins immédiats*, Ouagadougou: CILSS/Club du Sahel.

Eldin, M. and Milleville, P. (eds) (1989) *Le risque en agriculture*, Paris: ORSTOM.

Elias, N. (1985) *La société de cour*, Paris: Flammarion.

Falloux, F. (1992) 'Agriculture, population et environnement: leurs inter-relations en Afrique', *Journées scientifiques de l'Association Nature, Sciences, Sociétés*, Paris.

FAO (1991) *Mali Sud III: Ressources et développement agricole de la région de Bougouni*, report 101/90 CP–MLI 28 SR, Rome: FAO.

Faulkingham, R. (1977a) 'Ecological constraints and subsistence strategies: The impact of drought in a Hausa village', in D. Dalby, H. Church and F. Bezzaz (eds) *Drought in Africa*, vol. 2, London: International African Insititue, 148–58.

—— (1977b) 'Fertility in Tudu. An analysis of constraints on fertility in a village in Niger', in J. C. Caldwell (ed.) *The Persistence of High Fertility. Population Prospects in the Third World*, Canberra: Australian National University Press: 157–87.

—— (1980) 'West African fertility: Levels, trends and determinants', duplicated paper, Abidjan: USAID.

Fay, Cl. (1989) 'Systèmes halieutiques et espaces de pouvoirs: transformation des droits et des pratiques de pêche dans le delta central du Niger (Mali) 1920–1980', *Cahiers ORSTOM, Série Sciences Humaines*, XXVII, 1–2: 213–16.

Faye, J. (1980) 'Approche zonale des migrations dans le Bassin Arachidier Sénégalais', in L. G. Colvin (ed.) 'Les migrants et l'économie monétaire en Sénégambie', duplicated paper, University of Maryland.

Florent, Ch. and Pontanier, R. (1982) *L'aridité en Tunisie présaharienne*, 'Travaux et documents', no. 150, Paris: ORSTOM.

Floret, C., Pontanier, R. and Serpantié, G. (1993) *La jachère en Afrique tropicale*, Dossier MAB 16, Paris: UNESCO.

Floret, C. and Serpantié, G. (eds) (1993) *La jachère en Afrique de l'Ouest*, Collection Ateliers et Séminaires, Paris: ORSTOM.

FNUAP (1988) *Etat de la population mondiale*, New York: FNUAP.

Fortes, M. and Evans-Pritchard, E. E. (eds) (1963) *African Political Systems*, London: Oxford University Press.

Franke, R. W. and Chasin, B. H. (1980) *Seeds of Famine. Ecological Destruction and the Development Dilemma in the West African Sahel*, Montclair: Allanheld & Osmun.

Freund, B. (1981) 'Labour migrations to the northern Nigerian tin mines, 1903–45', *Journal of African History*, 22, 1: 73–84.

Freysson, S. (ed.) (1973) *Etude d'une campagne de vulgarisation de masse pour l'amélioration des techniques culturales de l'agriculture nigérienne*, Paris: MARCOMER.

Froelich, J.-C. (n.d.) *Cartes des populations de l'Afrique noire*, Paris: la Documentation française.

Funel, J.-M. and Laucoin, G. (1981) *Politiques d'aménagement hydro-agricole*, 'Collection techniques vivantes', Paris: PUF.

Gallais, J. (1965) 'Le paysan dogon', *Cahiers d'Outre-mer*, 18, 70: 123–43.

—— (1975) *Pasteurs et paysans du Gourma, la condition sahélienne*, Bordeaux: CEGET/CNRS.

—— (1984) *Hommes du Sahel. Espaces-temps et pouvoirs. Le delta intérieur du Niger 1960–1980*, Paris: Flammarion.

Garin, P. and Lericollais, A. (1990) *Evolution des pratiques agricoles depuis vingt ans et leur adaptation à la sécheresse dans un village du Sine au Sénégal*, Montpellier: Document de travail DSA/ISRA.

Gastellu, J.-M. (1978) Mais où sont donc ces unités économiques que nos amis cherchent tant en Afrique?, note no. 26, *Le choix d'une unité*, Paris: AMIRA/INSEE: 99–122.

—— (1981) *L'égalitarisme économique des Sereer du Sénégal*, 'Collection travaux et documents de l'ORSTOM' no. 128, Paris: ORSTOM.

—— (1988) 'Le paysan, l'état et les sécheresses (Ngohe, Sénégal, 1972–1982)', *Cahiers ORSTOM, Série Sciences Humaines*, XXI, 4: 413–32.

Gaston, A. (1981) 'La végétation du Tchad. Evolutions récentes sous des influences climatiques et humaines', unpublished thesis, University of Paris XII.

Gaston, A. and Dulieu, D. (1975) *Pâturages du Kanem, effet de la sécheresse sur les pâturages*, Maison-Alfort: IEMVT.

Giri, J. (1983) *Le Sahel demain, catastrophe ou renaissance?*, Paris: Karthala.

Godelier, M. (1984) *L'idéel et le matériel. Pensées, économies, sociétés*, Paris: Fayard.

Goody, J. (ed.) (1958) *The Development Cycle of Domestic Groups*, Cambridge: Cambridge University Press.

Grandidier, G. (1934) *Atlas des colonies françaises*, Paris: Société d'Editions Géographiques, Maritimes et Coloniales.

Grayzel, J. (1988) 'Land tenure and development in Mauritania: The causes and consequences of legal modernization in a national context', in R. E. Downs and S. P. Reyna (eds) *Land and Society in Contemporary Africa*, Hanover, NH: University Press of New England: 221–42.

Grégoire, E. (1980) 'Etude socio-économique du village de Gourjae (département de Maradi, Niger)', duplicated paper, Bordeaux: University of Bordeaux 2.

—— (1986) *Les Alhazai de Maradi, Histoire d'un groupe de riches marchands sahéliens*, 'Travaux et documents', no 187, Paris: ORSTOM (new edn 1990) (English version (1992) *The Alhazai of Maradi: Traditional Hausa Merchants in a Changing Sahelian City*, translated and edited by B. H. Hardy, Boulder, CO and London: Lynne Rienner Publishers).

—— (1993) 'Islam and Merchants' sense of identity in Maradi, Niger', in L. Brenner (ed.) *Muslim Identity and Social Change in Sub-Saharan Africa*, London: Hurst & Co. – Indiana University Press.

Grégoire, E. and Labazée, P. (eds) (1993) *Grands commerçants d'Afrique de l'Ouest, logiques et pratiques d'un groupe d'hommes d'affaires contemporains*, Paris: Karthala/ORSTOM.

Grégoire, E. and Raynaut, Cl. (1980) 'Présentation générale du département de Maradi (Niger)', duplicated paper, Bordeaux: University of Bordeaux 2.

Grouzis, M. (1984) *Problèmes de désertification en Haute-Volta*, 'Notes et documents voltaïques', 15 (1–2), Ouagadougou: IRSV.

—— (1988) *Structures, productivité et dynamique des systèmes écologiques sahéliens (Mare d'Oursi, Burkina Faso)*, 'Etudes et thèses', Paris: ORSTOM.

Hallaire, A. (1992) 'Les montagnards du Nord Cameroun et leur environnement', *Afrique contemporaine*, 161: 144–55.

Herbart, P. (1939) *Le chancre du Niger*, Paris: NRF/Gallimard.

Héritier, F. (1975) 'Des cauris et des hommes: production d'esclaves et accumulation de cauris chez les Samo (Haute-Volta)', in Cl. Meillassoux (ed.) *L'Esclavage en Afrique précoloniale*, Paris: Maspero: 477–507.

Herry, Cl. (1990) 'Croissance urbaine et santé à Maradi (Niger): caractéristiques démographiques, phénomènes migratoires', duplicated paper, Bordeaux: University of Bordeaux 2.

Hervouët, J.-P. (1977) 'Stratégies d'adaptation différenciées à une crise climatique. L'Exemple des éleveurs du Centre-Sud mauritanien', in J. Gallais (ed.) *Stratégies pastorales et agricoles des sahéliens durant la sécheresse 1969–1974*, 'Travaux et documents de géographie tropicale', Bordeaux: CEGET, CNRS.

—— (1992) 'Environnement et grandes endémies: le poids des hommes', in G. Pontié and M. Gaud (ed.) *L'Environnement africain*, special issue of *Afrique contemporaine*, 161: 155–67.

Hill, P. (1966) 'Landlords and brokers: A West African trading system', *Cahiers d'études Africaines*, VI, 23.

—— (1972) *Rural Hausa, a Village and a Setting*, Cambridge: Cambridge University Press.

—— (1976) 'From slavery to freedom: the case of farm slavery in Nigerian Hausaland', *Comparative Studies in Society and History*, 18, 3.

Huguenin, J. (1989) 'Evolution et situation actuelle du cheptel de trait dans les exploitations Sereer des terres neuves du Sénégal Oriental', *Cahiers de la recherche-développement*, no. 21: 30–43.

Hunter, J. H. (1977) *Zones d'exploration pour les terres neuves du Sahel*, Paris: CILSS/Club du Sahel.

IEMVT/CTA (1988) *Elevage et potentialités pastorales sahéliennes, synthèses cartographiques (Burkina Faso, Mali, Niger)*, Maison-Alfort.

IFAN (1959, 1960, 1963) *Cartes ethno-démographiques*, Dakar and Paris.

IGN (1975) *Cartes topographiques au 1/200.000: feuilles de l'Afrique de l'Ouest*, Paris: Institut de Géographie Nationale.

IUCN (1989) *Etudes de l'UICN sur le Sahel*, Gland: The World Conservation Union (IUCN).

Izard, M. (1971) 'Les Yarse et le commerce au Yatenga pré-colonial', in Cl. Meillassoux (ed.), *L'Evolution du commerce en Afrique de l'Ouest*, London: Oxford University Press: 214–27.

—— (1985) *Gens du pouvoir, gens de la terre. Les institutions politiques de l'ancien royaume du Yatenga (Bassin de la Volta Blanche)*, Cambridge: Cambridge University Press/ Paris: Editions de la Maison des Sciences de l'Homme.

Jean, S. (1975) *Les jachères en Afrique tropicale. Interprétation technique et foncière*, 'Mémoire' no. XIV, Paris: Institut d'Ethnologie/CNRS.

Johnson, M. (1970) 'The cowry currencies of West Africa', *Journal of African History*, xi, I: 17–49.

Jollivet, M. and Pavé A. (1993) 'L'Environnement: un champ de recherche en formation', *Natures, sciences, sociétés*, 1, 1: 6–20.

Jonckers, D. (1987) *La société Minyanka du Mali*, Paris: L'Harmattan.

Kane Diop, M. (1989) 'Les contraintes de la répartition de la population', *Historiens et géographes du Sénégal*, 4–5: 19–23.

Kintz, D. (1982) 'Pastoralisme, agro-pastoralisme et organisation foncière', in E. Le Bris, E. Le Roy and F. Leimdorfer (eds) *Enjeux fonciers en Afrique Noire*, Paris: ORSTOM/Karthala: 212–17.

Koechlin, J. (1980) *Rapport d'étude sur le milieu naturel et les systèmes de production*, Bordeaux: University of Bordeaux 2.

Lacoste, Y. (1984) *Unité & diversité du tiers monde*, Paris: Editions La Découverte/ Hérodote.

Lahuec, J.-P. and Marchal, J.-Y. (1979) *La mobilité du peuplement Bissa et Mossi*, 'Travaux et documents' no. 103, Paris: ORSTOM.

Lake, L.-A., and Touré el Hadj, S. N. (1985) *L'Expansion du bassin arachidier. Sénégal 1954–1979*, document no. 48, Paris: AMIRA/INSEE.

Lambert, A. and Egg, J. (1992) 'Réseaux commerciaux et marchés céréaliers en Afrique de l'Ouest: l'approche des marchés par les acteurs', unpublished document.

Lavigne Delville, Ph. (1988) *Soudure et différenciations sociales. Essai d'analyse au Sénégal oriental*, 'Abordages' no. 7, Paris: AMIRA/INSEE.

—— (1991) *La rizière et la valise. Irrigation, migration et stratégies paysannes dans la vallée du fleuve Sénégal*, Paris: GRET/Syros Alternatives.

—— (1994) 'Migrations internationales, restructurations agraires et dynamiques associatives en pays soninké et haalpular (1975–1990): essai d'anthropologie du changement social et du développement', unpublished PhD thesis, Marseilles: EHESS.

—— (1995) *Valorisation des aménagements de bas-fonds au Mali: logiques paysannes et enjeux agro-économiques*, Paris: GRET.

Le Borgne, J. (1990) 'La dégradation actuelle du climat en Afrique entre Sahara et Equateur', in J.-P. Richard (ed.) *La dégradation des paysages en Afrique de l'ouest*,

Dakar: Faculté des Lettres et Sciences Humaines/Paris: Ministère de la Co-opération: 17–36.

Le Bris, E., Le Roy, E. and Leimdorfer, F. (eds) (1982) *Enjeux fonciers en Afrique Noire*, Paris: ORSTOM/Karthala.

Le Bris, E., Le Roy, E. and Mathieu, P. (1991) *L'Appropriation de la terre en Afrique Noire. Manuel d'analyse, de décision et de gestion foncière*, Paris: Karthala.

Leprun, J.-C. (1989) 'Étude comparée des facteurs de l'érosion dans le Nord-est du Brésil et en Afrique de l'ouest', in B. Bret (ed.) *Les hommes face aux sécheresses*, 'Travaux et mémoires de l'IHEAL', no 42, Paris: 139–53.

Lericollais, A. (1972) *Sob. Etude géographique d'un terroir sérèr (Sénégal)*, 'Atlas des structures agraires au sud du Sahara' no. 7, Paris: ORSTOM.

—— (1990) 'La gestion du paysage. Sahélisation, surexploitation et délaissement des terroirs sereers au Sénégal', in J.-P. Richard (ed.) *La dégradation des paysages en Afrique de l'ouest*, Dakar: Faculté des Lettres et Sciences Humaines/Paris: Ministère de la Coopération: 151–69.

Lericollais, A. and Diallo, Y. (1980) *Peuplement et cultures de saison sèche dans la vallée du Sénégal*, Introduction and 7 1/10.000 maps and accompanying commentaries, global presentation, no. 81, Paris: ORSTOM/Dakar: OMVS.

Lericollais, A. and Schmitz, J. (1984) 'La calebasse et la houe. Techniques et outils de culture de décrues dans la vallée du Sénégal', *Cahiers ORSTOM, Série Sciences Humaines*, XX, 3–4: 427–52.

Lewis, Van D. (n.d) 'Domestic Labor Intensity and the Incorporation of Malian Peasant Farmers into Localized Descent Groups', duplicated document, n.p.

Leynaud, E. and Cissé, Y. (1978) *Paysans Malinké du Haut Niger*, Bamako: Imprimerie Populaire du Mali.

Lombard, J. (1993) 'Riz des villes, mil des champs en pays sérer, Sénégal', *Espaces tropicaux*, 6, Bordeaux: CEGET.

Loriaux, M. (1991) 'La peur du nombre ou les défis de la coissance démographique', *Politique Africaine*, 44: 15–36.

Lovejoy, P. E. (1978) *Caravans of Kola. the Hausa Kola Trade – 1700–1900*, Zaria: Ahmadu Bello University Press/Oxford: Oxford University Press.

—— (1986) 'Problems of slave control in Sokoto caliphate', in P. E. Lovejoy (ed.) *Africans in Bondage: Studies in Slavery and the Slave Trade*, Madison: University of Wisconsin Press: 235–72.

Luxereau, (Levy), A. (1972) *Etude ethno-zoologique du pays Hausa en République du Niger*, Société d'Etudes Ethno-zoologiques et Ethno-Botaniques, Paris.

—— (1987) 'Maîtrise et transmission du foncier dans un village rural du départe-ment de Maradi', in M. Gast (ed.) *Hériter en pays musulman*, Paris: CNRS: 231–44.

Manière, R. (1990) 'Inventaire des milieux naturels', unpublished report, Répub-lique du Tchad, Schéma Directeur d'Aménagement de la Préfecture du Lac, Mai.

Marchal, J.-Y. (1983) *Yatenga, Nord Haute-Volta. La dynamique d'un espace rural soudano-sahélen*, 'Travaux et documents' no. 167, Paris: ORSTOM.

—— (1984) 'Lorsque l'outil ne compte plus. Techniques agraires et entités sociales au Yatenga', *Cahiers ORSTOM, Série Sciences Humaines*, XX, 3–4: 461–9.

—— (1986a) 'Prémisses d'un état moderne? Les projets coloniaux dans le bassin des Volta, 1897–1960', *Cahiers d'études Africaines*, 103, XXVI-3: 403–20.

—— (1986b) 'Vingt ans de lutte anti-érosive au nord du Burkina Faso', *Cahiers ORSTOM, Série Sciences Pédologie*, 2.

—— (1987) 'En Afrique des savanes, le fractionnement des unités d'exploitation rurales ou le chacun pour soi. L'Exemple des Moose du Burkina Faso', *Cahiers ORSTOM, Série Sciences Humaines*, XXIII, 3–4: 445–54.

—— (1989) 'En Afrique soudano-sahélienne: la course contre le temps. Rythmes des averses et forces de travail disponibles', in M. Eldin and P. Milleville (ed.) *Le risque en agriculture*, Paris: ORSTOM.

Marie, J. (1977) 'Stratégies d'adaptation à la sécheresse chez les éleveurs sahéliens. Perte de bétail, mobilité, ethnie', in J. Gallais (ed.) *Stratégies pastorales et agricoles des sahéliens durant la sécheresse 1969–1974*, 'Travaux et documents de géographie tropicale', Bordeaux: CEGET, CNRS.

Marzouk-Schmitz, Y. (1984) 'Instruments aratoires, systèmes de cultures et différenciation intra-ethnique', *Cahiers ORSTOM, Série Sciences Humaines*, XX, 3–4: 399–425.

—— (ed.) (1989) *Sociétés rurales et techniques hydrauliques en Afrique*, special issue of *Etudes rurales*, 115–16: 9–36.

Mathieu, P. (1987) 'Agriculture irriguée, réforme foncière et stratégies paysannes dans la vallée du fleuve Sénégal, 1960–1985', 2 vols, unpublished thesis, Arlon: Fondation universitaire luxembourgeoise.

—— (1989) *Etude de faisabilité d'actions de développement rural intégré dans les provinces de la Houet et de la Kossi*, 2 vols, Ouagadougou: Ministère de l'Agriculture et de l'Elevage.

Mauny, R. (1967) *Tableau géographique de l'Ouest Africain au moyen âge d'après les sources écrites, la tradition et l'archéologie*, 'Mémoire de l'Institut Français d'Afrique Noire', no. 61, (edn) Amsterdam: Swets & Zeitlinger NV (first published 1961).

Meillassoux, Cl. (ed.) (1971) *L'Evolution du commerce en Afrique de l'Ouest/The Development of Indigeneous Trade and Markets in West Africa*, Oxford: International African Institute and Oxford University Press.

—— (ed.) (1975) *L'Esclavage en Afrique précoloniale*, Paris: Maspero.

—— (1982) *Femmes, greniers & capitaux*, Paris: Maspero.

Meyer, F. (1985) 'Le paradoxe démographique', *Science*, 1, (January–February).

Michel P. (1990) 'La dégradation des paysages au Sahel', in J.-P. Richard (ed.) *La dégradation des paysages en Afrique de l'ouest*, Dakar: Faculté des Lettres et Sciences Humaines/Paris: Ministère de la Coopération: 17–36.

Michelin (1980) *Cartes routières de l'Afrique Occidentale*, 1954, 1965 edns.

Milleville, P. (1985) 'Sécheresse et évolution des systèmes agraires sahéliens. Le cas de l'Oudalan', *Cahiers de la recherche développement* 6: 11–3.

Milleville, P. and Dubois, J.-P. (1979) 'Réponses paysannes à une opération de mise en valeur de terres neuves au Sénégal', in Collectif, *Maîtrise de l'espace agraire et développement en Afrique tropicale*, 'Mémoire' no. 89, Paris: ORSTOM: 513–18.

Minvielle, J.-P. (1985) *Paysans-migrants du Fouta Toro*, 'Collection travaux et documents' no. 191, Paris: ORSTOM.

Miranda, E. (de) (1980) 'Essai sur les déséquilibres écologiques et agricoles en zone tropicale semi-aride', Doctorate of Engineering thesis, University of Montpellier.

Monimart, M. (1989) *Femmes du Sahel, la désertification au quotidien*, Paris: Karthala–OCDE.

Monnier, Y. (1981) *La poussière et la cendre; paysages, dynamique des formations végétales et stratégies des sociétés en Afrique de l'Ouest*, Paris: ACCT.

Mortimore, M. (1989) *Adapting to Drought. Farmers, Famines & Desertification in West Africa*, Cambridge: Cambridge University Press.

Moussa Soumah, (1980) 'Les migrations régionales dans le sud-est du Sénégal', in L. G. Colvin (ed.) 'Les migrants et l'économie monétaire en Sénégambie', duplicated paper, University of Maryland.

Niasse, M. and Voncke, P. P. (1985) 'Perception de l'environnement et réactions des agriculteurs et éleveurs du Galodjina face aux modifications récentes de leurs espaces traditionnels', *Mondes en développement*, 13, 52.

Nicolas, G. (1962) 'Un village bouzou du Niger', *Les cahiers d'Outre-mer*, XV: 138–65.

—— (1968) 'Un système numérique symbolique: le quatre, le trois et le sept dans la cosmologie d'une société hausa (vallée de Maradi)', *Cahiers d'études Africaines*, VIII, 4: 566–627.

—— (1975) *Dynamique sociale et appréhension du monde au sein d'une société hausa*, 'Travaux et mémoires de l'Institut d'Ethnologie', LXXVIII, Paris: Institut d'Ethnologie.

Nicolas, G. and Mainet, G. (1964) *La vallée du Gulbi de Maradi. Enquête socio-économique*, 'Document des études Nigériennes' no. 16, Niamey.

Ninnin, B. (1993) 'L'Influence des marchés sur la distribution des populations rurales dans l'espace Ouest-Africain. Elements d'analyse et de modélisation, 1960–1990', duplicated working paper, Paris: Club du Sahel.

Olivier de Sardan, J.-P. (1984) *Les sociétés songhay-zarma (Niger, Mali). Chefs, guerriers, esclaves, paysans*, Paris: Karthala.

Painter, T. (1985) *Peasant Migrations and Rural Transformation in Niger: A Study of Incorporation within a West African Capitalist Regional Economy*, Ann Arbor: University Microfilms International.

Painter, T., Sumberg, J. and Price, T. (1994) 'Your "terroir" and my "action space": Implications of differentiation, mobility and diversification for the "approche terroir" in Sahelian West Africa', *Africa*, 64, 4: 447–64.

Patterson, J.-R. *et al.* (1973) 'Mission forestière anglo-française Nigéria–Niger, Déc. 1936 – Fév. 1937', *Bois et forêts des tropiques*, 148: 2–26 (March–April) (first published 1937).

Péhaut, Y. (1970) 'L'Arachide au Niger', *Etudes d'économie africaines*, 'Série Afrique Noire' no. 1, Paris: Editions Pédone: 1–103.

Pélissier, P. (1966) *Les paysans du Sénégal. Les civilisations agraires du Cayor à la Casamance*, Saint Yrieix: Imprimerie Fabrègue.

—— (1980a) 'L'Arbre dans les paysages agraires de l'Afrique noire', *Cahiers ORSTOM, Série Sciences Humaines*, XVI, 3–4: 131–6.

—— (1980b) 'L'Arbre en Afrique Tropicale. La fonction et le signe (préface)', *Cahiers ORSTOM, Série Sciences Humaines*, XVII, 3–4: 137 – 150.

—— (ed.) (1980c) *Atlas du Sénégal*, Paris: Editions Jeune Afrique.

Père, M. (1988) *Les Lobi – tradition et changement*, Laval: Siloë.

Péron, Y. and Zalacain, V. (eds) (1975) *Atlas de la Haute-Volta*, Paris: Editions Jeune Afrique.

Piéri, Ch. (1989) *Fertilité des terres de savane. Bilan de trente ans de recherche et de développement agricoles au Sud du Sahara*, Montpellier: CIRAD/IRAT/Paris: Ministère de la Coopération.

Pochtier, G. (1989) *Evolution des systèmes agraires sereer. La communauté rurale de N'Gaye Kheme*, Montpellier: CIRAD.

Polanyi, K. (1957) *Origins of Our Time: The Great Transformation*, Boston: Beacon Press (French edition (1983), *La grande transformation. Aux origines politiques et économiques de notre temps*, Paris: Gallimard).

Pollet, E. and Winter, G. (1972) *La société Soninké (Diahunu, Mali)*, Brussels: University of Brussels, Institute of Sociology.

Poncet, Y. (1973) *Cartes ethno-démographiques du Niger*, 'Etudes Nigériennes' no. 32, Niamey: CNRSH.

Pons, R. (1988a) *L'Elevage dans les pays sahéliens*, Paris: Club du Sahel–CILSS.

—— (1988b) *La télédétection satellitaire et le Sahel*, Paris: OCDE–Club du Sahel.

Portella, E. (ed.) (1992) *Entre savoirs. L'Interdisciplinarité en acte: enjeux, obstacles, résultats*, Paris: Erès.

Portères, R. (1950) 'Vieilles agricultures de l'Afrique intertropicale. Centres d'origine et de diversification variétale primaires et berceaux d'agriculture antérieure au XVIe siècle', *L'Agronomie tropicale*, 99: 9–10.

Pradeau, Ch. (1970) 'Kokolibou (Haute Volta) ou le pays Dagari à travers un terroir', *Etudes rurales* 37–38–39: 85–112.

Prygogine, I. and Stengers, I. (1991) *La nouvelle alliance. Métamorphose de la science*, Paris: Gallimard/Folio.

Quensière, J. (1993) De la modélisation halieutique à la gestion systémique des pêches, *Natures, sciences, sociétés*, 1, 3: 211–19.

—— (ed.) (1994) *La pêche dans le delta central du Niger*, 2 vols, Paris: ORSTOM/ Karthala.

Quiminal, C. (1991) *Gens d'ici, gens d'ailleurs; migrations soninké et transformations villageoises*, Paris: Christian Bourgois.

Raison, J.-P. (1988) 'Les "parcs" en Afrique. Etat des connaissances et perspectives de recherches', duplicated working paper, Paris: Centre d'Etudes Africaines.

Raulin, H. (1967) *La dynamique des techniques agraires en Afrique tropicale du Nord*, 'Etudes et documents', Paris: Institut d'Ethnologie.

—— (1984) 'Techniques agraires et instruments aratoires au sud du Sahara', *Cahiers ORSTOM, Série Sciences Humaines*, XX, 3–4: 339–58.

Raynaut, Cl. (1973a) 'La circulation marchande des céréales et les mécanismes d'inégalité économique', *Cahiers du Centre d'Etudes et de Recherches Ethnologiques*, 2, Bordeaux: University of Bordeaux 2: 1–48.

—— (1973b) *Structures normatives et relations électives. Etude d'une communauté villageoise haoussa*, Paris and The Hague: Mouton.

—— (1975) 'Le cas de la région de Maradi', in J. Copans (ed.) *Sécheresses et famines du Sahel*, 2 vols, Paris: Maspero: 5–43.

—— (1977a) 'Circulation monétaire et évolution des structures économiques chez les Haoussas du Niger', *Africa*, 47, 2.

—— (1977b) 'Lessons of a Crisis' ('Leçons d'une crise'), in D. Dalby, H. Church and F. Bezzaz (eds) *Drought in Africa*, African Environment Special Report, London: International African Institute: 17–32.

—— (1978) 'Aspects socio-économiques de la préparation et de la circulation de la nourriture en pays haoussa', *Cahiers d'études Africaines* XVII, 4.

—— (1980a) 'Collecte du combustible et équilibre des relations avec le milieu naturel dans les communautés villageoises du département de Maradi (Niger)', in proceedings of the Colloquium, *L'Energie dans les collectivités rurales du tiers-monde*, Bordeaux: UNU – CEGET.

—— (1980b) *Rapport de synthèse sur les études multidisciplinaires menées dans le département de Maradi (Niger)*, Bordeaux: University of Bordeaux 2.

—— (1983) 'La crise des systèmes de production agro-pastorale au Niger et en Mauritanie', in Cl. Raynaut (ed.) *Milieu naturel, techniques, rapports sociaux*, Paris: Editions du CNRS.

—— (1984) 'Outils agricoles de la région de Maradi (Niger)', *Cahiers ORSTOM, Série Sciences Humaines*, XX, 3–4: 505–36.

—— (1988a) 'Aspects of the problem of land concentration in Niger', in R. E. Downs and S. P. Reyna (eds) *Land and Society in Contemporary Africa*, Hanover, NH: University Press of New England: 221–42.

—— (1988b) 'Disparités économiques et inégalités devant la santé' in *Urbanisation et santé dans le tiers-monde*, 'Mémoire', Paris: ORSTOM: 477–503.

—— (1989a) 'La crise sahélo-soudanienne, un paradigme possible pour l'analyse des relations milieu/sociétés/techniques', in M. Bruneau and D. Dory (eds) *Les enjeux de la tropicalité*, Paris: Masson, 136–44.

—— (1989b) 'La culture irriguée en pays haoussa nigérien. Aspects historiques, sociaux et techniques', *Etudes rurales* 115–16: 105–28.

—— (1989c) 'Quelques réflexions sur la notion d'enclavement', in proceedings of the Colloquium Etat-Sahel, *Cahiers géographiques de Rouen*, collective, no. 32: 129–32.

—— (1991a) 'Rapport méthodologique en vue de la réalisation d'un zonage dans la région Mali–Sud', duplicated paper, University of Bordeaux 2/Bamako: CMDT.

—— (1991b) *The Process of Development and the Logics of Change: The Need for an Holistic Approach*, Stockholm: Stockholm Environment Institute (first published in French (1989) *Genève Afrique*, XXVII, 2).

Raynaut, Cl. and Abba Souleymane, (1990) 'Le Niger, trente ans d'indépendance; repères et tendances', *Politique Africaine*, 38: 3–29.

Raynaut, Cl. (ed.), Koechlin, J., Brasset, P., and Cheung, Ch. (1988) *Le développement rural, de la région au village*, Bordeaux: GRID – University of Bordeaux 2.

Rémy, G. (1967) *Yobri (Haute-Volta)*, 'Atlas des structures agraires au sud du Sahara' no. 1, Paris and The Hague: Mouton.

Retaillé, D. (1989) 'Le destin du pastoralisme nomade en Afrique', *L'Information géographique*, 53: 103–13.

Richard, J.-P. (ed.) (1990) *La dégradation des paysages en Afrique de l'Ouest*, Dakar: Faculté des Lettres et Sciences Humaines/Paris: Ministère de la Coopération.

Richard-Mollard, J. (1952) *Cartes ethno-démographiques de l'Ouest Africain*, Dakar: IFAN.

Richards, J. F. and Tucker, R. P. (eds) (1988) *World Deforestation in the Twentieth Century*, Durham, NC and London: Duke University Press.

Richards, P. (1985) *Indigeneous Agricultural Revolution*, London: Hutchinson.

Riddell, J. C. and Dickerman, C. (eds) (1986) *Country Profiles of Land Tenure: Africa 1986*, Madison: Land Tenure Center.

Rocheteau, G. (1979) 'Pionniers mourides: un exemple de colonisation agricole spontanée des terres neuves du Sénégal', in Collectif *Maîtrise de l'espace agraire et développement en Afrique tropicale*, 'Mémoire' no. 89, Paris: ORSTOM: 167–71.

Rochette, M. (1986) 'Les migrations et la colonisation des terres nouvelles', duplicated working paper, Atelier sur le contrôle de la désertification et la gestion des resources renouvelables dans les zones sahélienne et soudanienne, Oslo: World Bank.

—— (ed.) (1989) *Le Sahel en lutte contre la désertification. Leçons d'expériences*, Margraf: CILSS/GTZ-Weikersheim.

Ruthenberg, H. (1980) *Farming Systems in the Tropics*, 3rd edn, Oxford: Clarendon Press.

Sachs, I. (1980) *Stratégies de l'éco-développement*, Paris: Les Editions ouvrières.

Sahlins, M. (1972) *Stone-Age Economics*, Chicago: Aldine-Atherton.

Salifou, A. (1971) *Le Damagaram, ou sultanat de Zinder, au XIXe siècle*, 'Etudes Nigériennes' no. 27, Niamey: CNRSH.

Sanogo, B. (1989) *Le rôle des cultures commerciales dans l'évolution de la société sénoufo (Sud-Mali)*, Bordeaux: CRET/University of Bordeaux 3.

Santoir, Ch. (1977) 'Les sociétés pastorales du Sénégal face à la sécheresse (1972–1973). Réactions à la crise et degré de rétablissement deux après. Le cas des Peul du Galodjina', in J. Gallais (ed.) *Stratégies pastorales et agricoles des sahéliens durant la sécheresse 1969–1974*, 'Travaux et documents de géographie tropicale', Bordeaux: CEGET, CNRS.

—— (1983) *Raison pastorale et politique de développement: les Peuls sénégalais face aux aménagements*, Paris: ORSTOM.

Saul, M. (1988) 'Money and land tenure as factors in farm size differentiation in Burkina Faso', in R. E. Downs and S. P. Reyna (eds) *Land and Society in Contemporary Africa*, Hanover, NH: University Press of New England: 243–79.

Sautter, G. (1962) 'A propos de quelques terroirs d'Afrique Occidentale', *Etudes rurales*, 4: 24–86.

337

Savonnet, G. (1979) 'Structures sociales et organisation de l'espace (exemples empruntés à la Haute-Volta)', in Collectif *Maîtrise de l'espace agraire et développement en Afrique tropicale*, 'Mémoire' no. 89, Paris: ORSTOM: 39–44.

Schmitz, J. (1986) 'Agriculture de décrue, unités territoriales et irrigation dans la vallée du Sénégal', *Cahiers de la recherche-développement*, 12: 65–77.

Schmoll, P. (1993) 'Black stomachs, beautiful stones: soul-eating among Hausa in Niger', in J. Comaroff and J. Comaroff (eds) *Modernity and its Malcontents: Ritual and Power in Postcolonial Africa*, Chicago: University of Chicago Press: 193–220.

Schwartz, A. (1991) 'L'Exploitation agricole de l'aire cotonnière burkinabé: caractéristiques sociologiques, démographiques, économiques', duplicated working paper, Ouagadougou: ORSTOM.

—— (1992) 'Coton et développement au Burkina Faso – histoire d'une recherche', *Chroniques du Sud*, ORSTOM.

Schwartz, J. (1992) 'Le défi démographique', in G. Pontié and M. Gaud (eds) *L'environnement africain*, special issue of *Afrique contemporaine*, 161: 43–6.

Sebillotte, M. (1993) 'L'agronome face à la notion de fertilité', *Natures, sciences, sociétés*, 1, 2: 128–41.

Seck, S. M. and Lericollais, A. (1986) 'Aménagements hydro-agricoles et systèmes de production dans la vallée du Sénégal', *Cahiers de la recherche-développement*, 2: 3–9.

Seignebos, Ch. (1980) 'Fortifications végétales de la zone soudano-sahélienne (Tchad et Nord Cameroun)', *Cahiers ORSTOM, Série Sciences Humaines*, XVI, 3–4: 191–222.

—— (1984) 'Instruments aratoires du Tchad méridional et du Nord-Cameroun', *Cahiers ORSTOM, Série Sciences Humaines*, XX, 3–4: 537–73.

Serpantié, G., Mersadier, G. and Tezenas du Montcel, L. (1986) 'La dynamique des rapports agriculture-élevage en zone soudano-sahélienne du Burkina-Faso: diminution des ressources, organisation collective, et stratégies d'éleveurs-paysans au nord du Yatenga', *Cahiers de la recherche-développement*, 9–10: 40–50.

Shaikh, A. (1990) *The Segou Roundtable on Local Level Management in the Sahel*, Washington: USAID.

Sigaut, F. (1976) *L'Agriculture et le feu. Rôle et place du feu dans les techniques de préparation du champ de l'ancienne agriculture européenne*, Paris: Mouton.

—— (1985) 'Une discipline scientifique à développer: la technologie de l'agriculture', in Ch. Blanc-Pamard and A. Lericollais (eds) *A travers champs*, Paris: ORSTOM.

Sirven, P. (1987) 'Démographie et villes au Burkina Faso', *Cahiers d'Outre-mer*, 40, 159: 265–83.

Smith, M. (1954) *Baba of Karo: A Woman in the Muslim Hausa*, London: Faber & Faber (French edition (1969) *Baba de Karo. L'autobiographie d'une musulmane haoussa du Nigéria*, Paris: Plon).

Steinkamp-Ferrier, L. (1983) 'Sept villages du Guidimakha mauritanien face à un projet de développement: l'histoire d'une recherche', unpublished thesis, Ecole des Hautes Etudes en Sciences sociales, Paris.

Suret-Canale, J. (1968–72) *L'Afrique noire. Géographie, civilisation, histoire*, 3 vols, Paris: Editions Sociales.

Tchayanov, A. (1990) *L'organisation de l'économie paysanne*, Paris: Librairie du Regard (original edn 1924).

Teme, B. (1985) 'Système agraire villageois et développement rural dans cinq villages du Bélédougou au Mali', unpublished thesis INRA/SAD, IER, University of Dijon.

Thomson, J. T. (1988) 'Deforestation and desertification in twentieth-century arid

Sahelien Africa', in J. F. Richards and R. P. Tucker (eds) *World Deforestation in the Twentieth Century*, Durham, NC and London: Duke University Press: 71–90.

Thornton, J. (1992) *Africa and African in the Making of the Atlantic World*, Cambridge: Cambridge University Press.

Tiffen, M., Mortimore, M. and Gichuki, F. (1994) *More People, Less Erosion; Environmental Recovery in Kenya*, London: ODI/Wiley.

Toupet, Ch. (ed.) (1977) *Atlas de la République Islamique de Mauritanie*, Paris: Editions Jeune Afrique.

Toutain, B. and Dewispelaere, G. (1978) *Etude et cartographie des pâturages de l'ORD du Sahel et de la zone de délestage au N.E. ou Fada N'Gourma (Hte Volta)*, no. 31, 3 vols, Maison-Alfort: IEMVT.

Traoré, M. (ed.) (1981) *Atlas du Mali*, Paris: Editions Jeune Afrique.

UNESCO (1977) *Développement des régions arides et semi-arides, obstacles et perspectives*, Technical notes on the Man and Biosphere Programme, 6, Paris: United Nations Educational Scientific and Cultural Organization.

—— (1991) *Ecosystèmes pâturés tropicaux*, Paris: United Nations Educational Scientific and Cultural Organization.

Van der Pol, F. (1991) 'L'épuisement des terres, une source de revenus pour les paysans du Mali sud?', in *Savanes d'Afrique, terres fertiles?* Paris: Ministère de la Coopération, Coll. Focal Coop.

Villenave, D. (1983) 'Organisation de l'espace et schistosomiase urinaire dans 3 communautés mossi de la région de Kaya en Haute-Volta', unpublished thesis, University of Bordeaux 3.

Vlaar, J. C. J. (ed.) (1992) *Les techniques de conservation des eaux et des sols dans les pays du Sahel*, Ouagadougou/Wageningen: CIEH/UAW.

Weigel, J.-Y. (1982) *Migration et production domestique des Soninké du Sénégal*, Paris: ORSTOM.

White, F. (1986) *La végétation de l'Afrique* (note accompanying the African vegetation map), Paris: ORSTOM–UNESCO.

Yamba, B. (1993) 'Ressources ligneuses et problèmes d'aménagement forestier dans la zone agricole du Niger', unpublished thesis, University of Bordeaux 3.

Yung, J.-M. and Bosc, P.-M. (eds) (1992) *Le développement agricole au Sahel*, 5 vols, 'Documents systèmes agraires' no. 17, Montpellier: CIRAD.

Zachariah, K. C., Clairin, R. and Condé, J. (1977) *Aperçu sur les migrations en Afrique de l'Ouest*, Document CD/R (78) 30-CD, Paris: OCDE.

Zanoni, M. and Raynaut, Cl. (1994) 'Environnement et développement: quelle recherche, quelle formation?', in *Sociétés, développement, environnement*, special issue of *Cadernos de Desenvolvimento e Meio Ambiente*, 1, Bordeaux: GRID/Curitiba, Brazil: Universidade Federal do Parana: 167–90.

Zoungrana, I. (1991) 'Recherches sur les aires pâturées du Burkina-Faso', unpublished thesis, University of Bordeaux 3.

BIBLIOGRAFÍA

INDEX